ANALYZING RECORDED MUSIC

Analyzing Recorded Music: Collected Perspectives on Popular Music Tracks is a collection of essays dedicated to the study of recorded popular music, with the aim of exploring "how the record shapes the song" (Moylan, *Recording Analysis*, 2020) from a variety of perspectives. Introduced with a Foreword by Paul Théberge, the distinguished editorial team has brought together a group of reputable international contributors to write about a rich collection of recordings.

Examining a diverse set of songs from a range of genres and points in history (spanning the years 1936–2020), the authors herein illuminate unique attributes of the selected tracks and reveal how the recording develops the expressive content of song performance.

Analyzing Recorded Music will interest all those who study popular music, cultural studies, and the musicology of record production, as well as to popular music listeners.

William Moylan is Distinguished University Professor (2019–2022) at the University of Massachusetts Lowell, where he is a Professor of Music and Sound Recording Technology and served as Chairperson of the Department of Music. As a recording engineer and a producer, he has worked with emerging and leading artists across a broad spectrum of popular and classical genres. William Moylan is the author of *Understanding and Crafting the Mix, 3rd Edition* (Focal Press, Routledge, 2015) and *Recording Analysis: How the Record Shapes the Song* (Focal Press, Routledge, 2020).

Lori Burns is Professor of Music at the University of Ottawa. Burns' interdisciplinary research merges musical analysis and cultural theory to explore representations of gender in the lyrical, musical, and visual texts of popular music. Her monograph, *Disruptive Divas: Feminism, Identity, and Popular Music* (2002), won the Pauline Alderman Award from the International Alliance for Women in Music (2005). She is co-editor of two previous essay collections: *The Pop Palimpsest* (2018) and *The Bloomsbury Handbook of Popular Music Video Analysis* (2019). She is co-editor of the *Ashgate Popular and Folk Music Series* and Associate Editor of *Music Theory Spectrum*.

Mike Alleyne is Professor Emeritus with the Department of Recording Industry, Middle Tennessee State University (MTSU) and a Visiting Professor at the Pop Akademie in Germany. His work has been published in *Popular Music & Society, Rock Music Studies,* and *Billboard magazine*. His books include *The Essential Hendrix* (2020) and *The Encyclopedia of Reggae* (2012) and a co-edited collection entitled *Prince and Popular Music* (2020). He also writes and edits for the SAGE Business Case Series in Music Marketing and contributed liner notes to the groundbreaking 9-CD box set, *The Smithsonian Anthology of Hip-Hop and Rap* (2021).

ANALYZING RECORDED MUSIC

Collected Perspectives on Popular Music Tracks

Edited by William Moylan, Lori Burns, and Mike Alleyne

Routledge
Taylor & Francis Group
LONDON AND NEW YORK

Designed cover image: simonkr via Getty Images

First published 2023
by Routledge
4 Park Square, Milton Park, Abingdon, Oxon OX14 4RN

and by Routledge
605 Third Avenue, New York, NY 10158

Routledge is an imprint of the Taylor & Francis Group, an informa business

British Library Cataloguing-in-Publication Data
A catalogue record for this book is available from the British Library

Library of Congress Cataloging-in-Publication Data
Names: Moylan, William, edit | Burns, Lori, edit | Alleyne, Mike (Michael Randolph) edit
Title: Analyzing recorded music : collected perspectives on popular music tracks / edited by William Moylan, Lori Burns, and Mike Alleyne.
Description: Abingdon, Oxon ; New York : Routledge, 2023. | Includes bibliographical references and index. | Identifiers: LCCN 2022033137 (print) | LCCN 2022033 (ebook) | ISBN 9780367546328 (hardback) | ISBN 9780367546311 (paperback) | ISBN 9781003089926 (ebook)
Subjects: Classification: LCC MT146 .A57 2023 (print) |
LCC MT146 (ebook) | DDC 781.49--dc23/eng/20220720
LC record available at https://lccn.loc.gov/2022033137
LC ebook record available at https://lccn.loc.gov/2022033138

ISBN: 978-0-367-54632-8 (hbk)
ISBN: 978-0-367-54631-1 (pbk)
ISBN: 978-1-003-08992-6 (ebk)

DOI: 10.4324/9781003089926

Typeset in Bembo
by KnowledgeWorks Global Ltd.

Access the Support Material: www.routledge.com/9780367546311

CONTENTS

FIGURES AND EXAMPLES

Figures

Examples

TABLES

CONTRIBUTORS

Mike Alleyne is Professor Emeritus with the Department of Recording Industry, Middle Tennessee State University (MTSU) and a Visiting Professor at the Pop Akademie in Germany. His work focuses on aspects of popular music history and music production, with material published in *Popular Music & Society, Rock Music Studies, Ethnomusicology Forum, the Journal on the Art of Record Production, the Journal of Latin American and Caribbean Studies, Popular Music History* and *Billboard magazine.* His books include *The Essential Hendrix* (2020), *Prince and Popular Music* (2020—co-edited with Kirsty Fairclough), and *The Encyclopedia of Reggae* (2012). In addition to his numerous book chapters, he writes and edits for the SAGE Business Case Series in Music Marketing and contributed liner notes to the groundbreaking 9-CD box set, T*he Smithsonian Anthology of Hip-Hop and Rap* (2021).

Tom Attah is a musician, teacher, and published author who leads the BMus (Hons) Popular Music Performance degree at Leeds Arts University. His research and writing focus on the effects of technology on popular music and culture, particularly the blues. As a guitarist and singer, Tom performs solo, with acoustic duos, and as leader of an electric band. Tom's solo acoustic work includes his own original blues compositions and has led to international theatre tours and performances at major music festivals. Tom's media appearances as a subject matter expert in blues include multiple performances and documentaries for BBC Radio and Sky Arts.

Samantha Bennett is a sound recordist, guitarist, and academic from London, UK, Associate Dean Higher Degree Research, and Professor of Music at the Australian National University. Her research interests include sound recording, music production, music technology, and the manifestation of technology and process in recorded popular music. Her co-authored book with associate professor Eliot Bates (CUNY), *Gear: Cultures of Audio and Music Technologies*, is forthcoming from The MIT Press, and a short-form monograph, *Secrets and Revelatory Discourse in Music and Audio Technology Culture*, is forthcoming from Cambridge University Press. She is the author of two further monographs, *Modern Records, Maverick Methods: Technology and Process in Popular Music Record Production 1978–2000* (Bloomsbury Academic), and *Peepshow*, a 33 1/3 series edition on the album by Siouxsie and the Banshees (Bloomsbury Academic). She is also a co-editor *of Critical Approaches to the Production of Music and Sound* (Bloomsbury Academic) and *Popular Music, Stars and Stardom* (ANU Press).

Zack Bresler is a lecturer in music production and recording at the University of Stavanger, and a recent PhD graduate from the Department of Popular Music, University of Agder, Norway. Motivated by an obsession with immersive and interactive media such as surround and 3D sound, virtual reality, and interactive art, Zack is a musicologist studying popular music as it is presented in such media with an eye toward the aesthetics of music productions, compositional design, immersion, listener experience, and staging. In addition, Zack is also an active artist, working as a drummer and percussionist, a sound artist and composer working on immersive and interactive installations, performances, and collaborations.

Ragnhild Brøvig is Professor of Popular Music Studies at RITMO and the Department of Musicology, University of Oslo, Norway. Her interdisciplinary research explores the aesthetic and sociopolitical implications of experimenting with technology in popular music, including via sampling and remixing, glitches and musical fragmentation, sound and spatialization, and rhythm and groove. She is the author of *Parody in the Age of Remix: Mashup Creativity vs. the Takedown* (MIT Press, forthcoming) and *Digital Signatures: The Impact of Digitization on Popular Music Sound* (with Anne Danielsen, MIT Press, 2016). Her research has been funded by the Research Council of Norway and the Arts Council of Norway.

Lori Burns is Professor of Music at the University of Ottawa. Burns' interdisciplinary research, funded by the Social Sciences and Humanities Research Council of Canada, merges musical analysis and cultural theory to explore representations of gender in the lyrical, musical, and visual texts of popular music. Her monograph, *Disruptive Divas: Feminism, Identity, and Popular Music* (2002), won the Pauline Alderman Award from the International Alliance for Women in Music (2005). She is co-editor of *The Pop Palimpsest* with Serge Lacasse (2018) and The *Bloomsbury Handbook to Popular Music Video Analysis* with Stan Hawkins (2019). She was a founding Co-Editor of the *Tracking Pop Series* of the University of Michigan Press and is now Co-Editor of the *Ashgate Popular and Folk Music Series* and Associate Editor of *Music Theory Spectrum.*

Anne Danielsen is Professor of Musicology and Director of *RITMO Centre for Interdisciplinary Studies in Rhythm, Time and Motion* at the University of Oslo, Norway. She has published widely on theoretical and empirical aspects of rhythm, groove, and sound recording in post-war African American popular music and is the author of *Presence and Pleasure: The Funk Grooves of James Brown and Parliament* (2006) and *Digital Signatures: The Impact of Digitization on Popular Music Sound* (with Ragnhild Brøvig, 2016), and the editor of *Musical Rhythm in the Age of Digital Reproduction* (2010/2016).

Walter Everett is Professor Emeritus of Music at the University of Michigan. In addition to authoring 35 peer-reviewed articles and book chapters and presenting to more than 80 audiences for conferences or by invitation on various aspects of both common-practice repertoires and pop and rock music, he has published *The Beatles as Musicians* (two volumes), *The Foundations of Rock*, and—with co-author Tim Riley—*What Goes On: The Beatles, Their Music, and Their Time*, all for Oxford University Press, and is the editor of *Expression in Pop-Rock Music*, for Routledge. His current book project, for Bloomsbury Press, has the working title *Sex and Gender in Rock and Pop from the Beatles to Beyoncé.*

Michail Exarchos is Creative Director of Music Innovation company RT60 Ltd. Mike holds a doctorate in hip-hop musicology and is an award-winning rap artist (first MTV Best Greek Act winner), with nominations for multiple VMAs and an MTV Europe Music Award. He has led programs in recording, mixing, mastering, and audio production at various UK universities and has published with Routledge, Bloomsbury Academic, and the journals of *Popular Music*, *Popular Music Studies*, and *Popular Music Education*. Mike's self-produced album, *Xli3h*, has been included in the 30 best Greek hip-hop albums of all time and he is the first Greek artist to ever perform at South By Southwest. Mike's studio career began as a sound engineer at Vault Recording Studios in London, where he worked with pioneering British hip-hop acts, including Klashnekoff, Bury Crew, Skinnyman, and Taskforce. Under his Stereo Mike alias, he's produced three critically acclaimed solo albums and numerous singles for international artists on labels such as EMI, Sony, Universal, and Warner Music.

Andrew Flory is Associate Professor of Music at Carleton College. He has published widely on rock and R&B topics, including articles on Marvin Gaye, the Beach Boys, and the Grateful Dead. Flory is the author of *I Hear a Symphony: Motown and Crossover R&B* (Michigan, 2017) and the co-author of the industry-leading history of rock textbook *What's That Sound* (Norton, 2018). He also works as a consultant for Universal Music Enterprises and has co-produced many Motown-oriented reissues. He is currently writing a biography of Marvin Gaye.

Steven Gamble is a Marie Curie Research Fellow in Music at University College Cork and an incoming Research Fellow at the University of Bristol, specializing in the study of popular music and the Internet. He is the author of *How Music Empowers: Listening to Modern Rap and Metal* (Routledge, 2021) and *Digital Flows: Internet-Based Hip Hop Music and Culture* (Oxford University Press, 2023). He has published on empowerment, listening, digital music technology, and new media in journals including *Popular Music & Society*, *Journal on the Art of Record Production*, and *Metal Music Studies*. He is a board member and webmaster for the *International Society for Metal Music Studies* and the *Equality, Diversity and Inclusion in Music Studies* network.

Phil Harding started at the Marquee Studios, London, engineering for acts such as The Clash, Killing Joke, Toyah, and Matt Bianco. During the 1980s, Phil was chief mix engineer for Stock Aitken Waterman (PWL Studios), working with acts such as Dead or Alive, Mel & Kim, Bananarama, Rick Astley, Pet Shop Boys, and Kylie Minogue. In the 1990s, Phil and Ian Curnow ran P&E Music Studios from The Strongroom Studio complex, producing hits for East 17 (including "Stay Another Day"), Deuce, Boyzone, and 911. Phil's first book, *PWL from the Factory Floor* (Cherry Red Books, 2010), was followed by his PJS Music productions for Holly Johnson, Samantha Fox, Belinda Carlisle, Curiosity, and Mel & Kim. He completed his PhD in Music Production at Leeds Beckett University in April 2017. Phil's new book is *Pop Music Production* published by Routledge in July 2020.

Stan Hawkins is an Emeritus Professor at the University of Oslo and University Agder, Norway. He has published widely on popular music, culture, sexuality, and identity and is the author of *Settling the Pop Score* (2002/2017), *The British Pop Dandy* (2009/2016), *Prince: The Making of a Pop Music Phenomenon* (co-author Sarah Niblock, 2011/2013), and *Queerness in Pop Music* (2016). Edited volumes include *Music, Space and Place* (co-editors Sheila Whiteley and Andy Bennett) (2005/2017), *Essays on Sound and Vision* (co-editor John Richardson, 2007), *Pop Music & Easy Listening* (2011), *Critical Musicological Reflections* (2012/2016), *The Routledge Research Companion to Popular Music and Gender* (2017), and *The Bloomsbury Handbook of Popular Music Video Analysis* (co-editor Lori Burns, 2019). Hawkins has also edited special editions on Michael Jackson, *Popular Music and Society* (co-editor Susan Fast, 2012), David Bowie, *Contemporary Music Review* (2018), and Prince, *Popular Music and Society* (co-editor Anne Danielsen, 2020). In 2017, he was elected a life academy member of the *Norwegian Academy of Science and Letters.*

John Howland is Professor and Chair of Musicology at the Norwegian University of Science and Technology. He divides life between work in Trondheim, Norway, and home in Lund, Sweden. His research examines the interrelations among class hierarchy discourse and arranging, production, and entertainment aesthetics in popular music and jazz-related orchestral idioms. Howland is the author of the books *Ellington Uptown: Duke Ellington, James P. Johnson, and the Birth of Concert Jazz* (University of Michigan Press, 2009), *Duke Ellington Studies* (Cambridge University Press, 2017), and *Hearing Luxe Pop: Glorification, Glamour and the Middlebrow in American Popular Music* (University of California Press, 2021), among other writings and editorial work, and is the former editor/co-founder of the Routledge Press journal, *Jazz Perspectives*. He was formerly a professor of music and jazz studies at Rutgers University-Newark. Howland's latest book project (*Disco Gold!*) extends the themes of *Hearing Luxe Pop* through a study of mainstream discomania across 1973–1980.

Katie Kapurch is Associate Professor of English at Texas State University. She has authored many articles and chapters about visual and narrative representations of gender, sexuality, and race in popular culture. Her books include *Victorian Melodrama in the Twenty-First Century* (2016) and *New Critical Perspectives on the Beatles* (2016), the latter co-edited with Kenneth Womack. Forthcoming books include *Blackbird Singing*, which explores how Black musicians shaped and responded to the Beatles and is supported by the National Endowment for the Humanities; *Disney Plus Beatles* theorizes the impact of the title's two media empires on the production and reception of popular music in the 21st century. Katie's next co-edited collection, *The Beatles and Humor*, includes a chapter on the Beatles and Black music that she is co-authoring with Walter Everett and Mike Alleyne. Katie is co-editor of the journal *AMP: American Music Perspectives* (Penn State UP).

Timothy Koozin is Professor and Area Coordinator of Music Theory at the Moores School of Music, University of Houston. His research interests include music and meaning, popular music, film music, and music theory pedagogy. His analytical essays on musical gesture in popular music focused on artists including Nick Drake, Bob Dylan, Sarah McLachlan, and Jimi Hendrix and appeared in journal articles and in collected essays published by Cambridge University Press, Ashgate, and Routledge. Koozin is co-author of music theory textbooks published by Cengage and Oxford University Press. He is the former editor of the electronic journal of the Society for Music Theory, *Music Theory Online*.

Olivia R. Lucas is Assistant Professor of Music Theory at Louisiana State University. Her interdisciplinary research blends music theory with ethnography, sound studies, and ecomusicology, and up until the present chapter has been focused primarily on extreme metal music. Her work on issues of rhythm and meter and performance analysis has been published in *Music Theory Online* and *The Oxford Handbook of Public Music Theory*. An article on loudness appears in the *Journal of Sonic Studies*, and articles on environmental and identity issues have been published in *Popular Music*. She also served as a founding member of the editorial board for *Ground Works*, an interdisciplinary arts journal.

Michael Millner currently serves as the Nancy Donahue Professor of the Arts (2020–22) at the University of Massachusetts Lowell (UMass Lowell), where he is also an Associate Professor of English and American Studies. His scholarship often situates American culture within its political history. He is the author of *Fever Reading: Affect and Reading Badly in the Early American Public Sphere* and a number of scholarly essays, as well as writings for mainstream publications. He is Director of The Jack and Stella Kerouac Center for the Public Humanities and has served as the President of the New England American Studies Association. Before arriving at UMass Lowell, Millner was a Harper fellow in the Society of Fellows at the University of Chicago. He is currently working on a book about Dylan, Barbara Rubin, Andy Warhol, and the pop avant-garde in New York City circa 1965.

William Moylan is Distinguished University Professor (2019–2022) at the University of Massachusetts Lowell, where he is Professor of Music and Sound Recording Technology and served as Chairperson of the Department of Music and the Coordinator of Sound Recording Technology program. His research explores the intersection of the sound qualities of recordings, recording production, music theory/musicology, lyrics, psychoacoustics, aural perception, and cognition. He is particularly interested in the unique sonic dimensions and qualities generated by the recording process and how those elements transform recorded song. As a recording engineer and a producer, he has worked with emerging and leading artists across a broad spectrum of popular and classical genres. William Moylan is the author of *The Art of Recording* (Van Nostrand Reinhold, 1992), *Understanding and Crafting the Mix, 3rd Edition* (Focal Press, Routledge 2015), and *Recording Analysis: How the Record Shapes the Song* (Focal Press, Routledge 2020).

Mark Mynett, in addition to being a record producer, engineer, mix, and mastering engineer, is a senior lecturer in Music Technology and Production at Huddersfield University. His principal genre of focus, production-wise, is rock and metal music, an area where he is also a front of house engineer for a number of prominent acts. This focus is reflected in his area of research, where, following a PhD in contemporary metal music production and numerous authored chapters in edited collections by Bloomsbury, Routledge, and Ashgate, Mark authored the world's first book on the subject *Metal Music Manual: Producing, Engineering, Mixing and Mastering Contemporary Heavy Music*, published by Focal Press (Taylor and Francis).

Jocelyn R. Neal is the Bowman and Gordon Gray Distinguished Term Professor of Music at the University of North Carolina at Chapel Hill. She is the author of *The Songs of Jimmie Rodgers: A Legacy in Country Music* (Indiana University Press) and *Country Music: A Cultural and Stylistic History* (Oxford University Press). Her articles on country music have appeared in numerous collected editions and journals, including the *Journal of Popular Music Studies*, *Musical Quarterly*, and *Music Theory Spectrum*. Her research addresses songwriting, music and dance, and copyright within American popular music. She also founded and directs the UNC Bluegrass Initiative.

Thea Sørli Paulsrud works as a Research Assistant (2020–present) at RITMO Center for Interdisciplinary Studies in Rhythm, Time and Motion. She holds an MA in Musicology from the University of Oslo (2021), with a master thesis on freelance musicians' organization of work and their working rights. Her academic interests span from popular music, music analysis, and music and media to cultural entrepreneurship, music business, and the creative industries. She is also a renowned blues, soul, and R&B vocalist with her bands Soft City and a self-named solo project.

Stephen Spencer is a doctoral candidate in Composition and Music Theory at the CUNY Graduate Center. His research concerns the relationship between auditory perception and aesthetics in a broad range of repertoire, from Edgard Varèse to the Jesus and Mary Chain. His dissertation, supported by an SSHRC Doctoral Fellowship and an Art–Science Connect Research Award, focuses on the analysis and visualization of timbral structures in modernist and contemporary music.

Mark Spicer is Professor of Music at Hunter College and the Graduate Center, City University of New York, where he was the 2015 recipient of Hunter's Presidential Award for Excellence in Teaching. He specializes in the reception history and analysis of popular music, particularly British pop and rock, since the 1960s, and his writings have appeared widely in scholarly journals and essay collections. He coedited *Sounding Out Pop* (with John Covach), edited the volume on *Rock Music* for Ashgate's *Library of Essays on Popular Music*, and served as Associate Editor of *Music Theory Spectrum* from 2013 to 2015. Spicer's article "Fragile, Emergent, and Absent Tonics in Pop and Rock Songs" (*Music Theory Online*, 2017) won the 2020 Outstanding Publication Award from the Society for Music Theory. He is also a professional keyboardist and vocalist and continues to take the stage most weekends with his "electric R&B" group, the Bernadettes.

Paul Théberge is a Professor at Carleton University, Ottawa, Canada. He has published widely on issues concerning music, technology, and culture, and his research addresses the digital musical instrument industry, the history of sound recording, the relationship between music and the Internet, and the role of sound and music in visual media. Théberge is the author of *Any Sound You Can Imagine: Making Music/Consuming Technology* (Wesleyan University Press, 1997) and co-editor of *Living Stereo: Histories and Cultures of Multichannel Sound* (Bloomsbury Academic, 2015). In 2012, he produced and engineered a set of experimental recordings—*Glenn Gould: The Acoustic Orchestrations*—for Sony Classical. His recent work focuses on music technologies and practices (in contributions to anthologies such as *The Relentless Pursuit of Tone: Timbre and Popular Music*, 2019, and *The Bloomsbury Handbook of Music Production*, 2020); and on issues of celebrity and large-scale economic and structural change in the music industries (*Contemporary Music Review*, 2021).

Steve Waksman is the Elsie Irwin Sweeney Professor of Music at Smith College. His publications include the books *Instruments of Desire: The Electric Guitar and the Shaping of Musical Experience* (1999), and *This Ain't the Summer of Love: Conflict and Crossover in Heavy Metal and Punk* (2009). With Reebee Garofalo, he co-authored the sixth edition of the rock history textbook, *Rockin' Out: Popular Music in the U.S.A.* (2014), and with Andy Bennett, he co-edited the *SAGE Handbook of Popular Music* (2015). His latest book is *Live Music in America: A History from Jenny Lind to Beyoncé* (2022), published by Oxford University Press.

Jada Watson is an Assistant Professor of Digital Humanities in the School of Information Studies at the University of Ottawa. Principal investigator of the SSHRC-funded SongData project (www.SongData.ca), Watson's research explores the role of market data in the formation and evolution of genre categories, focusing on the impact of structural racism and sexism embedded within industrial structures on representation in country music culture. This research has been published in *Popular Music & Society, Popular Music History,* and *American Music Perspectives.* It has also been cited as a major source in a brief submitted to the US Federal Communications Commission in response to the National Association of Broadcasters' proposal for ownership deregulation, as well as in the Grammy Recording Academy's Report on Inclusion and Diversity. She is co-editor of *Whose Country Music? Genre, Identity, and Belonging in Twenty-First Century Country Music*, a collection of essays forthcoming with Cambridge University Press.

Justin Williams is an Associate Professor in Music at the University of Bristol (UK), the author of *Rhymin and Stealin: Musical Borrowing in Hip-Hop* (2013), and editor of *The Cambridge Companion to Hip-Hop* (Cambridge, 2015). His latest book, *Brithop* (2020), looks at the intersections of UK hip-hop and politics. As a musician in California, he led a jazz piano trio and played trumpet with the award-winning band Bucho!.

Albin Zak is a composer, songwriter, record producer, and musicologist. He holds degrees in composition and performance from the New England Conservatory and in musicology from the City University of New York. His articles and reviews appear in the *Journal of the American Musicological Society, Journal of the Society for American Music, Current Musicology, Journal of American History,* and several volumes of collected essays. He is the author of *The Poetics of Rock: Cutting Tracks, Making Records* and *I Don't Sound Like Nobody: Remaking Music in 1950s America*; and editor of *The Velvet Underground Companion: Four Decades of Commentary*. He has recently published a music appreciation text—*Rock on Record* (SUNY Press)—which takes a guided listening approach to rock music styles and genres. His recordings include the albums *An Average Day, Waywardness and Inspiration, Villa Maria Road*, and *Homefront*. He is Professor Emeritus of Music at the University at Albany (SUNY).

FOREWORD

As I write this during the Spring of 2022, the world is still emerging, in fits, and starts from a two-year hiatus in daily life brought on by the COVID-19 pandemic. Throughout this period, recorded music has offered some degree of distraction and solace from the long months of isolation. In the early days of the pandemic, it was the more spontaneous outbursts of music, often recorded by design or by chance on cell phones, that caught our attention: the sound of Italians singing from balconies, in harmony with other singers, across an empty street during the interminable days and nights of lockdown, or the sight of TikTok fans trying to outdo one another, performing improvised dance moves to the sound of pop music recordings in their bedrooms. Over time, well-known artists added their own contributions to the daily mix, dropping new songs and videos on YouTube. While the preparation, purpose, and quality of these recordings varied enormously, they nevertheless each bore the marks of their making, exhibiting distinct musical, sonic, and spatial signatures.

Taking a somewhat longer historical perspective, it is worth noting that these recordings and videos largely came to us via streaming platforms, and, although music services such as Spotify had already been well established by the time of the onset of the pandemic, digital media companies benefitted greatly as subscriptions soared to new heights: with restrictions on mobility and deprived of opportunities to hear live music, fans turned increasingly to digital media. Indeed, we are now two decades into a new century of recorded music, and, aided by digital distribution, recordings are now more ubiquitous than at any time in the history of sound recording.

The record industry, once on the verge of collapse due to digital file sharing, has found (or perhaps created) a new ally during the past decade, opening its vaults of recorded music to streaming platforms. Together, these industry partners claim to have made virtually every commercial recording made in the past

120 years available at the tap of a finger. Tens of millions of "songs" (a generic term used loosely by streaming services to cover everything from pop songs to symphonies) reside on distant servers waiting to be streamed or downloaded, one at a time or, more likely, subsumed within endless playlists. In this regard, digital distribution may have become a victim of its own recent success: we live within a vast continuum of recorded music, but it is music that may be as much overheard as listened to, and hence, the allegiance of subscribers appears to be already waning as other opportunities for experiencing music begin to emerge after the long hiatus.

Significantly, one of the goals of the present volume, *Analyzing Recorded Music*, is to encourage us to actually *listen* to recordings again—to listen closely to the sounds and spaces created by recordings and to listen outwardly to situate those recordings within broader contexts. This is a laudable goal; and one that is not encouraged by the streaming medium, neither in the proliferation of its algorithm-generated playlists nor in its reliance on MP3 files as the primary vehicle for music consumption (as Jonathan Sterne has argued, from the outset the MP3 format was designed to counterbalance the quality of the listening experience against ease of transmission, 2012). But one of the things we become aware of when we listen carefully to recordings that have been made at different times over the past 50 years or so is that their character is enormously variable, each responding to different aesthetic, generic, historical, and technical possibilities. And at a time when recordings of the past and present are constantly juxtaposed with one another on playlists and in everyday listening, developing an ear for the subtleties found in recorded sound is an important skill.

Indeed, as described by the editors, the initial impetus behind this volume was, in part, to explore the history of popular music by analyzing a series of significant recordings utilizing, in the first instance, a focused mode of listening. Observational, perhaps even empirical in character, the listening approach developed by William Moylan (2020) places musical, timbral, spatial, and temporal elements of a recording within a balanced analytic framework. But while the recorded object is central to the analysis, Moylan's approach is not simply textual in nature: Moylan argues that the interpretation of a recording is informed by a wider engagement with social, historical, technical, and cultural information. Thus, observation quickly opens up to a much broader hermeneutic.

The strength of Moylan's approach is evident in the diverse contributions present in this anthology: each of the essays focuses on a range of material evident both within and beyond the confines of a chosen recording and employs a variety of methodologies in coming to terms with the significance of its sounds. This marks the volume as fundamentally interdisciplinary in character: musical, social, and technical analyses alternate with interviews, ethnographies, and broader theoretical considerations of issues such as gender and race. It seems to me that the scope and diversity of approach found here work against the idea of referring to this observational and interpretive enterprise as a "musicology of record production," a term used by Simon Zagorski-Thomas (2014), among others. While there are, no doubt, legitimate reasons for adopting such an expression, it should be noted that musicology as a discipline has come rather late to considering the significance of sound recording in musical culture: studies by individuals such as Robert Philip (2004) and, more generally, the work associated with the Centre for the History and Analysis of Recorded Music (CHARM), are exceptions, and even in these instances there has been as much emphasis placed on using records as little more than historical documents—as vehicles for studying performance practices of the past—as in addressing sound recordings as musical objects in their own right.

But if we consider the recorded object itself as a point of departure, it is important to recognize that sound recording is not simply something that happens *to* music: within popular music especially, sound recording is a precondition, a significant element in the creation and performance of music from the outset. This point is made abundantly clear in the first essay in this volume: as Tom Attah argues, as early as the 1930s, blues musician Robert Johnson sounded different from his contemporaries because he had already learned to adapt his style of playing and singing to the medium.

The majority of chapters in this volume, however, address recordings that were made under a vastly different set of preconditions than those that existed for Johnson: from the 1960s onward, multitrack recording came to define popular music production, first in the analog realm and then in the digital. The ability to isolate, process, and mix sounds on separate tracks (including even the individual drums and cymbals in a drum kit) made it possible to hear recorded musical performances in a new way kind of way: as a series of events within an enhanced, technologically constructed multidimensional space—a "sound stage." With overdubbing, the musical "performance" itself became part of the construction—a by-product of a technological process—and as pointed out often within this volume, recorded performances thus cannot be taken for granted, they need to be understood as part of an elaborate "staging" process (c.f., Lacasse 2000).

To some extent, the framework for analyzing recordings proposed by Moylan and adopted by others within this volume might also be understood as a by-product of studio recording methods: that is, the observational and interpretive form of analysis employed here has grown out of a profound engagement with sound recording as a technological practice. Moylan's own work as a recording engineer and producer, author, and educator affords him with unique insights into recorded sounds and the techniques utilized in creating them. Similarly, while most of the contributors to this volume are, at least at the present time, academics by profession, many have also worked as professional engineers and producers, guitarists and DJs, and these experiences inform their analyses along with a diverse set of theoretical and cultural concerns.

These diverse experiences and perspectives are matched by an equally diverse (and sometimes eclectic) repertoire of recordings spanning several decades of pop music history. In terms of musical styles and genres alone, it is rare to find soul music, rock and pop, hip-hop, reggae, doom metal, and country music all covered within the same volume. Among artists, there are, of course, some of the usual suspects: there are chapters devoted to well-known recordings by Marvin Gaye, The Beatles, and U2, for example, but in each instance, the analyses reveal new details and offer fresh insights into the music and the recording process. And in the case of Bob Dylan, where one might expect to find an analysis of one of his "classic" recordings of the 1960s, we are presented instead with a detailed discussion of a track released in 2020, the most recent track covered in the entire volume. It is also refreshing to see the number of recordings by female artists (and essays by female musicians, engineers, and scholars) represented within this volume: female artists are seldom given their due in discussions of technology and recording production practices, so it is significant that artists such as The Judds, Nicki Minaj, Anneke van Giersbergen, and others are given serious consideration here and understood in terms of the agency they exhibit both on stage and in the studio.

As befits a collection of essays devoted to music recordings, celebrity artists are not the only individuals to be taken into consideration: the myriad contributions of arrangers and session musicians, engineers, and producers are also addressed with considerable detail in a number of the chapters. This is important, given the commercial imperative to market star performers and the resulting tendency for backing musicians and production personnel to drift in and out of public awareness. Liner notes were one way in which they could achieve some degree of public recognition in the past and, on rare occasions, such as in the case of collaborations between Brian Eno, Daniel Lanois, and U2 during the mid-1980s, Eno and Lanois were given prominent credit on the LP covers, liner sleeves, and on the center labels of the discs themselves. Ironically, in our contemporary, information-rich society, it is probably more difficult for fans to obtain this kind of information as it is not typically included directly in the streaming experience, one has to search elsewhere for it.

This is not to suggest that there is some kind of "golden age" of recording that existed sometime in the past—there isn't one—and neither does this volume present the history of recorded music as in decline nor as a story of endless progress. Indeed, some of the recordings discussed in the volume are not even "definitive" commercial releases: some are demos, remixes, alternate versions, or recordings within music videos. And some are not exclusively studio recordings, *per se*: for example, Steve Waksman's analysis of Prince's "Purple Rain" grapples with the artificial divide between studio and live recording, arguing, ultimately, that studio processes are as much responsible for the staging of "liveness" as the "live" filmed performance itself. And in the case of Mike Alleyne's account of "Concrete Jungle," we have a salient example of how commercial pressures led reggae recordings, initially made in Jamaica, to be adapted and overdubbed by British engineers and producers for an international market. His insights into this process say much about commercial and colonial relationships and find resonance in earlier work by Jocelyne Guilbault (1993), where recordings of zouk, a Creole form of Caribbean popular music, were similarly filtered through the sensibilities of engineers and producers in France to situate them within the market for "world music."

These comments, brief as they may be, attest to the fact that *Analyzing Recorded Music* is a diverse and rich resource. Each recording presented in it is understood within the unique conditions that gave rise to it and within the musical, social, and economic contingencies that influenced its completion. Equally important, the authors attempt to find evidence of these conditions within the musical, sonic, and spatial characteristics of the recordings themselves; and in doing so, they help us to *listen* to recordings in new ways.

Paul Théberge
Carleton University, Ottawa, Canada

References

Guilbault, Jocelyne. 1993. *Zouk: World Music in the West Indies*. Chicago: University of Chicago Press.

Lacasse, Serge. 2000. *Listen to My Voice: The Evocative Power of Vocal Staging in Recorded Rock Music and Other Forms of Vocal Expression*. Doctoral Dissertation, University of Liverpool.

Moylan, William. 2020. *Recording Analysis: How the Record Shapes the Song*. New York: Routledge.

Philip, Robert. 2004. *Early Recordings and Musical Style: Changing Tastes in Instrumental Performance, 1900–1950*. Cambridge: Cambridge University Press.

Sterne, Jonathan. 2012. *MP3: The Meaning of a Format*. Durham, NC: Duke University Press.

Zagorski-Thomas, Simon. 2014. *The Musicology of Record Production*. Cambridge: Cambridge University Press.

DISCLAIMER AND CREDITS

Every effort has been made to contact copyright-holders. Please advise the publisher of any errors or omissions, and these will be corrected in subsequent editions.

INTRODUCTION

Analyzing and Interpreting Recorded Song Tracks

Lori Burns, Mike Alleyne, and William Moylan

This chapter is supported by a webpage.

This volume of essays contributes to a growing body of scholarly work that is dedicated to the study of recorded popular music. The pursuit to understand the creative, technological, and aesthetic aspects of recordings has led to the emergence of a scholarly field of inquiry, for which Simon Zagorski-Thomas has suggested the topic names *Record Production Studies* (Frith and Zagorski-Thomas 2012) and the *Musicology of Record Production* (2014). Inviting the participation of engineers, historians, theorists, and analysts, the field is interdisciplinary in scope. William Moylan has been a leader in this field, with his foundational study of the creative processes that are engaged within *The Art of Recording* (1992), followed by *Understanding and Crafting of the Mix* in 2007 and the recent *Recording Analysis* (2020).

Authors such as Paul Théberge (1997), Mark Cunningham (1998), Mark Katz (2004), and Paul Théberge, Kyle Devine, and Tom Everrett (2015) have adopted historical approaches to record production, while theoretically based and process-based studies of record production are to be found in the studies by Albin Zak (2001), Richard James Burgess (2005), Jay Hodgson (2010), Mark Mynett (2017), and Samantha Bennett (2019). The aesthetic potential of the creative process emerges in these writings, as well as specifically in the analytic writings of Serge Lacasse (2000); Robert Fink, Melinda Latour, and Zachary Wallmark (2018); Zagorski-Thomas, Isakoff, Stévance, and Lacasse (2019); Paul Thompson (2019); and Paula Wolfe (2019). Music scholars study not only the history of the recording process, but also think about ways in which particular records have had an impact upon popular music history (Smith 2009; Zak 2022).

The goal of this collection is to explore "how the record shapes the song" from a variety of diverse perspectives. Collectively, the group of 28 authors represents various specializations with the disciplines of popular music studies and

DOI: 10.4324/9781003089926-1

recording technologies, including specialists in song lyrics, production practice, popular music history, cultural studies, sociology, feminism, and gender theory. We invited authors to choose a specific track for its unique content and significance in the history of recorded popular music. Resulting in a diverse set of musical works from a range of genres and points in history (spanning the years 1936–2020), the analyses presented herein offer different perspectives while weaving common threads from Moylan's analytic framework in *Recording Analysis: How the Record Shapes the Song* (2020). Each chapter illuminates the unique attributes of the individual track under consideration and accounts for how attributes of the recording develop the inherent character and content of the selected song performance.

Moylan's foundational yet malleable framework and principles are meant to illuminate all aspects of a recorded song—a comprehensive set of musical elements and material attributes, from formal song organization to lyrical meanings and musical persona, to structural elements of rhythm, pitch, and harmony, to the content and expression of captured performances, song lyrics, and to stereophonic space, timbre, dynamics, and performance intensity. *Recording Analysis* (2020) establishes an observation- and evaluation-based listening process that encourages analysts to gather a host of analytic data upon which interpretive conclusions can be drawn. Guided by a few core principles, Moylan's analytic method promotes a "frame of mind" for exploring each unique record. These principles assume the equivalence of all ***elements*** within the track, the importance of an ***observational perspective*** that fosters attentive and intentional listening, and the value of drawing ***interpretive conclusions*** based on the thorough gathering of analytic data.

Elements of the Recording

Moylan's framework (2020) for the elements of a recording offers many listening pathways through a recorded popular song. By distinguishing these five primary elements—space, timbre, pitch/frequency, loudness, and rhythm/time—it is possible for an analyst to gather discrete points of musical content and expression that can then become the basis for an interpretation (see Table 0.1).

The recording elements related to *spatial properties* are of particular significance to tracks. Every track establishes its own unique world—its own set of physical relationships between and among sounds—and most of these defy the physics of our natural world. Space is redefined in every track to serve artistic intentions. Stereophonic space allows sound sources to be of apparent size from a single needlepoint in space to a sound that occupies the entire width of the sound stage; further, sounds can be placed anywhere along the width of the sound stage, change locations and become larger or smaller in width. The illusion of distance between the listener's location and the voice and instruments is integral to the track; sounds may be localized anywhere from within the listener's sense of self to an unfathomably far place. Each instrument and voice will have qualities of a

TABLE 0.1 The elements of recording, with a summary of their states and functions at all levels of perspective

Element	*States, manifestations, qualities*
Space	Sound stage dimensions (breadth and depth) Sound source positions and relationships (stereo image, distance) Environments (individual source, sound stage) Ambience, echo, reverberation
Timbre	Timbral balance Pitch density Crafted sound qualities and timbres of instruments and voices Performance intensity, expression, and performance techniques Component parts of timbre/sound quality Timbral qualities of distance cues and environments
Pitch/frequency	Pitch registers and frequency placement Pitch areas and range of sounds/sound sources Pitch density Timbral balance
Loudness	Dynamic contours (overall, individual) Loudness balance of voices and instruments Contour/shape of dynamic and spectral envelopes Dynamic contours within reverberant energy
Rhythm/time	Tempo, meter, and metric durations Patterns of durations (surface rhythms) in all recording elements Environments, timbre, reflections, and density in relation to time units Time patterns in relation to song form

space blended with its inherent timbral qualities; the space or host environment within which the sound source appears adds coloration to the sound as well as a sense of depth and ambience. The multidimensionality of records is further enhanced by the relationships of environments in tracks; the entire track exists within an all-encompassing holistic environment, within which the instruments and voices (situated within their individual host environments) are contained. A sense of place is established for each instrument and voice and for the complete, overall experience of the track.

Timbre manifests on many levels in records. The overall sound (texture) of the track has a timbre that can be explored as "timbral balance." The composite texture can be conceptualized as interacting pitch densities of materials unfolding over time. Timbre at the basic level relates to individual instruments and voices; the timbres of these sources are the result of the performance of the parts as much as the inherent characteristics of the instruments and voices that produced them; this can be explored through performance intensity as well as timbre analysis and typologies. At the immediate level of an individual sound, timbral content can be analyzed for the inner workings of dynamic contour, spectral content, and spectral envelope. At all levels, timbre may be examined and evaluated for both timbral character (what it communicates) and timbral content (its sonic characteristics and components). Concepts of timbre may also be applied to environments within the recording. Further, the level of detail within a source's timbre is central to distance location (from the listener's position) of the instrument or voice.

Pitch/frequency as a recording element differs significantly from its manifestation in music performance, especially since it does not always align with equal temperament. The psychological phenomenon of pitch and the sensations of frequency levels are entwined, with pitch and frequency levels both used in observing this aspect of recording content. Instrumental and vocal lines are reframed as pitch density for understanding timbral balance. Pitch registers function as a means of recognizing and notating these frequency/pitch levels. Registers form a continuum for calculating the bandwidths of pitch areas within sounds and the bandwidths of sound source pitch density boundaries. These attributes of sounds and sound sources are integral to timbral balance and also to the character of individual sounds and lines. Pitch areas and pitch density both establish linkages for observing and examining timbral balance and evaluating the spectral content of timbres.

Loudness also manifests as a recording element at many levels of listening and analysis. Through the dynamic contour of a track's changing loudness levels, the loudness contour often articulates the unique shape of a track and supports its unfolding drama. The impression of each sound's loudness (reflected in the content and character of its timbre) and the resulting loudness balance can have a significant impact on the interpretation of the drama that unfolds between the lead vocal and the ensemble. Loudness is manifest in the dynamic contours of individual sounds and in the loudness contours of individual harmonics and overtones within spectra; further, the changing loudness levels of reverberation and reflection can provide environments with characteristic qualities. In all its apparent simplicity, loudness is perhaps the most often misperceived of the elements; an individual listener's focus of attention can provide sounds with a level of prominence that is very often misunderstood as being loudness-related.

In contrast with musical rhythm, *rhythm/time* appear with different attributes and musical syntax and can interact in interesting ways with other musical elements. Rhythm (metric-related) patterning and durations occur within the materials of each recording element; however, it is important to distinguish the recording treatments of rhythm/time from the surface rhythms of the song. In this regard, it is productive to identify how this element serves to articulate macrorhythmic relationships and to characterize microrhythmic patterns. The recording treatments of rhythm/time tend to reflect qualities that remain consistent within a structural section; shifts between sections (such as verses and choruses) can establish a macrorhythm of unfolding form. Time is important for certain central qualities of recording elements; time durations and relationships (functioning outside the metric grid of meter and tempo) establish important microrhythmic characterizations, such as the patterning of reflections in an early time field or the density of reverb. A microrhythm of similar events progressing too rapidly to be perceived individually can influence a perception of timbre.

Ideal Conditions for an Observational Perspective: The Impact of Playback

Listening is the basis of recording analysis. The playback system of the analyst establishes the sound qualities from which the analysis will emerge, thus, the system itself is part of the analytic methodology. Each variable of the playback system has the potential to shape what is heard within the track (the character and qualities of sounds and their balances). As Moylan states,

> It is possible for even an experienced listener to be fully unaware of the veil of lost detail, of acoustic shadows creating dead zones absent of spatial cues, of unbalanced colorations of spectrum tinting timbral balance (or percepts in other dimensions), of playback formats storing and delivering an incomplete reproduction of the original sound, of unbalanced or mismatched loudspeakers, or a misaligned listening position—and this is just the beginning, as very much more might distort the record.
>
> *Moylan 2020, 220*

The variables of an analyst's playback system might be reduced to several large categories: storage media and formats, audio source reproduction; loudspeakers and power amplifiers; listening room considerations and loudspeaker placements; and loudness level of playback. Each of these playback categories impacts the reproduced sound of the track. Quality of devices and level of performance will vary significantly between various options—as will cost. These are topics worthy of an analyst's attention; one does not need to be an audiophile or recording professional to understand the fundamental concerns of these matters. Putting together a system, setting it up in a room, and calibrating the system are all significant concerns.

It can be tempting to use headphones as a means to simplify a listening experience. Headphone listening, though, is not an equal substitute for the open-air playback system. Headphones and earphones and earbuds produce a different experience entirely; some qualities of the track are physically transformed, and others will be perceived differently with near-ear or in-ear listening. It is important to recognize that "the vast majority of tracks were created (tracked, mixed, assembled, mastered) while listening over loudspeakers" (Moylan 2020, 227).

The content and quality of each listening experience are critical to accurate listening and analysis. The playback system—the analyst's only access to the sounds of the track—might be likened to aural sunglasses, tinting all you hear and coloring all you seek to study and analyze. The issue, then, is the extent to which one's system transforms the sound of a track. A professional quality system is not needed to establish an acceptably accurate listening condition; one can seek a system that produces only minimal changes in the sounds of the track. It is possible to assemble a listening system and physical arrangement within modest financial means and typical household logistical constraints. The goal should be to reproduce the track with as few alterations as is practical for one's situation—and, importantly, to learn the sonic consequences of one's listening devices and conditions.[1]

On the webpage for this introductory chapter, the reader will find further information and directions for establishing accurate and consistent track playback, with guidance and resources for evaluating and calibrating playback systems. While cursory, the provided materials and information will allow the reader a solid introduction to pertinent issues and some resources for exploring their own system. Further research on track playback, listening conditions, and playback systems is encouraged, and links to the web resources of *Recording Analysis* (Moylan 2020) and *Understanding and Crafting the Mix* (Moylan 2015) are included in the endnotes.[2] This involved topic is ever shifting; technologies, equipment, and storage/streaming media are in continual development. This involved topic is ever shifting; technologies, equipment, and storage/streaming media are in continual development.

Leading to Interpretive Conclusions

Once analytic data has been gathered and collated, the interpretive phase allows for a great deal of creativity and thoughtful reflection on the meanings and messages that emerge from the record. Our collective of authors has adopted a variety of perspectives in order to situate their chosen track within cultural, social, political, economic, and historical contexts that are pertinent to the artist/band and song. To this end, the essays in this collection lead to significant interpretive conclusions, from which we can ascertain the following implicit goals that are valued collectively by the scholars within these pages:

- to appreciate the art of song performance and record production by discovering how songs are captured and processed to become the expressive artworks that endure across the decades;
- to hear music in slow motion by presenting written analyses that can be read and received at a rate of reflection and thoughtful understanding of the expressive elements being explored;
- to demonstrate how sonic analysis can be mobilized to reveal historical contexts, cultural messages, and human stories;
- to decenter and decolonize the musical canon by considering the contexts and circumstances that were relevant to the crafting of the record and/or the artist's career within the frameworks of genre and style;
- to reveal how assumptions of creative agency in recorded works are too often restricted by perceptions of gender, race, and place (socially, culturally, and geographically);
- to elucidate relationships between performance, production, and listener reception, demonstrating possible readings/hearings that are rooted in engagements with the texts and their respective surrounding contexts, the latter of which may expand over time.

Mapping the Chapters

The collection is organized into seven sections, based on the following themes:

i. Musical Genre, Culture, and Technology
ii. Track Revelations and Aural Mirrors
iii. Layers Make the Record
iv. Sonic Journeys
v. Sampling and Reframing
vi. Deconstructing the Mix and Production Process
vii. Voicing Identity through Genre

The subsequent paragraphs offer a broad perspective on this gathering of analyses, with the purpose of illustrating how the chapters in each section work together as a whole.

Section I: Musical Genre, Culture, and Technology

The two essays in this opening section apply music analysis to raise questions about the mythologies attached to race and racism in America. Set at the extreme ends of the temporal spectrum for popular music recording (1936 and 2019), the Robert Johnson record, with its rudimentary technology and strong cultural associations that have endured for over 80 years, can be understood as a contextual backdrop for Injury Reserve's track, with its extreme reliance upon digital technologies to convey its cultural and societal commentary. These essays implicitly discuss forms of performative authenticity that have informed many other recordings in the anthology (even as new technologically driven authenticities were established), while others have radically exceeded many of the perceived creative boundaries manifested in vintage blues recordings.

We begin (in Chapter 1) with **Tom Attah**'s analysis of "Preachin' Blues (Up Jumped the Devil)" by Robert Johnson (recorded in 1936). Attah explores the legacy of this recording, reflecting on its cultural impacts and its implications for attributions of authenticity, as well as questions about the treatment of blues as a genre, demythologizing the common beliefs around this lasting musical contribution. Temporal and harmonic content are revealed to animate the organic nature of the music and its function as an invitation to dance. Considering the longevity of this recording across decades of new technology, the author illuminates the sonic and sociocultural texts embedded in monophonic recordings without the spatial expanse or dimensional presence characterizing many works after the late 20th century. While reading this essay, readers can appreciate the depth and layers of musical, cultural, and technological elements that went into its creation as well as its lasting effects. Factors of race arise here and similarly permeate the essays in virtually every section of the anthology, applicable across genres.

We place this foundational blues track into dialogue with contemporary hip-hop practices by chasing it (in Chapter 2) with **Steven Gamble**'s and **Justin Williams**' analysis of a 2019 track by Injury Reserve, "Jailbreak the Tesla (featuring Aminé)." This essay raises (along with others in the anthology) the roles of samples as historical conduits transporting sonic features of the times, places, and conditions within which the source recordings were made, inclusive of the technologies used in their construction. It also expresses ways in which sonic and vehicular technologies (and their associated social and material values) have impacted collective and individual perceptions of popular music recordings. Gamble and Williams carefully break down the mbira sample that opens the track and develop an argument about the sounds of the track as points of reference that interact with the rap lyrics. The opening ostinato pattern has many elements that deserve detailed consideration. We discover here the principle of depth beneath apparent simplicity and are made aware of the risks of reductive listening that leads to unfounded conclusions. Instead, we are pointed to attentive listening linked to the cultural meanings of genre expression. These analytic implications are linked to the cultural implications of racism, specifically linked to the policing of black bodies in America.

Section II: Track Revelations and Aural Mirrors

The gathering of chapters featured in this section explores the significance of the recording contexts for three leading figures in popular music history: Marvin Gaye, Barry White, and Prince. Through these analyses, we reflect on how the record shapes the artist's career as well as how the recording strategies and choices ultimately deliver hit songs that have endured over time.

Andrew Flory's essay (in Chapter 3) on Marvin Gaye's "I Heard It Through the Grapevine" (1968) addresses the transformative, revivifying effect that the success of this track had on the seemingly stalled career of Marvin Gaye, referencing numerous preceding Motown versions that ultimately helped determine the nature of his hit recording. "Grapevine" helped redefine Gaye's popular music presence, reestablishing his solo identity after his series of hit duets with Tammi Terrell. The darker musical tone of the "Grapevine" recording significantly contrasted the generally lighter character of his earlier singles, effectively transitioning his focus toward less idealized and arguably more adult realms of real-world experience, characterized in this case by sinister intimations of the imminent demise of a love relationship. Although Flory does not explicitly state this, his narrative accurately implies that the growth of Marvin Gaye into an iconic solo artist with a unique identity found one of its catalysts in the recording of "Grapevine." While Gaye was neither the writer nor the producer of the record, his vocal interpretation perhaps revealed the possibilities of a more independently shaped creative output, subsequently resulting in records such as the classic album *What's Going On* (1971) and the vastly underrated, compositionally ambitious soundtrack for *Trouble Man* (1972), a release for which Flory wrote detailed liner notes in its 40th Anniversary Deluxe Edition. The essay explores what the recording's other central participants—such as producer and co-writer Norman Whitfield—brought to the process of creating Gaye's version, and Flory describes the song's multiple evolutionary paths in examining the preceding versions by other Motown artists. The presence of multitrack technology and the physical characteristics of the studio space were intrinsic, defining features affecting the mood and texture of the piece, creating a unique aural imprint. The essay establishes that Motown remained entirely uncertain about the commercial viability of Gaye's "Grapevine" recording and that the label's preparation of a stereo mix in an industry era still transitioning from mono created what Flory describes as a "spatial dialogue" between the recorded components.

With his analysis of Barry White's 1976 song, "Let the Music Play" (in Chapter 4), **John Howland** asks how listeners can *hear* a hit track historically, treating this song to a careful analysis of the musical expression and recording qualities while at the same time situating the work in its cultural contexts. In order to address four different versions of the recorded track (with alterations in mixing and editing), this process of historical listening allows for reflections on the song's fit within the industry standard for mainstream disco. Howland's analytic attention to differences in the four versions illuminates the significance of timbre and texture for the hit-making process. The examination of multiple mixes illuminates the sonic aesthetics of the targeted marketing practices being probed by the author. The analysis engages with the multitrack mechanics, editing, and their relationships to the record's various iterations. His analysis also interrogates ways in which the manifestation of recorded sound, especially when heard in a historical context, may be directly informed by commercial considerations resulting in identifiable revisions of structure and form.

Also concerned with the contexts of the recording, **Steve Waksman** (in Chapter 5) addresses the hybrid nature—the interwoven live and studio performances—of Prince's "Purple Rain" (1984). This aspect of the final track draws attention to the intentional construction of the record, encompassing multiple edits and overdubs to ironically achieve what many audiences came to view as the "authentic" collective sonic artifact. Drawing attention to the evidence of a live audience in the recording, the author considers the interplay between the audience sounds, the recorded live guitar, and the post-recording studio elements. In such aural contexts, the performances in the live and studio environments may often imitate each other, with the live rendition attempting to reflect a modicum of studio-like coherence, while the studio overdubs aim to transcend the clinical space in which they are created by injecting the electric energy of spontaneity and response to audience presence into the finished composite work. Waksman also considers how the crossover imperatives of the 1980s—particularly for black artists—informed the sonic contours, dynamics, and subsequent marketability of recorded works. He analyzes structural features such as harmony and phrase as these are projected by the guitar and drums through performative as well as recording strategies. Exploring the significance of this song for Prince's career, Waksman interprets the placement of the song within the narrative of the album *Purple Rain* as well as the film *Purple Rain*. Waksman contributes further to the layers of meaning that emerge from this iconic track by underscoring the racial implications of this song's appeal for both white and black audiences.

Section III: Layers Make the Record

The authors in this section provide thoughtful reflections on the multi-layered content that comprises a recorded track, including musical gestures and formal content, layers of musical and extra-musical meanings, as well as the multiple layers of recorded "tape" that are ultimately mixed and balanced ("bumped down") to the final product.

Mike Alleyne opens this section (in Chapter 6) with his analysis of Seal's debut solo single "Crazy" (1990), focusing on its enhancement in its extended album version by Trevor Horn's cinematic production approach and examining how the recording's multiple spatialities and aural contrasts paradoxically create holistic cohesion. Alleyne emphasizes the accretive role of overdubbing on the record and the degree to which those layering processes help define the overall sonic presence. Both the single and its accompanying album are placed within the context of Seal's total career, thereby highlighting the uniqueness of that initial creative moment and the sonic energies embedded in the recordings. The critical importance of the mix is valorized in the analysis as a vital vehicle for Horn's staging mechanisms and the ways in which they literally dramatize a recording. Although only briefly addressed, the chapter also raises issues surrounding race, market perception, and creative fixity, with Seal representing numerous international Black artists whose material has cross-genre appeal that is not defined by their ethnicity but by their cultural experience and creative synergy with the producer and musicians. The recorded sound communicates with multiple audiences.

Walter Everett and **Katie Kapurch** offer here (in Chapter 7) a multi-layered analysis of an impactful track, "Come Together," from The Beatles' album, *Abbey Road* (1969). The recording process is unveiled through the careful description of the eight layers of recorded tape tracks that served as the raw material for the final mix. The authors also reveal the layers of intertextuality that occur within the song expression, including the influence of Chuck Berry and Muddy Waters, whose musical "voices" are present in Lennon's song design and gestures. Layers of lyrical and cultural meanings are paramount to the analytic emphasis, as the authors reveal the song's polyvalent social and cultural meanings. We are also brought into the stratified layers of melody and harmony as these are conveyed through the instrumental and vocal texture. Its lyrical and instrumental presence ambiguously ranges from the sociopolitical to the sexual, simultaneously revealing and obscuring meaning amidst a maze of both intertwined and seemingly contradictory musical gestures discussed in the chapter. In past and present contexts, it is the record and its post-Beatles interpretations that "Come Together" in the studio and on stage, rather than fulfillment of the song's intimations of global unity. The numerous notable cover versions partially manifest the song's utopian social integration through music, perhaps fractionally counterbalancing the fractured racial and recording industry power dynamics highlighted in discussing the song's roots in a Chuck Berry composition and the overall debt that The Beatles owed to black influences. Ultimately, the song is shown to endure in a unique way through musical versioning, as the covers by several Black American singers are declared to be "reclamations" in response to The Beatles' borrowings from their Black forebearers.

The Beatles' influence is directly connected to the creative eclecticism of the10cc track in "I'm Not in Love" (1975), that is analyzed (in Chapter 8) by **Mark Spicer** and **Stephen Spencer**. The analysis presented here places the band's studio at the center of its creative activity, establishing a rich technological and artistic familiarity with the available tools that culminate in the song expression. The authors explore the record's "harmonic design," demonstrating how its compositional complexities exceed pop song norms. They also highlight the recording's numerous sonic peculiarities, for instance, the exclusion of the drum kit, which is replaced with synthesized percussive punctuations, and the creation of choral vocal layers as integral instrumental components. These are key examples of artistic influence catalyzing creative intuition, consequently resulting in unique soundscapes and applications of technology. Fundamentally, the analysis focuses on how the group harnessed (pre-digital) multitrack studio technology to create one of the most distinctive pop records of all time. Taken from the album *The Original Soundtrack*, "I'm Not in Love" truly conveys the title's audiovisual import and its potential allusions to creative innovation.

Another iconic and multi-layered track is examined (in Chapter 9) by **Timothy Koozin**, whose analysis of U2's "Gloria" (1981) grapples with the sonic challenges posed by 80s arena rock. The author examines the centrality of The Edge's guitar gestures and expressive effects in a contrapuntal relationship with Bono's passionate vocalizations, which demanded clarity of musical gesture in a vast musical space, whether live at the arena or effected through the recorded track. Through this analytic focus on the spatial and temporal relationships between guitar and voice, the reader experiences the gestural tensions and expressive dynamics of a complex musical texture as the song form proceeds. Koozin's analysis engages questions of compositional design and performativity, illuminating the significance of the gestural and sonic elements of a recorded song that conveyed its epic quality into the performing arenas of this hugely influential stadium rock band.

Section IV: Sonic Journeys

The three essays in this section aim to draw the listener into the process of shaping and sculpting a sonic narrative. These authors invite the listener to attend to the particulars of musical nuance as they trace pathways through the selected song tracks. By illuminating the subtlest of details, the authors here are successful in connecting musical expression to the extra-musical narratives on display.

Michael Millner and **William Moylan** focus (in Chapter 10) on a Bob Dylan track, "I've Made Up My Mind to Give Myself to You" (2020), which represents a late stage in the artist's career. Interdisciplinary in scope, this study brings literary interpretation and musical interpretation into a cohesive approach to song expression and meanings. The nuances of Dylan's vocal expression and intonations and the sparse recording's overall timbral and spatial subtleties are investigated and illuminated in an essay demonstrating how the often-elusive element of feeling is embedded within a rendition. Furthermore, the exploration of the record's structure underscores its vital role in communicating meaning in popular song, even in the more recent work of a veteran artist such as Dylan, who, according to the discussion, is still finding fresh means of articulating his subject matter.

Returning to *Abbey Road*, **Albin Zak** takes us on a sonic journey (in Chapter 11) through the medley track, "You Never Give Me Your Money." Focusing on the *Abbey Road* (1969) medley, Zak illustrates how this unusual sequence of musical materials can be understood to convey a narrative—a montage—through motivic connections at the level of musical content and recording processes. Zak's harmonic analysis reveals one of the primary means that musical narrative and coherence operate across the song sequence. However, connection and cohesion also emerge through instrumental and timbral effects that Zak brings forward in some detail. Seemingly disparate in vocal character, Zak examines the vocal expression of Lennon and McCartney to illustrate the journey that the listener travels in attending to this medley. He also directly considers the roles of improved recording studio technologies in adding fresh sonic textures to the mixes compared to previous Beatles albums. As a final consideration, Zak breaks down the stereophonic staging of the song sequence, wherein a dynamic treatment of left and right sonic fields conveys its own dramatic unfolding.

With attentive listening at the forefront, **Jocelyn Neal**'s analysis (in Chapter 12) of The Judds' "John Deere Tractor" (1984) invites readers to discover a narrative of sonic gestures and structures as these are expressed in three different recorded versions. Not remaining solely within the realm of sonic analysis, Neal situates these sounds and structures within larger cultural and music-historical contexts. As she includes the lyrics and their interpretive ambiguities in her analysis of the song, she gives her musical analysis a greater expressive purpose. Consequently, the female protagonist of the song is shown to convey an intimate level of gendered subjectivity that agitates and resonates throughout the song's narrative. Her perceptive comparison to the original song, written by Lawrence John Hammond, illuminates subtle genre attributes and harmonies that Neal suggests are implicitly present in the Judds' version. Herein we can appreciate the incredible power of musical performance to embed historical content and contexts.

Section V: Sampling and Reframing

Sampling embodies sonic paradoxes that only begin with the digital replication of pre-existing recordings. The process (re)consumes produced recorded material while simultaneously subsuming elements of the original creative direction as the "new" artist and/or production team recontextualizes the work to support their own purposes, reframing the sometimes familiar or retrofitting the less familiar to enhance commercial viability. Inevitably, hip-hop, in particular, has faced a range of copyright infringement issues as well as more aesthetic concerns about artistic authenticity. Despite their validity, such considerations often overlook or undervalue sampling's potential for sonic reinvention as tones, timbres, and textures from a variety of sources are often placed in seemingly contradictory collages to facilitate the creation of new aural mosaics. As the essays in this section demonstrate, sampling provides foundational elements, points of sonic departure coexisting with new aural composites to reframe the pre-recorded moments, giving them fresh spatial, rhythmic, and even historical perspectives.

Opening this section, **Michail Exarchos** (in Chapter 13) challenges the scope of hip-hop scholarship and its inadequate relationships to mapping what the author describes as "the material mechanics" engaged in the genre's inherent sonic layering characteristics, with a notable emphasis on spatialities. It sets out to explore the multiple spatial dimensions in hip-hop recording aesthetics with specific reference to MC. Conway's "Biscotti Biscuit" (*Blakk Tape*, 2018) providing figurative food for thought regarding overlooked aspects of sample-based staging techniques. Exarchos indicates that both the work under primary consideration and the genre, in general, are involved in the phonographic reframing of sampled material, often redefining its sonic emphases through additions that affect the nature of the aural gestures and punctuations. The author focuses on the instrumental text, aiming to counterbalance the emphasis in scholarship on the auralities of oralities that often foreground the voice at the expense of its crucial rhythmic accompaniment. In so doing, Exarchos discusses timbral nuances that collectively influence the spatiality and sonic texture of hip-hop works which, as the practitioner author readily recognizes, are typically re-recordings imbued with new life and performative meaning under ideal circumstances. The analysis of "Biscotti Biscuit" is directly informed by clinical structural analysis that reveals both the articulation of hip-hop phonographic authenticities and production idiosyncrasies. The discussion meaningfully demonstrates ways in which the recorded past becomes aurally integrated into the present, aesthetically retrofitted through production purpose and performance rendering. Like most of the anthology's essays, it is only through repeated listening and revisitation of the sonic texts (both sampled sources and reframed recorded expansions) that underlying aesthetic strategies emerge.

Ragnhild Brøvig (in Chapter 14) discusses intersections between improvised application of sampling technology and the aesthetics of specific sampled blues works in Kid Koala's "1 Bit Blues" (2012). Here too, structural analysis helps to reveal key aspects of the compositional and timbral content, thoroughly recontextualizing vintage blues recordings by blending otherwise unrelated works. The author's assessment implies that sonic layering of old and new technology in the Kid Koala recording creates aesthetic peculiarities that only potentially assume anomalous significance through the rigor of recording analysis and the presence of historical contexts. There's a compelling point about the blues as a medium for the incorporation of pre-existing songs and the implication that sampling extends a roughly similar aesthetic approach, although sampling offers and delivers the possibility of exact performative replication, challenging concepts of phonographic authenticity. The author considers the nature of the recording medium as a means of capturing sonic performances, noting that theoretical and listener perspectives sometimes create oppositional recording paradigms. The Kid Koala recording embodies a merger of the traditional performance continuities (structural and sonic) with technologically facilitated means of textual reorganization.

Continuing with sampling issues, **Anne Danielsen** and **Thea Sørli Paulsrud** (in Chapter 15) choose Azealia Banks's "212" (2011) for a study that focuses on the elemental assets of the track. The analysis underscores the intertextuality of modern pop forms, demonstrating how the source influences may become masked by sampling and other means of rhythmic and melodic overlay. The work in question is a result of stylistic fusion, creating an innumerable variety of hybrids resisting straightforward categorization, as theoretically disparate sonic environments collide on a single record informed by audio collage techniques. Consequently, the chapter considers the admixture of Afro-Caribbean (dancehall) and Latin (reggaetón) influences on an otherwise ostensibly typical western pop text. As a secondary point, this analysis also raises the controversial legal potential inherent in replicating unique pre-existing genre rhythm formulae (dancehall reggae) in a purportedly new form (reggaetón), with both styles subsumed within house, rap, and pop frameworks in "212."

Section VI: Deconstructing the Mix and Production Process

Recording engineers listen to tracks with a unique expertise, whether they were involved in the song recording or not. Here we feature three analyses written by expert recordists and producers in order to excavate the technical aspects of recording capture and production treatment. As we are given this access to a specific kind of listening experience, we might be surprised to learn how recordists and producers hear popular recordings.

Samantha Bennett (in Chapter 16) brings her considerable engineering experience to the track "Cirice" (2015) by Swedish metal band Ghost. Her primary goal in analyzing the track is to demonstrate how the technical processes were mobilized to achieve the sonic effects that build the spatial environment and emotional affect of the song. Adopting Moylan's notion of the sound stage, she analyzes the use of depth and breadth to position instrumental gestures and effects. Not offering her analysis merely as dry description, these sonic attributes are always tied to the meaning of the song, as they suggest the particular environments of a cathedral and a storm. Taking those environmental evocations further, she then illustrates how these sonic stagings construct emotional connotations for the listener, linking her interpretation of affect to the song meanings that emerge through the lyrical expression.

Responsible for the production, engineering, mixing, and mastering of the album *The Ghost of Orion* (2020) by British doom-metal band, My Dying Bride, **Mark Mynett** holds a unique perspective on the track he chooses to analyze, "Your Broken Shore" (in Chapter 17). Specifically taking up the challenge of illustrating how the sonic content and expression of doom metal calls for a unique technological response, Mynett explores how he captured and processed the expressive content of guitars, bass, kit, and vocals within the intensely slow tempi and heavily distorted timbres of doom metal. With this analysis, we are drawn into the temporal factors, through attention to sonic attack and decay, as well as spatial aesthetics, as these emerge through room ambience, pitch intensity, and atmospheric effects.

Also responsible for his chosen song study, Taz Vegas's "Am I Dreaming" (2019), **Phil Harding** (in Chapter 18) deconstructs a commissioned remix that he supervised, providing valuable practitioner insights into how and why the sonic transformation of the record was achieved. The distinguished pop producer explores the structural and gestural revision process that emphasizes the remixed record's sonic punctuations that maximize commercial audience impact. Harding's analytical methodology also includes post-production interviews with the client and arrangers/performers on the revamped interpretation. His discussion capitalizes on his enormous global commercial success in the 1980s as part of the prolific PWL production team, illustrating how the focused application of sonic dynamics influences a record's emotional and psychological connectivity with audiences. The chapter, emphasizing the remixer's access to multitracked digitally captured performances, also implicitly highlights a stark contrast with the technically confining immediacy of the recording techniques characterizing the blues works discussed at the beginning of this anthology. Both approaches engage with the musical text in different ways—one by requiring holistic pre-production to be translated into usually unaltered performance, while the other invites post-production application to recontextualize existing performances with latitude to add parts and alter structure, thereby affecting audience experience.

Section VII: Voicing Identity through Genre

Is there space within genre-based forms for musical artists to extend boundaries and carve out new modalities of musical subjectivity and identity? The authors in this section would answer with a resounding affirmative response to this question. Their song analyses operate at the edges of genre exploration, with examples that ask us to reflect on the meanings of metal, country, pop, rock, and reggae. As genres evolve, they must and do diversify to expand notions of inclusion and break down hegemonic barriers. All authors here demonstrate that the production features of the recording are essential to the communication of the cultural negotiations on display.

Lori Burns (in Chapter 19) places the focus on Anneke van Giersbergen, a female metal and progressive rock vocalist who has made a significant contribution to the doom metal scene and to progressive rock and metal music over the course of her 30-year career. Although never part of the popular mainstream, The Gathering holds a significant place in metal music history as a doom metal band that featured a strong female vocal presence, carving through the gloomy effects of heavy guitars and kit. With the track "Strange Machines" (1995) from her debut album with the band, Anneke van Giersbergen stands out in the history of female-fronted metal for offering a new way to hear female vocals: not vulnerable and ethereal in her delivery and pitch register, Van Giersbergen's voice carves out a new space for the female voice and thus influenced female vocalists from the late 1990s to the present. Taking a holistic approach to metal music composition and performance expression, Burns explores the sonic parameters (registral, spatial elements, instrumental effects) and structural parameters of the song (overall form, riff and phrase structure, modality, and harmonic progression). Placing the sonic and structural (compositional) parameters into dialogue, what emerges is an exploration of how these parameters are mutually shaped and staged, with particular attention to Van Giersbergen's role in leading the song.

Stan Hawkins and **Zack Bresler** collaborate (in Chapter 20) to explore Orville Peck's "Hope to Die" (2019). The authors explore challenges to the country music genre through aesthetics, production, and musical expression in Orville Peck's country-pop track, "Hope to Die." Showing Peck to be a queer icon, they analyze sound and image in the official music video, revealing how details of the musical performance and sound production contribute to his signature style. That style is presented in the context of popular artists and is shown to resignify country music traditions. Taking up the elements of style and technique, the authors consider the parameters of structure, space, rhythm, timbre, and production, finding compositional and production nuances throughout that show Peck to be manipulating conventions and disrupting expectations. Turning to the music video, musical elements are enhanced by visual developments and levels of intensity. The queering of the audiovisual content is illuminated through intertextual references and specific choreographic and vogueing cues. Ultimately, the song/video is revealed to be a multidimensional, multimodal, and highly referential cultural expression that impacts substantially upon genre identity.

Genre identity is a significant aspect of **Jada Watson**'s study (in Chapter 21) of Mickey Guyton's 2020 song, "What Are You Going to Tell Her." Watson receives this song as an emotional appeal to the country music industry, calling for a response to the marginalizing practices that exclude women and, more specifically, black women. Identifying the song as an instance of a feminist "breaking point," Watson adopts an ethnographic as well as analytic approach. Interviews with the artist reveal multiple contexts for the writing and performing of this song, and industry data show how Guyton's career evolved in Nashville. The theme of "breaking point" emerges clearly in Watson's account of Guyton's impetus for writing a song about the fraught legacy that professional songwriters, operating within this industry, will leave for their daughters. This breakdown transfers into a musical expression that Watson explores through a range of analytic parameters, but especially through her finely crafted attention to the gestures, recording capture, and production of Guyton's voice. Illustrating clearly the communication of emotion in this song, the author reveals remarkably how genre is a powerful force to navigate in musical culture.

With the aim of illustrating how a rapper can draw upon both rap and pop conventions in ways that challenge performance and compositional norms, **Olivia Lucas** (in Chapter 22) analyzes Nicki Minaj's song, "Stupid Hoe" (2011). Presenting clearly the formal, instrumental, and vocal elements of this song, Lucas brings forward a myriad of ways in which Minaj resists and rejects genre conventions. A subtle sensitivity to stylistic elements allows Lucas to illustrate Minaj's transformation of the "diss track" through integration with the genre of bounce. The analysis ultimately focuses on Minaj's vocal flow and the performative implications of her recording, harnessing what Lucas describes as "sonic brashness" coupled with numerous structural ambiguities and the fluid dynamics of the instrumentation.

To close the volume, **Mike Alleyne** (in Chapter 23) analyzes the Wailers' "Concrete Jungle" (1973). Alleyne explores the recording's sound sources with direct reference to the studio track sheet to demonstrate how it was crafted to enhance its commercial appeal to rock audiences. Detail and perspective are further supported by two interviews with Tony Platt, the overdub and mixing engineer, underlining the creative and commercial realities informing the recording process. Essential historical contexts help to explain how some aspects of the recording's sound crystallized as expressions of postcolonial conditions, stylistically and philosophically resonating into the 21st century. The precise commercial thinking at Island Records is outlined as a foundation shaping the sonic transformational exercise of overdubbing the original 4- and 8-track recordings from Jamaica on 16 tracks in London in 1972. Overall, the discussion points toward a blend of both calculated and more organic circumstances creating one of reggae's most memorable, influential, and historically significant recordings.

Although this collection focuses on specific recordings, there are also fascinating record production connections between individual examples and across genre boundaries. For instance, one studio space discretely connects two records of acutely contrasting sonic identity discussed in this anthology. Parts of both The Wailers (Ch. 23) and Seal (Ch. 6) recordings took place in Island's Basing Street complex in West London, subsequently taken over by Trevor Horn and renamed SARM West. The now converted building location is currently designated as SARM Residences and the multiple apartment structure now bears musical remnants of its creative past including framed album covers of commercially successful and influential releases crafted on the site, including both The Wailers and Seal recordings. As an additional relationship, Island Records and ZTT co-owner, producer Trevor Horn briefly conducted a successful joint label operation in the 1980s that included Grace Jones's *Slave to the Rhythm* album (1985) (Blackwell and Morley 2022, 258) that is referenced in Ch. 6. Such multilayered and sometimes unseen linkages related to both sonic production and historical contexts connect many of the analyzed recorded works in this volume.

Concluding Remarks

As the chapters unfold, readers will encounter analyses of tracks representing a range of musical styles, genres, and production techniques. Authors present different vantage points as they explore the content of their chosen recording, while the discussion of unreleased and/or alternative versions permeates several chapters. Readers will be able to observe a great variety of approaches in how each author uniquely engages the principles and framework of *Recording Analysis,* and we hope that the reader will find such diversity informative. Ultimately, the goal of each chapter is to directly engage the qualities of the recording. How recording elements shape and contribute to the character and content of these tracks is explored in a substantive manner, adjusted according to each author's perspective and to the unique qualities of each individual track. In effect, each author is creating his or her own analytical mix, applying forms of critical attenuation to reveal the dimensions that they have found most compelling and noteworthy.

Notes

1 Adapted from *Recording Analysis* (Moylan 2020) website, page: https://routledgetextbooks.com/textbooks/9781138667068/playback_system.php.
2 For further information on establishing accurate playback, see *Recording Analysis* (Moylan 2020, 4–5, 223–29) and its companion website, pages: https://routledgetextbooks.com/textbooks/9781138667068/default.php and https://routledgetextbooks.com/textbooks/9781138667068/playback_system.php; and for evaluating and calibrating systems, see *Understanding and Crafting the Mix,* 3rd ed. (Moylan 2015, xxxix–xl, 360–72, 437–38) and in the companion website, pages: https://routledgetextbooks.com/textbooks/9780415842815/default.php and https://routledgetextbooks.com/textbooks/9780415842815/instruction.php.

References

Bennett, Samantha. 2019. *Modern Records, Maverick Methods: Technology and Process in Popular Music Record Production, 1978–2000.* New York: Bloomsbury Academic, Bloomsbury Publishing.

Burgess, Richard James. 2005. *The Art of Music Production.* London: Omnibus Press.

Cunningham, Mark. 1998. *Good Vibrations: A History of Record Production*. London: Sanctuary Publishing Ltd.

Fink, Robert Wallace, Melinda Latour, and Zachary Wallmark, eds. 2018. *The Relentless Pursuit of Tone: Timbre in Popular Music*. New York: Oxford University Press.

Frith, Simon, and Simon Zagorski-Thomas, eds. 2012. *The Art of Record Production: An Introductory Reader for a New Academic Field*. Farnham: Ashgate Publishing.

Hodgson, Jay. 2010. *Understanding Records: A Field Guide to Recording Practice*. New York: Continuum.

Lacasse, Serge. 2000. "'Listen to my Voice': The Evocative Power of Vocal Staging in Recorded Rock Music and Other Forms of Vocal Expression." PhD diss., University of Liverpool.

Katz, Mark. 2004. *Capturing Sound: How Technology Has Changed Music*. Berkeley: University of California Press.

Moylan, William. 1992. *The Art of Recording: The Creative Resources of Music Production and Audio*. New York: Van Nostrand Reinhold.

———. 2015. *Understanding and Crafting the Mix: The Art of Recording*, 3rd edition. Burlington: Focal Press.

———. 2020. *Recording Analysis, How the Record Shapes the Song*. New York: Routledge, Focal Press.

Mynett, Mark. 2017. *Metal Music Manual: Producing, Engineering, Mixing, and Mastering Contemporary Heavy Music*. New York: Routledge.

Smith, Chris. 2009. *101 Albums That Changed Popular Music*. Oxford: Oxford University Press.

Théberge, Paul. 1997. *Any Sound You Can Imagine: Making Music/Consuming Technology*. Hanover: University Press of New England.

Théberge, Paul, Kyle Devine, and Tom Everrett, eds. 2015. *Living Stereo: Histories and Cultures of Multichannel Sound*. New York: Bloomsbury Academic.

Thompson, Paul. 2019. *Creativity in the Recording Studio: Alternative Takes*. Cham: Palgrave Macmillan.

Wolfe, Paula. 2019. *Women in the Studio: Creativity, Control and Gender in Popular Music Production*. Abingdon: Routledge.

Zagorski-Thomas, Simon. 2014. *The Musicology of Record Production*. Cambridge: Cambridge University Press.

Zagorski-Thomas, Simon, Katia Isakoff, Sophie Stévance, and Serge Lacasse, eds. 2019. *The Art of Record Production: Creative Practice in the Studio, Vol. 2*. Abingdon: Routledge.

Zak, Albin. 2001. *The Poetics of Rock Cutting Tracks, Making Records*. Berkeley: University of California Press.

———. 2022. *Rock on Record*. Albany: State University of New York Press.

SECTION I

Musical Genre, Culture, and Technology

1

"I BEEN STUDYING RAIN"

What do We Hear When We Listen to the Blues of Robert Johnson? ("Preachin' Blues," 1936)

Tom Attah

Introduction

The track to be analyzed is "Preachin' Blues (Up Jumped the Devil)" performed by Robert Johnson, recorded in the improvised recording studio of room 414 of the Gunter Hotel, San Antonio, Texas, on Friday November 27, 1936 (Conforth and Wardlow 2019, 181). Recordists were Vincent Liebler engineering for the American Recording Company (ARC), with Don Law producing. Capture equipment for this original session was a single microphone feeding a portable record lathe which cut Johnson's solo guitar and vocal blues performance direct to acetate.

This essay considers the music, recording, and lyrics of the track alongside its perceived emotional, energetic, and expressive elements in order to explore what is represented by this performance and its performer in successive eras of contemporary popular music culture. Specifically, in what ways does this recording enable and constrain notions of authenticity, race, creativity, and myth-making within the discourse on popular music? How has this recording transcended its origins and several rounds of technological mediation? In eras of socio-economic crisis and Civil Rights activism, what particular truths do creators, consumers, and cultural workers seek—and find—in this recorded text?

The analysis considers the ways that the original mono recording affects listeners' perception of the performance's rhythms and notes that while Johnson's emotional intensity is without question, the clarity of some of his words is lost to time—particularly the phrase from the 4th verse which is often parsed as "I been studyin' rain." The critical perspective does not offer any new, hitherto undiscovered information concerning the recording sessions or the life of Robert Johnson. This chapter is concerned with a brief analysis of the music, lyrics, and recording process of "Preachin' Blues (Up Jumped the Devil)" based on the

DOI: 10.4324/9781003089926-3

framework described by Moylan (2020). The analysis is then used to support a discussion concerning the ways that these elements have combined to shape the song and its position in popular culture since the song's initial release in 1939.

The recording used for this discussion is from *The Centennial Collection* (2011), remastered from acetates and the best-quality existing original pressings and issued by Legacy/Sony BMG as part of a boxed set. Playback equipment for this analysis is an SSL AWS924 recording console feeding a pair of Genelec 1237A speakers and a pair of Avantone studio monitors in the context of an isolated recording studio console room.

Robert Johnson and the Blues

Robert Johnson (1911–1938) was a pre-war black American singer and guitarist whose recorded works, biography, and iconography have exerted a significant cultural influence on the development and marketing of Western popular music, particularly guitar-based meta-genres of blues and rock & roll (Hayne, Mee, and Rumble 2005). There are two persistent cultural controversies concerning the artist's life and work. The first is that he reputedly "… sold his soul to the devil" at a rural crossroads in return for his prodigious talent, a fact supported in myth by his death at the age of 27 through poisoning. The second is that his recordings had their speed mechanically accelerated by ARC prior to their original release (Asante 2010; Wilde 2010), although this claim has been robustly refuted by blues scholars (Wald 2010b). Both of these controversies point to the subaltern position of black Americans during the 1930s and have the potential to impact how we hear and understand Johnson's recorded blues.

The definition of "blues music" is fluid and contested. In other words, "blues" in reference to music and culture means different things to different people at different times and in different places. For some scholars, the blues is a site of ideological incorporation and resistance (Baker 1984; Jones 1963), yet for others, including musicians and cultural workers, blues is a purely commercial popular musical enterprise (Grazian 2004; King 2011; Ryan 2011). Certain researchers and writers regard the blues as an essential musicological forebear of Jazz (Blesh 1946; Murray 1978), and others variously consider blues as: socio-political evidence of the existence of the negro proletariat (Gellert and Siegmeister 1936); a barometer of and map for the changing fortunes of black Americans through the 20th century (Barlow 1989; Levine 2007); the wellspring of American popular musical identity and practice during the 20th century (Small 1987); a revisionist hegemonic construction (Filene 2000; Hamilton 2007); and a site of myth-making (Morrison 1992; Schroeder 2004). Additionally, for some, the blues represent "… the strangest, most moving, and most thrilling songs I had ever heard" (Oliver 1989, 59) and, for others, simply "… good-time dance music … group and joy music" (Shaw 1979, xvi). With the above in mind, it may be that the definition of blues is itself a matter of discourse (Förnas 1995).

Muir (2010, 1) identifies the music that emerges from "… the black subculture … into the American mainstream in the second decade of the twentieth century" as "popular blues," while Davis (1995, 59) identifies music from the same period as "… blues-in-name-only." Similarly, in a 2013 survey concerning "the relevance of blues for a modern audience,"[1] a researcher listed no fewer than 34 different styles of blues, including acoustic blues, blues-rock, blues shouter, British blues, Chicago blues, classic female blues, contemporary blues, country blues, Delta blues, Detroit blues, electric blues, harmonica blues, hill country blues, hokum blues, jazz blues, jump blues, piano blues, Piedmont blues, rhythm & blues, slide guitar blues, soul blues, swamp blues, Texas blues—and still added a box for "other"; in the early 1960s, Charles Keil referred to "phony" blues (Keil 1966, 221).

The aim of the above examples is to illustrate the multiplicity of definitions that have constituted the blues from various cultural perspectives. These definitions do not encroach on the purely musicological, which typically include reference to: numbers of bars/measures; tempo; chord types; the use of I7–IV7–V7 chords; blues form; blues scales; cyclic form; A-A-B lyric verse patterns and triplet swing. Reference to these elements prevails in very many blues texts (e.g. Barlow 1989, 8; Davis 1995, 4; Gioia 2008, 14; Moore 2012, 126–127), but strict adherence to "twelve-bar-blues" or "the tradition of call-and-response," (Wald 2010a, 3–4) as a marker for what constitutes a piece of blues music excludes almost as many compositions as it embraces. To further complicate matters, from a modern, capitalist perspective, "blues" has also been described as "… a broad range of popular music primarily created by and for black Americans … thus it is more a trade category than a genre" (Ripani 2006, 6). In summary, diverse and competing views of what-counts-as-blues are facilitated and exacerbated by technological mediation and discourse. Thus, in agreement with blues scholar Elijah Wald, "… I have no problem with people using whatever definition [of blues] they like, as long as they grant that it is not the only one" (Wald 2004, 4).

Initially marketed at the black community, blues music was disseminated by public performance and subsequently sheet music, waxed cylinders, and phonograph recordings (Muir 2010, 15). Katz (2010, 9–46) observes that capturing sound in media brings "phonograph effects," which include tangibility, portability, invisibility, repeatability, temporality, receptivity, and manipulability. In other words, as soon as a recording is made, it exists as an artifact which may be commodified beyond the original performance and may be touched, held, stored, traded, and collected in a way that an ephemeral performance may not; it may be moved beyond its point of origin to other locations in space and time; early audio recordings made no concession to visual representation beyond the packaging of the record and so the performer was largely invisible; the recording could be played repeatedly with the only limitation being the condition of the media under repeated use; the media placed a limit on the amount of time that could be recorded—early discs had a recording time of a little over four minutes per side (Katz 2010, 31); early limitations of the recording process and format affected which sounds could and could not be accurately represented; and once captured, it is possible to alter recordings with post-production effects and by adding or correcting musical parts. Rothenbuhler (2007) discusses phonograph effects in specific relation to blues music. Discussing the work of blues musician Robert Johnson, he writes:

> Johnson learned his craft from records and radio, and polished his songs to be recorded, he effectively developed a 'for-the-record' aesthetic that made his music sound different to that of his Delta contemporaries and many others who used musical techniques honed in performance for an audience. Decades later, when a 'for-the-record' aesthetic was the taken-for-granted standard in popular musical culture, Robert Johnson's records sounded better than those of his contemporaries, and the audience from the 1960s to today has had a reason to think that he and his music were special.
>
> *Rothenbuhler 2007*

This locates the musician's work firmly in the era of phonographic modernity.

For many blues fans and scholars during the 1960s blues revival, their first contact and most enduring relationship with the music was through long-playing (LP) phonograph records (Schwartz 2007) which included the work of Robert Johnson. The collection and trade of these discs afforded the recordings a certain value beyond performance, as well as bestowing status on the collector as an expert (Katz 2010, 171). The collation of complete discographies for certain artists and recording companies was an early and intense project for enthusiasts, first in the United Kingdom and then in the United States (Titon 1993, 226), while other collectors went to extreme—in many cases, expensive—lengths in order to acquire specific examples of blues music on record (Wardlow and Komara 1998). In an environment where not all available material was recorded, and not all recorded material was released (Sanjek 2009), some recordings acquired mythical status. In addition, physical limitations on the pressing and storage of vinyl records meant that, in many cases, certain titles were only available for a short time, increasing their rarity value in later years.

Few histories of the blues in the 20th century omit mention of the alleged Faustian pact in which Robert Johnson received supernatural musical gifts in return for the sale of his soul. The perpetuation of this myth illustrates the learning and enculturation process and its intersection with technology in blues music and blues culture. There is also a racial dimension to the retelling of this story. The implication is that a black man may not improve his talent, skills, and ability purely by application, concentration, and practice without the influence of some external, supernatural force. According to Eddie "Son" House (1902–1988), himself a blues artist of stature, Johnson would regularly attend the dances at which he and his fellow guitarist and singer Willie Brown (1900–1952) performed. Johnson was at this time in his mid-to-late teens and

> … could blow harmonica real good. Everybody liked it. But he just got the idea that he wanted to play guitar … He used to sit down between me and Willie. And we'd be sitting about this distance apart, and [Robert] would come and sit right on the floor, with his legs up like that, between us.

> So when we'd get to a rest period or something, we'd set the guitars up and go out – it would be hot I the summertime, so we'd go out and get in the cool, cool off some. While we're out, Robert, he'd get the guitar and go bamming with it, you know? Just keeping noise, and the people didn't like that. They'd come out and they'd tell us, "Why don't you or Willie or one go in there and stop that boy? He's driving everybody nuts."
>
> I'd go in there and I'd say "Robert, don't do that, you'll drive the people home." I'd say "… just leave the guitars alone."
>
> But as quick as we're out there again, and get to laughing and talking and drinking, here we'd hear the guitar again, making all kind of tunes "BLOO-WAH, BOOM-WAH" – a dog wouldn't want to hear it!
>
> *House, quoted in Wald 2004, 109–110*

Here, House reports evidence of the young Johnson using imitation to learn (Green 2001; Merriam 1964). Similarly, House alludes to Johnson's harmonica playing and indicates that the younger musician received praise from his community for his performance on this instrument, in line with Merriam's comments concerning social acceptance in which "… the individual must learn behaviour which is specified in the culture as being correct or best" (Merriam 1964). The learning musician is then rewarded by the community with approval and/or valorization. Additionally, Johnson receives "… scolding, warning, ridiculing, [and] the use of sarcasm" (Merriam 1964, 151) at the hands of House and the audience for a performance which is not deemed satisfactory or conducive to the social context.

Further brief examination of the Johnson narrative provides evidence of education, schooling, and mentorship. Following the death of his wife in childbirth in 1930, Johnson left Robinsonville, where he had encountered House and Brown. Wald (2004) and Rothenbuhler (2007) each report that Johnson spent the following two years in Hazlehurst, where he encountered guitarist Ike Zimmerman—a less well-known figure in the Delta who never recorded for a label but who "… claimed to have learned to play by visiting graveyards at midnight" (Palmer 1981, 113). Zimmerman reportedly guided Johnson's learning throughout this period. Additionally, at this time, Johnson "… must have had access to Victrolas … on which he could closely listen to his contemporaries on records" (Wardlow and Komara 1998, 199). Specifically, many of his musical influences "… came from phonograph records, and during the next few years, while Robert was cementing his reputation as the most formidable young bluesmen in the Delta, he borrowed heavily from recordings by … Kokomo Arnold, Peetie Wheatstraw, and Skip James" (Palmer 1981, 114).

Johnson, therefore, had primary oral and mimetic 1:1 teaching from Zimmerman, as well as using secondary oral materials available via new media in the form of radio and 78 rpm records. This facilitated—but did not *determine*—the development of Johnson's unique style, which is not formed solely by local influences but draws on sources outside the region which are clearly audible when comparing his *Sweet Home Chicago* (Johnson 1990b) with Kokomo Arnold's "Old Original Kokomo Blues" (Decca 7026, 1934), Charlie McCoy's "Baltimore Blues" (Decca 7007, 1934)—or indeed, any of the recordings included on 2007's *The Road to Robert Johnson and Beyond* CD (JSP 7795, 2007) with the 29 pieces available on *Robert Johnson—The Complete Recordings* (Johnson 1990a). Then, as now, it appears that talent borrows, genius steals (Eliot 1920). As Rothenbuhler (2007) also indicates, as Johnson learned, in part, from phonograph records, this also influenced his own compositions in terms of song structure, length, and content. In other words, Johnson's work featured a for-the-record aesthetic rather than the open-ended improvisatory form which characterized many early Delta Blues recordings. For Rothenbuhler, this standardization of form in line with the affordances and limitations of 78 rpm media contributed to Johnson's accessibility and musical intelligibility to a secondary audience separated from his original context by time and space—a secondary audience which included British Blues Boom artists such as Eric Clapton and Keith Richards (Clapton and Sykes 2008; Richards and Fox 2011).

Johnson's narrative illustrates several ideas proposed by Merriam and Herskovits with regard to enculturation, education, and schooling. Further, even if the logical confluence of Johnson's talent and technology is abandoned in favor of the "crossroads" myth—there is still evidence of mentoring and tutelage in the form of the "big black man"—who may be the devil, the Yoruba God Orisha, or the voodoo intermediary Papa Legba—interacting with the supplicant Johnson at the crossroads, and on whom supernatural skills are conferred.

In summary, Johnson is significant to popular music culture for several reasons. His performance and recording practice make manifest the influence of recording technology in terms of his repertoire, which draws on recorded sources for influence. In turn, his own compositions are commodified for distribution and consumption beyond their point of origin, and these exert influence over popular musicians and critics who are attracted to the style. Simply put, Johnson is one of the major artists who represents the bridge between the oral and technological transmission of blues music and blues culture, particularly in the way that (a) he is influenced by phonograph records and passes on his own influence through the same media and (b) he was valorized and mythologized by a secondary audience decades after his passing in 1938. Further, in terms of the "… sold his soul at the crossroads" myth, it is possible to trace the racial tension in the hegemonic relationship between black Americans and the dominant white culture.

As indicated above, Johnson was an aspiring guitar player who suffered a traumatic life event in the shape of the death of his young family. He then spent time being tutored by Ike Zimmerman and, with access to phonograph records, spent two years developing his musicianship. In short, he put in the hours and learned to play the guitar. In Western culture, written references to a deal with the devil in return for diabolical favors are found from the 6th century; Theophilus of Ardana apparently sold his soul for ecclesiastical power. Christopher Marlowe's 1604 play Doctor Faustus also features this central plot point, and 19th-century classical violinist Niccolò Paganini is subject to the same claim. The repetition and persistence of the story indicate that for some, it is easier to believe that a black American from Johnson's time could only acquire his incredible talent via a deal with the devil, rather than the more prosaic option that Johnson sat with widely available phonograph records over many hours patiently practicing and honing his craft. This version of the story is rather less marketable—and preserves the essentially racist idea that black Americans such as Johnson have only a limited cognitive ability and are unable to develop skills over time without the intervention of the supernatural.

The final bond between Johnson and the phonograph record in popular culture comes with the *From Spirituals to Swing* concert organized by record producer John Hammond in December 1938. Hammond had originally intended to include Johnson in the concert held at New York's prestigious Carnegie Hall, which would have cemented Johnson's position as a living exponent of the delta blues before a white audience, an early example of a "crossover" artist (Dugan and Hammond 1974). However, as Hammond wrote in the program notes for the concert:

> Robert Johnson was going to be the big surprise of the evening for this audience at Carnegie Hall. I knew him only from his Vocalion blues records and from the tall, exciting tales the recording engineers and supervisors used to bring about him from the improvised studios in Dallas and San Antonio. I don't believe Johnson ever worked as a professional musician anywhere, and it still knocks me over when I think of how lucky it is that a talent like his ever found its way onto phonograph records. [Tonight] we will have to be content with playing two of his records, the old *Walkin' Blues*, and the new unreleased *Preachin' Blues*, because Robert Johnson died last week at the precise moment when Vocalion scouts reached him and told him that he was booked to appear at Carnegie Hall on December 23. He was in his middle twenties and nobody seems to know what caused his death.
>
> *Hammond, quoted in Conforth and Wardlow 2019, 264*

Here, we see that Johnson's first performance to a mainstream crossover audience is not made in person but posthumously and mediated by a phonograph record. Additionally, Hammond's own myth-making concerning the uncertain death of the young musician lays foundational stones in the legend of the artist. In other words, the very moment Johnson steps onto the national stage and into the public consciousness—he is a phantom.

The Recording

Significant value is placed on perceived authenticity in blues music and blues culture. While it is beyond the scope of this essay to explore connotations regarding content, context, race, and gender in the performance of popular music in general and the blues in particular, "... the creation of the *effect* of authenticity ... [as] a matter of culturally determined convention, not an expression of essence" (Auslander 1998, 6 italics mine) is of importance when considering the semiotic resonance of a black man singing and playing the blues in pre-war America. As Allan F. Moore indicates, "… [a]uthenticity' is a matter of interpretation which is made and fought for from within a cultural, and thus, historicised position. It is ascribed, not inscribed" (Moore 2002, 210). With this in mind, we now turn to an examination of the music, lyrics, performance, and recording of the reference track.

Capture and Release

"Preachin' Blues (Up Jumped the Devil)" was recorded on Friday November 27, 1936. The environment was an improvised studio set up in room 414 of the Gunter Hotel, San Antonio, Texas. Johnson was not the only musician time-tabled to be recording, and his sessions were spread through the week from the preceding Monday (Miller 2010, 317). Despite the apparent risk of ambient sounds made by other guests at the hotel being captured alongside the performance, only Johnson's voice and guitar are audible on the recording. The session was arranged by Art Satherley for ARC from New York. The recording supervisor in the room was Don Law, and the engineer was Vincent Liebler. Capture was a single microphone set in front of the performer. Signal from the microphone was then sent to the cutting head of a portable recording lathe which was located in an adjoining room. The sound was then direct cut directly to an acetate blank, producing the "master" recording. The specific cutting machine is unknown. A small picture article posted in the San Antonio Light on November 11, 1936, shows Vincent Liebler "… inspecting a new thin type of master-record being used for recordings … for the first time in the United States." The scene has been notably depicted twice in popular cultural history. Firstly, the scene is illustrated by Don Wilson on the sleeve of *King of the Delta Blues Singers Volume II*, where Johnson is shown performing into the microphone placed in the corner of the room with his back to the viewer above the caption "Robert Johnson first records in a makeshift studio in a San Antonio hotel room – November 1936." On the back of the album, Law and Liebler are illustrated monitoring the lathe in the adjoining room. Secondly, the scene is recreated in the opening minutes of the *Crossroads* motion picture (dir. Hill 1986); Johnson (played by Tim Russ) is met at the hotel room by Don Law (actor uncredited), and the session for "Crossroad Blues" commences while Vincent Liebler (actor uncredited) operates the lathe, clearing acetate cuttings from the blank as it turns.

For many years, researchers believed that the microphone was set in the corner of the room and that Johnson played into the corner, effectively facing the wall (Clapton and Sykes 2008, 40). This notion is repeated in the two illustrations of the session mentioned above. Several theories for this positioning were suggested over time: (i) that Johnson was shy and wished to perform without distraction, (ii) that he was "corner loading" the sound—playing into a confined space to enhance the pressure of soundwaves being reflected, (iii) that he was hiding his performance techniques from other musicians who may have been observing the session. As indicated above, this myth was repeated and passed into cultural history. This long-standing notion has since been attributed to a misunderstanding, reinforced by films and documentaries, and Don Law subsequently confirmed that Johnson, in fact, performed with the microphone in the center of the room (Conforth and Wardlow 2019, 158–160).

Two takes of the song were recorded. This was standard practice at the time, allowing for redundancy in the capture, given the fragility of the recording media. Because there was no facility for overdubbing and correcting minor technical errors, performances would have to be played and sung correctly at each recording pass. This also meant that the recording process could be taxing for the performer both physically and mentally. While the artist is clearly acknowledged in the sessions, recordist Vincent Liebler is of equal importance in the session since, as we acknowledge—the recording shapes the sound. At the time of recording, Liebler had built up experience within the Brunswick record company prior to its purchase by Columbia. Writing about Liebler, Susan Schmidt-Horning comments that:

> Like the recording artist, the recordist either had it, or didn't; even experienced recorders might lack … the necessary knack of recording. That knack involved quick thinking, patience with often temperamental people, and the ability to remain unflappable in the face of trying situations, particularly while on location.
>
> *Schmidt-Horning 2013, 24*

While Johnson's recordings made history, it is Liebler's ability to capture the sound that is of equal value. As Schmidt-Horning indicates, "… these men incorporated intuitive knowledge, craft skills, musical sensibility, engineering expertise, and the ability to adapt to unforeseen circumstances" (Schmidt-Horning 2013, 8). As with recordings made using equipment of this type from this time, the primary challenges with the recording are speed, equalization, and surface noise.

The first take, ARC SA2632-1, is the version that was retained for the issue. ARC SA2632-2 was not issued and is assumed to be lost. Table 1.1 indicates the major releases of take SA2632-1.

TABLE 1.1 Releases of "Preachin' blues (Up Jumped the Devil)"

Year	*Label*	*Issue title*	*Format*	*Catalog #*
1939	Vocalion	*Preachin' Blues "Up Jumped the Devil"*	10″ 78 rpm	04630
1959	Folkways	*The Country Blues*	LP	RF1
1961	Columbia	*King of the Delta Blues Singers*	LP	CL1654
1970	Columbia	*King of the Delta Blues Singers Vol. II*	LP	30034
1990	CBS/Sony	*The Complete Recordings*	CD, LP, Cassette	CK46222
2011	Legacy/Sony	*The Centennial Collection*	CD, LP	88985374761

TABLE 1.2 Differences between musical transcriptions

	Mann	*Billmann*
Transcription	Intro + first verse (0:00–0:47)	Entire song
Key	D	Eb
Tempo	None	122 BPM
Crotchet timing for intro	1+4+6	5+4+2

I consulted two published transcriptions to assist with the analysis of the music: Mann (1981, 68–70) and Billmann (1999, 103–110). The differences between the two illustrate the tendency of popular music to resist conformity to Western notation and to wilfully defy it (Moylan 2020, 12). While there is clear cohesion and internal logic to what Johnson is playing in terms of pulse and backbeat as represented by the audio recording, there is a disagreement between the two transcriptions concerning the key, tempo, and the ways to count the opening bars of the song. See Table 1.2.

Further, both transcriptions are referring to the same issued take of the song—ARC SA2632-1. The aural analysis undertaken for this essay disagrees with both transcriptions for the opening bars, reading the song as being in the key of E, the tempo starting at around 130 BPM and rising to 146 by the song's climax, and counting the intro bars as 1+4+4+2. What can account for this discrepancy? The link between key and tempo in the transcriptions is significant, given that a lower playback speed will result in a lower pitch for musical tones and so a perceived change in key. Despite the issue of *The Complete Recordings* on CD in 1990, it is possible that Mann was using either a tape or phonograph copy that was playing back at a speed which meant he was hearing the music in the key of D. It is unclear why Billmann hears the track in Eb, as analysis of the track from several different digital sources accessed via streaming services places the tonality of the music in E. As indicated above—the variations are all based on the single issued take of SA2632-1, so the only accounting for the difference in pitch can come from variables arising from the playback mechanism(s).

Music and Lyrics

"Preachin' Blues (Up Jumped the Devil)" is a single-take recording of a solo male vocal, self-accompanied by a steel-string acoustic guitar. The song is indicative of the oral blues tradition, in that while the song contains original material by Johnson, the title of the song is taken from a previously issued Son House tune, and one of the verses is lifted from House's *Walking Blues*. The tempo starts at 130 BPM and accelerates to 146 BPM by the song's climax.

Despite the static nature of the harmony—the music is a modal vamp rooted around the open E chord tuning of the guitar—Johnson's voice and the inventive, percussive nature of his playing create propulsion, excitement, variety, and texture through the use of (i) a heavy backbeat and syncopated melodic lines in the guitar, and (ii) cries and muttered asides which accompany the lead vocal. The vocal itself is, in turn, bellowed, whispered, and sung, presenting a variety of lyrical and emotional scenes to the listener. The form of the piece is rooted in its function; this is dance music, designed to provide a steady, unrelenting pulse for dancers attending social occasions. As indicated above, the pulse remains constant, but the meter varies throughout the song; extra beats which appear throughout the performance mean that bars can be counted as 5/4 and 2/4 as well as the standard 4/4. This contributes to the organic nature of the music; the feeling is of something alive in the performance, something developing and growing in both intensity and emotion. The surface noise of the recording also adds a sense of distance to the recording; a covering of aural dust that highlights that the performance is coming to us as listeners from a place and time removed from the present.

As Moore (2012) explains, the artist has a tripartite engagement with their audience in the construction of an "authentic" performance of popular music. Firstly, the artist is a performer—the actual person who walks on stage or into the studio in order to record. Secondly, the artist has a persona—which is the mode in which they conduct themselves on stage while they are performing the material. Thirdly, the artist is the protagonist—the person who experiences and/or narrates the events of the song from a first-person perspective. While the three elements are separate, for the audience and indeed some artists, the lines between them are blurred. Certainly, in the case of Robert Johnson, there is a persistent sense that Johnson, the performer, is, in fact, Johnson the persona and protagonist. In other words, a certain community of blues fans and marketers have long held that the events Johnson sings about are autobiographical rather than fictional, and this has contributed to his legend and iconic status. The lyrics of "Preachin' Blues (Up Jumped the Devil)" give weight to this legend; Johnson sings excitedly and passionately about meeting the blues and being trapped in a diabolical whirlwind of emotional turmoil. The confessional and seemingly unmediated nature of the recording—one voice, one guitar, one microphone—adds to the sense of intimate truth-telling in the performance.

The lyrics are split into five verses and make extensive use of metaphor, particularly personification. In the first verse, Johnson explains he woke up that morning and encountered the blues—which was walking like a man. This nightmarish image suggests that from the moment the singer opened his eyes that day, the living blues was haunting him. In the second verse, the blues transforms and "falls," with the visceral power to "[tear] me all upside down," highlighting the power of the personified blues to engage with Johnson physically. In the third verse, Johnson likens the blues to a "… chill," "… like consumption, killing me by degrees." This description of blues as a cancerous "heart disease" further illustrates the slippage between metaphysical and physical torment. Finally, by the third verse, the song has accelerated by 15 BPM, and Johnson delivers the verse lyrics ins syncopated gasps, seemingly imploring the listener or other external agency to "... drive my blues away." This verse also includes lines which parse phonetically as "I been studying rain." While writer Steve LaVere commented in his notes to *The Complete Recordings* that the lines made were "... phonetically correct, but meaningless," others have argued that the description of "studying rain" is poetic and have likened them to Shakespeare (Marcus 2008, 189). Arguably, without access to the dialect and/or idiolect that Johnson, the performer and protagonist, was working with in the context of his primary audience, it is impossible to confirm precisely what these words mean. In the context of the music however, there is a sense of catharsis by this point of the song; as Johnson sings the first line of this verse, the song is rocketing along at breakneck speed. As the artist sings "... drive my blues away" the first time, the rhythm and meter collapse like a river that has burst its banks. For the second statement of the line, the rhythm and tempo resume their pulse and pressure. As the song climaxes and Johnson sings that he will "... go out to the distillery and stay there all day," there is a ritardando, and the music decelerates abruptly before coming to rest, illustrating in sound the sanctuary from the blues that might be sought in the oblivion of alcohol.

Conclusion

In the case of Robert Johnson's "Preachin' Blues (Up Jumped the Devil)," not only did the recordings shape the song, but it also helped to shape the artist in the minds of a secondary audience who have only known the musician through his recordings. While this could be said to be true of any artist who has passed away leaving only their music, the lack of biographical detail concerning Johnson and the comparative lightness of his back catalog have combined to position him as a cultural icon. Johnson, more than any other modern artist, is a cipher; a space into which notions of blackness and blues culture may be projected without refraction in order to construct a legend (Schroeder 2004). What do we hear when we listen to the blues of Robert Johnson? We hear our own notions of the past, our own sense of authenticity as constructed by our engagement with a long-passed performer, and our own imagined ideas of the pre-war blues as enacted by a skilled singer and musician. In other words, the recording and the music absorb our expectations and deliver them back to us as part of a cathartic process. While Johnson's primary audience of black Americans during the pre-war years afforded him limited success during his brief lifetime, it is through his recorded works over the subsequent 80 years that the myth of Johnson has grown and continued to influence a secondary audience removed from his artistry by space and time. Denied the opportunity to appear at Carnegie Hall, a performance which would have surely altered his life, this essay closes with the words of musicologist Albert Murray:

> The phonograph record has served as the blues musician's equivalent to the concert hall almost from the outset. It has been in effect his hall without walls ...
>
> *Murray 1978*

Note

1 https://www.surveymonkey.com/s/BluesFanSurvey

References

Asante, John. 2010. "Robert Johnson, We Heard You All Wrong." NPR, accessed June 12, 2021, https://www.npr.org/sections/talk/2010/06/03/127404458/robert-johnson-we-heard-you-all-wrong?t=1627938136409.

Auslander, Philip. 1998. "Seeing Is Believing: Live Performance and the Discourse of Authenticity in Rock Culture." *Literature and Psychology* 44, no. 4: 1–26.

Baker, Houston A., Jr. 1984. *Blues, Ideology, and Afro-American Literature: A Vernacular Theory*. Chicago, IL; London: University of Chicago Press.

Barlow, William. 1989. *"Looking up at Down": The Emergence of Blues Culture*. Philadelphia, PA: Temple University Press.

Billmann, Pete. 1999. *Robert Johnson: The New Transcriptions*. Milwaukee: Hal Leonard.

Blesh, Rudi. 1946. *Shining Trumpets: A History of Jazz*. 4th ed. London: Cassell.

Clapton, Eric, and Christopher Simon Sykes. 2008. *Eric Clapton: The Autobiography*. London: Arrow Books.

Conforth, Bruce M., and Gayle Wardlow. 2019. *Up Jumped the Devil: The Real Life of Robert Johnson*. Chicago, IL: Chicago Review Press.

Davis, Francis. 1995. *The History of the Blues*. London: Secker & Warburg.

Dugan, James, and John Hammond. 1974. "An Early Black-Music Concert from Spirituals to Swing." *The Black Perspective in Music* 2, no. 2: 191–207. https://doi.org/10.2307/1214236. http://www.jstor.org/stable/1214236.

Eliot, T. S. 1920. *The Sacred Wood. Essays on Poetry and Criticism*. London: Methuen & Co.

Filene, Benjamin. 2000. *Romancing the Folk: Public Memory & American Roots Music*. Chapel Hill, NC; London: University of North Carolina Press.

Förnas, Johan. 1995. "The Future of Rock: Discourses That Struggle to Define a Genre." *Popular Music* 14, no. 1: 111–25. http://www.jstor.org/stable/853345.

Gellert, Lawrence, and Elie Siegmeister. 1936. *Negro Songs of Protest*. New York: American Music League.

Gioia, Ted. 2008. *Delta Blues: The Life and Times of the Mississippi Masters Who Revolutionized American Music*. New York; London: W. W. Norton.

Grazian, David. 2004. "The Production of Popular Music as a Confidence Game: The Case of the Chicago Blues." *Qualitative Sociology* 27, no. 2 (Summer): 137–58. http://search.ebscohost.com/login.aspx?direct=true&db=aph&AN=12611131&site=ehost-live.

Green, Lucy. 2001. *How Popular Musicians Learn: A Way Ahead for Music Education*. Aldershot: Ashgate.

Hamilton, Marybeth. 2007. *In Search of the Blues: Black Voices, White Visions*. London: Jonathan Cape.

Hayne, Michael, David Mee, and Ron Rumble. 2005. "Influence of Music Genre and Composition on Entertainment Noise Limits." Paper presented at the Annual Conference of the Australian Acoustical Society: Acoustics in a Changing Environment.

Jones, Le Roi. 1963. *Blues People: Negro Music in White America*. New York: Morrow.

Katz, Mark. 2010. *Capturing Sound: How Technology Has Changed Music*. Rev. ed. Berkeley, CA; London: University of California Press.

Keil, Charles. 1966. *Urban Blues*. Chicago, IL; London: University of Chicago Press.

King, Stephen A. 2011. *I'm Feeling the Blues Right Now: Blues Tourism and the Mississippi Delta*. American Made Music Series. Jackson, MS: University Press of Mississippi.

Levine, Lawrence W. 2007. *Black Culture and Black Consciousness: Afro-American Folk Thought from Slavery to Freedom*. 30th anniversary ed. Oxford: Oxford University Press.
Mann, Woody. 1981. *The Complete Robert Johnson (Guitar Tablature Edition)*. New York: Oak Publications.
Marcus, Greil. 2008. *Mystery Train: Images of America in Rock "N" Roll Music*. 5th ed. New York: Plume.
Merriam, Alan P. 1964. *The Anthropology of Music*. Evanston, IL: Northwestern University Press.
Miller, Karl Hagstrom. 2010. *Segregating Sound: Inventing Folk and Pop Music in the Age of Jim Crow*. Durham, NC: Duke University Press.
Moore, Allan F. 2002. "Authenticity as Authentication." *Popular Music* 21, no. 2: 209–23. http://www.jstor.org/stable/853683.
———. 2012. *Song Means: Analysing and Interpreting Recorded Popular Song*. Farnham, Surrey; Burlington, VT: Ashgate.
Morrison, Toni. 1992. *Playing in the Dark: Whiteness and the Literary Imagination*. Cambridge, MA; London: Harvard University Press.
Moylan, William. 2020. *Recording Analysis: How the Record Shapes the Song*. 1st ed. Oxford: Routledge.
Muir, Peter C. 2010. *Long Lost Blues: Popular Blues in America, 1850–1920*. Urbana, IL: University of Illinois Press.
Murray, Albert. 1978. *Stomping the Blues*. London: Quartet Books.
Oliver, Paul. 1989. "Overseas Blues: Europeans and the Blues." Chap. 57–73. In *Sounds of the South: A Report and Selected Papers from a Conference on the Collecting and Collections of Southern Traditional Music, Held in Chapel Hill, April 6–8, 1989, to Celebrate the Opening of the Southern Folklife Collection with the John Edwards Memorial Collection in the Manuscripts Department of the Academic Affairs Library, University of North Carolina*, edited by Daniel W. Patterson. Chapel Hill, NC: Southern Folklife Collection.
Palmer, R. 1981. *Deep Blues: A Musical and Cultural History from the Mississippi Delta to Chicago's South Side to the World*. New York: Penguin Books.
Richards, Keith, and James Fox. 2011. *Life*. London: Phoenix.
Ripani, Richard J. 2006. *The New Blue Music: Changes in Rhythm & Blues, 1950–1999*. American Made Music Series. 1st ed. Jackson, MS: University Press of Mississippi. Table of contents http://www.loc.gov/catdir/toc/ecip067/2006002022.html.
Rothenbuhler, Eric W. 2007. "For-the-Record Aesthetics and Robert Johnson's Blues Style as a Product of Recorded Culture." *Popular Music* 26, no. 1: 65–81. https://doi.org/10.2307/4500300. http://www.jstor.org/stable/4500300.
Ryan, Jennifer. 2011. "Beale Street Blues? Tourism, Musical Labor, and the Fetishization of Poverty in Blues." *Ethnomusicology* 55, no. 3: 473–503.
Sanjek, David. 2009. "Following Harry's Footsteps: Contemporary Strategies of Reissue and Reassessment of American Vernacular Music." Experience Music Project Pop Music Conference, Seattle, Washington.
Schmidt-Horning, Susan. 2013. *Chasing Sound: Technology, Culture, and the Art of Studio Recording from Edison to the LP*. Baltimore, MD: Johns Hopkins University Press.
Schroeder, Patricia R. 2004. *Robert Johnson, Mythmaking, and Contemporary American Culture*. Urbana, IL: University of Illinois Press.
Schwartz, Roberta Freund. 2007. *How Britain Got the Blues: The Transmission and Reception of American Blues Style in the United Kingdom*. Aldershot: Ashgate. Table of contents only http://www.loc.gov/catdir/toc/ecip077/2006103148.html.
Shaw, Arnold. 1979. *Honkers and Shouters: The Golden Years of Rhythm and Blues*. New York: Collier.

Small, Christopher. 1987. *Music of the Common Tongue: Survival and Celebration in Afro-American Music.* London: Calder.

Titon, Jeff Todd. 1993. "Reconstructing the Blues: Reflections on the 1960s Blues Revival." In *Transforming Tradition: Folk Music Revivals Examined*, edited by Neil V. Rosenberg, Urbana, IL: University of Illinois Press: 220–40.

Wald, Elijah. 2004. *Escaping the Delta: Robert Johnson and the Invention of the Blues.* 1st ed. New York: Amistad.

———. 2010a. *The Blues: A Very Short Introduction.* Oxford: Oxford University Press.

———. 2010b. "The Robert Johnson Recording Speed Controversy." *Elijahwild.com.* (Online), accessed September 4, 2020, https://elijahwald.com/johnsonspeed.html.

Wardlow, Gayle, and Edward M. Komara. 1998. *Chasin' That Devil Music: Searching for the Blues.* San Francisco, CA: Miller Freeman Books.

Wilde, Jon. 2010. "Robert Johnson Revelation Tells Us to Put the Brakes on the Blues." *The Guardian* (Online). https://www.theguardian.com/music/musicblog/2010/may/27/robert-johnson-blues.

2

ANALYZING HIP-HOP HACKTIVISM AND AUTOMOBILITY IN INJURY RESERVE'S (2019) "JAILBREAK THE TESLA" (FEAT. AMINÉ)

Steven Gamble and Justin Williams

Introduction

The mbira sample that begins Injury Reserve's (2019) "Jailbreak the Tesla (feat. Aminé)" sounds like a straightforward three-note ostinato: an ascending step and a leap on repeat. However, all is not what it seems at first encounter. Like the track as a whole, what appears quite simple on the surface conceals significant depth and complexity. Even in the opening few seconds, there is more than initially meets the ear: there are in fact two separate layered samples in this mbira loop. At a high frequency around 2 kHz, one melodic line ascends B♭–C♭–E♭, referencing the well-known Neptunes production, Teriyaki Boyz's "Tokyo Drift (Fast & Furious)."[1] The lower line of the two makes an E♭–F♭–A♭ run in unison with this pattern. As such, the timbral design of this loop has a powerful sonic identity and leaves space in the mix for key vocal frequencies in-between each layer, even though it can be difficult to hear what, exactly, is going on. After a few repeats of this mbira sample (represented by a spectrogram and simplified score notation in Figure 2.1),[2] a booming bass pitched around E♭1 (39 Hz) enters, and we are speeding off.

The sounds of the track are engaging, dynamic, and richly referential, and they interact with the lyrics of the rap verses, which follow in deeply meaningful (and sometimes contradictory) ways. Critics praised the production of "Jailbreak the Tesla" for turning "a supercar garage into a strobing nightclub" (Pearce 2019) and characterized Injury Reserve as "staking a claim to the role of hip hop's next great innovators" (Vinti 2019) through this experimental modern hip-hop song. This chapter, therefore, takes the recorded track seriously as a text for analysis, interested in how its sounds afford interpretations of innovation and automobility. As we'll go on to argue, a superficial reading of the track might simply bring to mind the thrill of joyriding, but just as the

DOI: 10.4324/9781003089926-4

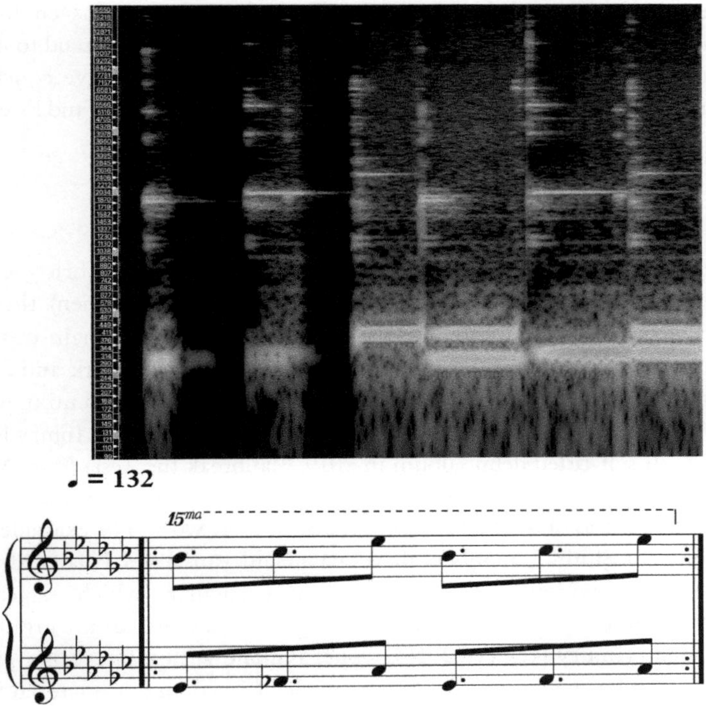

FIGURE 2.1 Mbira loop spectrogram and score reduction.

subtly layered mbira teaches us, there is more going on in the song than a quick glance will reveal.

William Moylan's (2002, 3) original approach urges analysts to consider how recordings "shape or create sound as an artistic statement." The ecological approach to perception holds that a major function of the human listening apparatus is to assess what the sounds we hear are the sounds *of* (Clarke 2005, 3). In the case of "Jailbreak," various audio sources specify automobiles in motion (engine noises, tires scraping, digital interfaces) alongside vocal performances which lyrically spotlight the cultural role of the car and technological customization. But there are complications. While the revving car samples and rattling metallic timbre of the mbira resemble the straining sounds of an internal combustion engine, they are at odds with the electric, famously near-silent "Tesla" referenced in the title and the lyrics. The vocal production is also conspicuously digital and highly modern (using heavy compression, pitch correction, and abrupt rhythmic edits) juxtaposed with somewhat outdated, nostalgic references to "Myspace," "Limewire," and "West Coast Customs" (the car customization shop of *Pimp My Ride* fame). How might listeners make sense of these tensions?

After an introduction to the group, we analyze the rhythm and timbre of the "Jailbreak" beat. We next consider the significance of such analysis with reference to listening contexts and the lyrics, identifying how the persona-environment relation refers to cultures of automobility and jailbreaking. We attend to the production of vocal timbre over the course of the track. Finally, we conclude by addressing how the track implicates driving, hacking technology, and listening to hip-hop as racialized practices.

Injury Reserve

Injury Reserve is a hip-hop collective described by Robert Christgau (2017) as "the most unpretentious hip-hop you ever heard." The (then) three-man Phoenix, Arizona group broke into the fringes of the hip-hop mainstream during the late 2010s, using online platforms to promote their work and grow an audience of digitally connected fans. Following two self-released mixtapes, *Live from the Dentist Office* and *Floss,* and the EP *Drive It Like It's Stolen*, Injury Reserve released their self-titled debut album in 2019. "Jailbreak the Tesla (feat. Aminé)" is the second promotional single and their most viewed music video on YouTube to date.[3] The automobile is a recurring motif in their work. For example, in the video for "Oh Shit!!!," members drive a U-Haul van and a Polaris Slingshot. The group has previously referenced Tesla in particular,[4] which suggests how clearly the brand symbolizes the pinnacle of personal success for the group. In "S on Ya Chest" (Injury Reserve 2016), rapper Ritchie with a T reveals aspirations to leave his "blue-collar" job and gain financial freedom and material wealth: "I can't wait 'till I can get to stunting, get a Tesla and take it to West Coast Customs."

By 2019, with a string of successful releases and tours under their belts, Injury Reserve had made good on this goal, as the music video for "Jailbreak"—shot in and around a Tesla Model X—makes clear (see Figure 2.2). However, their thinking about how to customize the Tesla had progressed. No longer content simply to tweak and repaint, they cite a particular culture of technology (lyrically recalled from an adolescence spent online) known as "jailbreaking." Most often used in the context of Apple iOS devices, this term refers to prosumer

FIGURES 2.2 (a and b) Stills from the music video for "Jailbreak the Tesla (feat. Aminé)," showing Injury Reserve members Ritchie with a T, Stepa J. Groggs, and Parker Corey (left-to-right) in their tricked-out Tesla.

hacking practices intending to subvert in-built software constraints (Magaudda 2012). To jailbreak the Tesla is to maximize the potential of computer equipment: in this case, the electric car. This might not be obvious from an unfamiliar listener's initial encounter with the track, instead potentially calling to mind images of vehicle theft (hotwiring?) or joyriding common in hip-hop music.[5] There is much more ingenuity at play than first appears, drawing on a discourse of creatively refashioning technology in Black popular music that Injury Reserve invokes through the recorded sounds and lyrics of "Jailbreak."[6]

Analysis of the "Jailbreak" Beat

Long since hip-hop DJs looped funk and rock breaks on vinyl records, contemporary hip-hop production has drawn from a vast range of sounds to create danceable, pleasurable grooves. The sonic lineage that "Jailbreak" references is clearly based in the car (and perhaps the club), moving loud blocks of abrasive sound and forceful bass around the recorded environment. Various sonic events are used in the main loop of the track: some of these approximate acoustic instruments (or, in ecological language, *specify* their source), many of which directly implicate the car (engine, tire, electric functions), whereas others are more digitally treated and obscure noises for which we have invented descriptive labels ("bubble," "fizz"). The annotated spectrogram and score transcription of non-vocal sounds in the first passage of the full beat (0:14–0:22) shows how these layers typically interact in the song (Figure 2.3).

The depicted section is indicative of the main looped beat, but it is rarely repeated in an identical fashion, as the carefully constructed layers of the beat change throughout different sections. Partway through the first verse (0:50), for example, the beat features the "white noise" sound prominently but omits the high-frequency "hi-hats," while the tire scrape sound is positioned later in the four-bar loop. We have represented the track structure, applying our own section labels, in Figure 2.4.

Only the bass and the mbira fall within conventional ranges for harmonic or melodic expression. Even then, the repetitive mbira pattern and single bass note are extremely limited in these domains. This allows the timbral and rhythmic interplay of the beat to become the main perceptual emphasis of the track, even though these are rarely "primary elements" of recorded music (Moylan 2020, 44–45). Nonetheless, as the nature and organization of the recorded sounds form the energizing force of "Jailbreak," they warrant analytical attention.

The dynamic envelope of most sounds comprises short attack, sustain, and release times, meaning listeners will characterize them as sudden, brief, and short-lived. Abrupt dynamic changes bring a sense of rhythmic vitality, as visible in the thin white spaces of the waveform in Figure 2.4. Few dynamic events linger—they come and go rapidly—which provide the beat a sense of constantly moving on. Furthermore, this locomotion is literalized by the timbres that resemble car sounds: tires skidding on tarmac, engines roaring, and electric whirring (1:01).

FIGURE 2.3 Annotated spectrogram and score reduction of non-vocal sounds (0:14–0:22).

Rather than using acoustic drum sounds, "Jailbreak" builds a complex sound stage out of automobile sounds that invite direct timbral interpretation of sound character and create an exciting rhythmic interplay.

The short, snappy rhythmic events contrast with the bass, which has longer decay and release periods. It, therefore, produces a frequent rumble—imagined

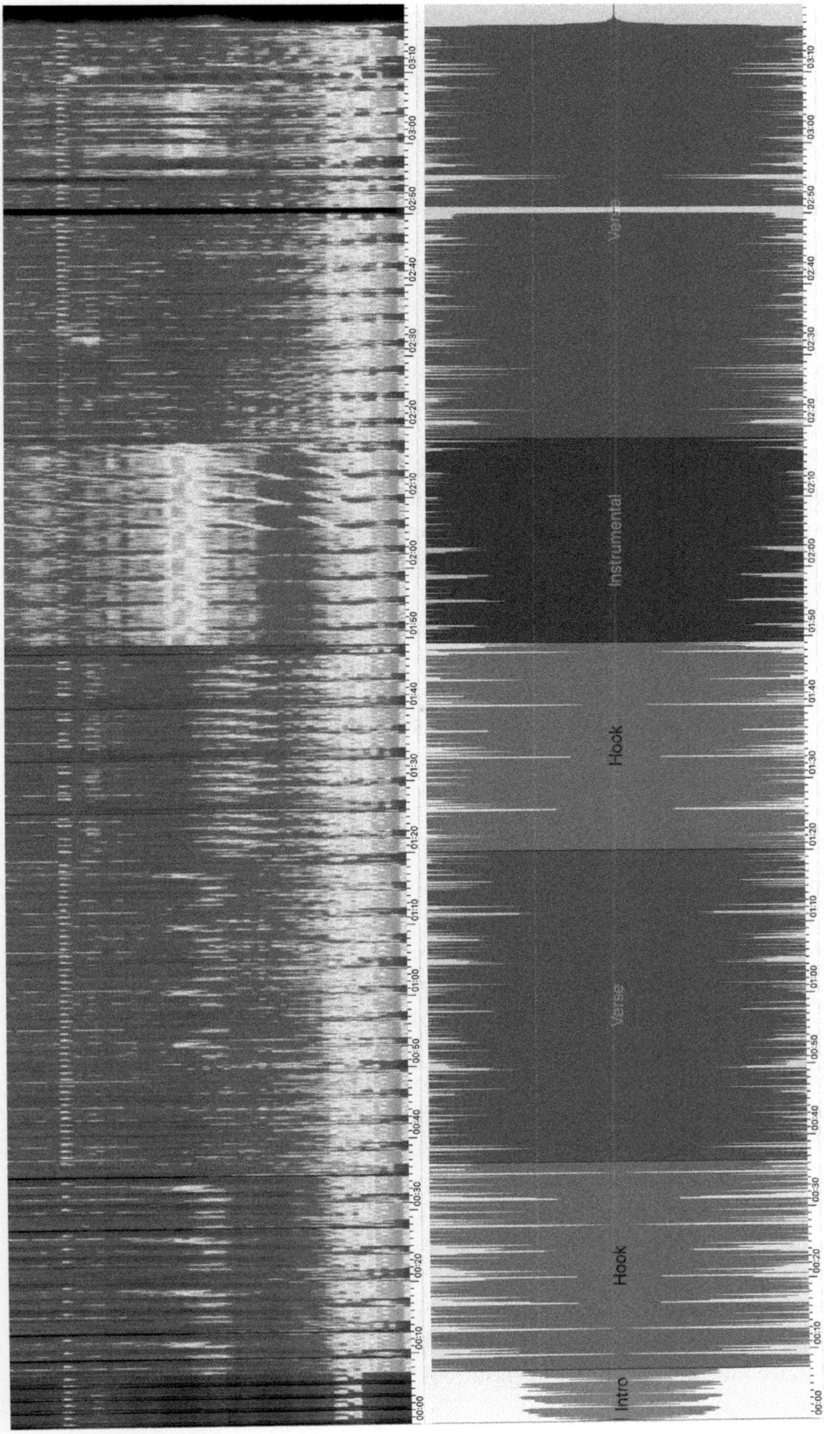

FIGURE 2.4 Full track spectrogram (top) and waveform (bottom) with overlaid structural labels.

like the road beneath the wheels, perhaps—within the overall timbral environment. As shown in the spectrogram of Figure 2.4, the booming bass occupies much of the low frequency range of the mix. It carries significant dynamic energy and is distorted, magnifying the overtones above its E♭1 fundamental. On each bass hit, a kick drum creates a snappy attack with significant force, then leaves a pitched note to decay and slide down slightly in pitch. It may be perceived as a powerful and invigorating sound which imitates the roar and "tail-off" after the revving of a car engine.

Despite the many cuts and ruptures in the beat, the metric pulse is always preserved. Any disruption is never so significant as to disorient the listener's sense of consistent motion, enabling them to follow along and move their body in time. The densely layered rhythmic organization of the beat provides ample opportunity for sympathetic bodily motion (Gamble 2021, 24–26). It is, therefore, clear that the song places "emphasis on formal qualities of texture and density," with a "vertical, layered" sound stage producing "a tactile, expanding dimensionality of sound" (King 2001, 433). In doing so, "Jailbreak" draws heavily on established stylistic traits of Black electronic popular music, especially those developed by the influential producer Timbaland. Jason King's (ibid.) writing highlights that "how the beat might not only bang against the eardrum, but seductively take hold of the body through sonic inspiration in various spaces of/for listening—the club, the jeep, the home—has become a major issue in black popular compositional practices." Given the recorded sounds and their layering, it is, therefore, necessary to consider the specific listening contexts of rap music.

Listening in (and to) the Car

By sonically evoking the car and automobile motion, "Jailbreak" raises important questions about the "accurate playback" (Moylan 2020, 4) environment for recordings. The cultural conventions of automotive listening (and the widespread use of headphones and earphones in hip-hop) also sit in tension with what equipment is "suitable" (ibid.) for listening. Our analysis of the mix, especially the prominence of the bass, encourages us to hear this as a track designed for the car, with additional application to the club, as ideal listening spaces. In many ways, then, the imagined context for listening to "Jailbreak" differs from the "stereo triangle" position presupposed by most academic texts and music production textbooks describing accurate playback conditions (Savage 2011, 14–18; Huber and Runstein 2017, 493–513; Moylan 2020, 230–32). This has previously been addressed by Simon Zagorski-Thomas (2014, 77), who notes that the "culturally constructed 'ideal' listening position and mental state" of record production involves a listener located in the producer's chair with a specifically arranged speaker setup, and trained, unimpaired listening ability.[7] "Jailbreak" provides an interesting analytical challenge to such conventions.

Furthermore, car audio technicians have highlighted the inadequacy of the automobile as an accurate environment for listening. According to Joe Pettitt (2003, 20), "car interiors are about the worst listening environments imaginable. Your location in the car is set. Listening seating is to the side of acoustic space in relation to the speakers." In a practical sense, therefore, the detailed rhythmic construction of "Jailbreak" might go entirely unnoticed in an automobile listening situation, where the rumble of the road, engine sounds, wind, and other environmental noises obscure the recording. The view of Stewart Copeland (of The Police), who describes the car stereo as "the first and best minisystem" and a "cocoon where you can have a really big sound in an enclosed environment" (quoted in Baber 1997), and many other producers who perform the infamous "L.A. Car Test" (Seabrook 2015, 36), are uncommon exceptions.[8] Why, then, spotlight the sounds and motion of the car?

It may be that an emphasis on "accurate" or "ideal," studio-style playback somewhat misses the point—the function, utility, and pleasure—of car audio systems. The car as a listening environment emphasizes "auditory privacy [in] public" (Bijsterveld et al. 2013, 2) and thereby acts more like "a technology of control through which drivers construct their privatized soundworlds, immersing themselves in a perfect sound booth" (Prior 2018, 106). As Will Moylan (2020, 235) reminds us, "along with the acoustic variables of automobile seating [...] there are conceptual matters where some listeners are prone to projecting or placing themselves *within* the contexts and activities of songs." Because of the sound staging in "Jailbreak," in the process of listening individuals may imagine themselves situated in the titular Tesla. This also helps make sense of the personal proxemic position of the vocal, which spans a wide space in the perceptual center of the mix. Compare, for example, the vocal positioning in the verse to that of the hook, which has a much narrower spatial profile (1:18–1:22). The sound stage affords an interpretation of sharing a small space with the vocalist, whereas the stereo width of the verse vocals also ensures the rapping remains prominent even using a less-than-ideal audio playback system (like a car). The stereo field may be split imperfectly between door speakers, yet the vocal will still project reasonably well as a perceptually central sound source.

Automobility in the Persona-Environment Relation

Conspicuous car sounds have long been used in popular music to various ends (Williams 2014, 118; McLeod 2020, 193), such as:

- flaunting an impressive engine (Beach Boys' "409," Simon and Garfunkel's "Baby Driver")
- celebrating particular manufacturers or models (Wilson Pickett's "Mustang Sally," Rush's "Red Barchetta," Jackie Brenston and his Delta Cats' "Rocket 88," ZZ Top, "Chevrolet")

- as sexual metaphor (Robert Johnson's "Terraplane Blues," The Beatles' "Drive My Car," Prince's "Little Red Corvette"), and
- evoking the experience of driving (Kraftwerk's "Autobahn," The Shangri-Las' "Leader of the Pack," Bobby Troup's "Route 66"), whether alone or communally.[9]

With recorded samples of car engines, tire screeches, horn noises, and so on, many tracks directly feature car sounds, a form of referentiality Serge Lacasse (2000) called "autosonic quotation." The synth bass in "Jailbreak" is not exactly the sound of vehicular acceleration, but it does resemble engine noise in timbre and dynamic shape, hence our potential hearing of it as "revving." Other elements—the metallic pops, scrapes, and especially the whooshing timbres that feature prominently in the instrumental section (1:49)—may be actual car sounds reproduced autosonically.[10] Because the recording features these sounds so conspicuously, the track evokes a tricked-out vehicle in spectacular motion. Reinforcing and even enacting the car-centric lyrics, we may hear the beat as "interventionist," drawing a term from Allan Moore's (2012, 191) typology of recorded environments. However, there is also a degree of contradiction at play.

The beat jostles with automobile references, but the lyrics—produced by the characters of the narrative, the personas—emphasize tuning up and hacking Teslas. Cars in the Tesla range are popularly known for their extremely quiet electric engines, so the persona and environment contradict one another, and their relationship might therefore be understood as "oppositional" (ibid., 191). But this friction is not lost on the artists. Indeed, the contrast between the whirring, straining, and straight-up loud beat and the supposed soundlessness of electric cars is spotlighted for a humorous lyrical punchline. In the second verse, the featured artist Aminé raps, "your engine goes 'vroom'; my engine goes …" then leaves a silence to boast about his quiet engine. The beat mutes altogether to reinforce his claim and drive the point home (2:50). The message is clear. Move over Ferrari, Lamborghini: the silent electric vehicle, not the raucous speedster, is now the prime object of automobile fascination, at least for these hip-hop artists.[11]

The aspirational car purchase is a well-known staple of hip-hop lyricism and a broader indicator of the significance of automobiles to urban US cultural experiences. The high status (and accompanying high price tags) associated with particular manufacturers and models makes the car an especially important feature of success discourses in hip-hop culture. Paul Gilroy (2001, 84) has described how the popularity of cars and the material wealth that they represent may be a reaction to racialized histories of economic marginalization (or the perception thereof) in the public sphere.[12] Automobiles and their affordances of personal mobility and spatial autonomy fit hand in hand with hip-hop music's sonic power, offering the ability to occupy space and express Black culture in the public domain (Rose 1994, 62–64). "Jailbreak" clearly draws on this lineage not only through the car sounds in the beat but also through lyrical references to other hip-hop songs about cars. The line mentioned in the introduction to this chapter, drawn from an earlier Injury Reserve track—"get a Tesla, take it to West Coast Customs"—also appears in the first verse of "Jailbreak." This self-quotation itself is quite striking, but the new context demonstrates a development from an abstract wish ("I can't wait 'till I can …") to actually making it happen (as demonstrated in the video, where the performers rap and drive around in the Tesla X series).

Jailbreaking, Lyrics, and Vocal Timbre

The track is obviously a love letter to the Tesla X and customization, but it also heavily references jailbreaking culture. In the first verse, rapper Ritchie with a T recalls "ninth grade," when he was being "raised by the internet" (another self-quotation of an earlier Injury Reserve lyric). His cultural education online includes engagement with prosumer hacking communities, name-dropping the software tools "redsn0w" and "Cydia," which significantly expanded the potential uses of the iPod touch.

Though technically a form of user privilege escalation specific to Apple Inc. devices, jailbreaking has become a widespread term for any kind of manipulation of factory-set software limitations on consumer electronics. Various tools for customizing handheld devices exist, adding new features that are not included in the basic product (nor downloadable as an add-on) and removing in-built bloatware (software deemed unnecessary which takes up space and slows the device down). Jailbreaking enables the user to gain more value from their device and gain more control over how it functions, making it work more precisely to their desired specification (Lee and Soon 2017, 355–57). Indeed, jailbreaking can be seen as a form of resistance to Apple, or at least the restrictions of the pre-installed operating systems on such devices (ibid., 358). Modifying a personal device can be viewed as an act of achieving greater technological autonomy, perhaps comparable to the liberating self-expressions witnessed in custom car culture, and there is an element of jailbreaking culture aligned with hacktivism and anti-authoritarian ethics. Nonetheless, many developers and users maintain brand loyalty toward Apple (ibid., 359), and although jailbreaking is seen as a potential violation of the warranty, the company evidently monitors jailbreaking software closely, as popular features have subsequently been released in official software updates.

With all of the above in mind, when Injury Reserve's Ritchie with a T name-checks his jailbroken iPod Touch, it brings with it a range of associated cultural memories: the excitement of purchasing a new device, flaunting it among friends, and pursuing a nerdy obsession with expanding its capacities. Recalling the impressive ability of the device to "hack security cams," his attention now turns to maximizing the potential of the automobile. Given that the Tesla is fully electric and its functions are controlled by a computer, why not apply the same techniques to your car and "turn the [Series] X into a Batmobile rip"?[13] Finally, the first verse clarifies the advantages of the Tesla over older exemplars of automobile customization in hip-hop. The Ferrari has severely limited customization options, predominantly aesthetic decisions since its functions are not fundamentally managed by a software operating system like the Tesla: "all you can change is paint and some f*** sh**."[14] This newer manifestation of aspirational hip-hop automobility and prosumer customization focuses on jailbreaking the Tesla, applying the same technological ingenuity that informed an adolescence of handheld device optimization to the latest developments in electric vehicle production.

The hook (from 1:20), delivered by Injury Reserve rapper Stepa J. Groggs,[15] repeats a straightforward phrase which can be characterized somewhere between a web search query and an instructional video title (like the deluge of online tutorials on "how to jailbreak your phone"). The voice is pitch-shifted down and highly compressed: like the Tesla and the iPod touch mentioned before it, it is conspicuously altered using technology. Voices are produced similarly, though not identically, in the two verses, where prominent processing gives the rapped vocals an unnaturally consistent timbral quality (Table 2.1). The vocals, like the car itself, are modified using digital technology.

Although it somewhat resembles an older, analogue practice of hip-hop vinyl scratching, the precise looping and pitch manipulation of the hook vocal (1:20) is the most clearly "digital" vocal effect. This "hacked" voice acts more like another rhythmic layer in the beat than a traditional chorus melody, repeating a catchy phrase that contrasts with the varying lyrical material in the preceding verse. The instrumental section, which follows the hook, with additional samples approximating acceleration and skidding tires, narratively imply that the Tesla has been successfully jailbroken, its settings altered: the instrumental presents the sounds of the car hacked to its full potential.

Finally, the guest verse by Aminé recalls another personal history of wishing for luxurious cars, from an "ice cream paint job" (a reference to Dorrough's track of the same name) to modern accessories like the blackout "Maybach curtains." Later, he alludes to the wild imagination of Tesla founder Elon Musk and the artist Grimes, joking that the latter's voice could be used as the GPS voice. Immediately after—as if in response, like a virtual audio assistant—a heavily computerized voice (albeit obviously not Grimes) begins issuing commands. This GPS voice narratively manifests the audio communication system of the titular Tesla directly in the recorded sound. Samples of acceleration and skidding are then reintroduced, alternating with Aminé's final lines, featuring jarringly artificial cuts to the sound envelopes. The last line is interrupted with an earlier vocal sample and one more tire scrape sound effect. In the closing moments, therefore, the track is at its most intensely digitally ruptured, as conspicuously edited samples jostle with computerized voices in frantic rhythmic interplay.

TABLE 2.1 Typology tables comparing the timbres of the recorded voices in the verse and hook of "Jailbreak the Tesla (feat. Aminé)"

Variable	*Verse (Ritchie with a T)*	*Hook (Stepa J. Groggs)*
Context		
Sound origin	Medium-register male African American voice	Deep (pitch-lowered) male African American voice
Performance technique	Rapped	Rapped, sampled
Formal function	Verse, referential, and narrative	Repetitive hook introducing minimal new content
Language	African American Vernacular English, Southwestern dialect	American English, ambiguous
Character		
Performance intensity	Loud, gently varying to match poetic contours	Loud, stable
Exertion level	Moderate level of exertion	Moderate level of exertion
Energy level	High, with some peaks and troughs	Medium, static
Expression	Heavily processed but natural rapped expression	Quasi-robotic, matter-of-fact; sampled repetition
Clarity/distortion	Clear diction, heavy compression with slight peak clipping in low-mid frequency area	Moderate clarity, processing slightly obscuring diction
Content		
Conspicuous digital processing	Compression providing extreme consistency in dynamic level, pitch correction with moderate glide time accentuating pitch	Dynamic consistency, lowered pitch, and shifted formants; stark edits for loop
Dynamic envelope	Short attack and release times, minimal decay between peak and sustain loudness levels	Short attack, long release times, and notable reverb; interrupted by edited loop
Spectrum	Wide bandwidth, moderate resonance; reduced high-frequency content through sibilance filtering	Wide bandwidth, low and low-medium frequencies; upper partials limited
Spectral envelope	Consistent density	Consistent density
Definition of fundamental frequency	Pitch occasionally defined, with longer sustained vowels more clearly pitched	Pitch moderately defined in sampled vocal
Formant frequencies and characteristics	Standard vocal formant production	Shifted pitch and formants, lowered and time-stretched
Space influences	High dynamic level, breaths, and stereo width places voice close to listener	High dynamic level but reduced stereo width, remains nearby but slightly narrower position

Conclusion: The "So What" of Analysis

Analyzing recorded sound means paying attention to what strikes listeners most significantly about recorded music. It challenges the primacy of formal structure in traditional approaches (designed for "common practice" Western European art music), such as Schenkerian analysis (Moylan 2020, 66–67), making it a preferable starting point for the analysis of popular music. Because it operates from the principle that it is the sound of a recording that takes precedence, recording analysis can fundamentally work in service of how listeners experience and understand songs. What, then, are we to make of "Jailbreak" and its contradictory qualities, spanning between various shades of youthful, nostalgic, geekish, cool, materialistic, resistant, consumerist, anti-corporate, liberating, and joyful?

The track illuminates aspirational uses of technology through rhythmically busy sound staging, sonic and lyrical automotive references, and conspicuously digital processing. Even though hacking and hip-hop are an effective combination, there is perhaps no other rap song so uniquely "about" jailbreaking, which indicates Injury Reserve's committed engagement with this aspect of popular digital culture. By tying their hacking practices to familiar themes of car customization (e.g., West Coast lowriding), however, the track develops upon a long-standing commitment to strategic enhancements of technology in hip-hop music (Weheliye 2005; McLeod 2020, 59). It repurposes found sounds and technologies to their own ends, with the intricate beat production emulating the pace and soundscapes of modern urban environments.

Like the track's innovative use of "black techno-interventions" (Rose 1994, 64), automobiles and jailbreaking are also implicated in racial discourses. The latter, and hacking communities more broadly, may be stereotypically (if inaccurately) viewed as the sole domain of white male nerdiness. In particular, Lori Kendall's (2011, 514) study of the hyperwhite racial identity of the nerd argues that the connections between nerd identity and whiteness can be portrayed as "directly contrasting [...] with the stereotypical black identity of the gangsta rapper." The track offers us another contradiction, then, between the African American cultural tradition of hip-hop car customization and the perceived whiteness of jailbreaking communities. However, it resolves this tension with clever nuance, drawing upon jailbreaking in relation to a familiar hip-hop success story, the rags-to-riches ascent (Haupt 2017). It takes this traditional expressive mode ("never would have thought I'd have a ride like mine") and upgrades the narrative itself, patching in the ingenuity of prosumer hacking tools to maximize the technological potential of the high-end car.

The automobile is therefore celebrated by the track as an emancipatory space of pleasure, one rife with "automotive emotions" (Sheller 2004). In sonically reproducing the car, "Jailbreak" draws upon it as a broader symbol of freedom in Black popular culture. Joanna Demers (2003, 78) argues that modern musical references to the car build on the metaphor of the Biblical chariot. Tracing a line from spirituals to hip-hop, she notes that "the chariot has long been an image of liberation and salvation, and at times a code word for the Underground Railroad or other anti-slavery movements" (ibid.). Recorded rap texts like Dr Dre's "Let Me Ride," as well as our analysis of "Jailbreak," shows that hip-hop carries the lineage of spirituals and funk music, updating from the chariot to the car—and now the electric car—as a source of potential liberation. The car is thus understood as "a sonic, visual, and mobile means of announcing and projecting cultural identity, as well as a means of resisting white corporate control" (McLeod 2020, 51). Although the track's apparent celebration of Tesla cars does not obviously challenge the concentration of power by white-owned megacorporations, it extends a Black cultural tradition of technological subversion and revision (Weheliye 2005).

However, such resistance is under threat in the contemporary context of "Jailbreak." While the track celebrates the car as a site of pleasure, of conspicuous consumption, and freedom of mobility, it also draws attention to the policing of Black drivers in America. The rapper might "tell my mom safety features ain't adjusted," but mechanical control over the car is far from the real threat that Black drivers face. It would be naïve not to hear the track in the contemporary context of political action against police violence, reproducing both the pleasures and the dangers of "driving while black" (Seo 2019). African Americans are more likely to be pulled over than their white peers and they are more likely to be shot and killed by police during the interaction (Edgar 2020). Rappers have commented in lyrics that they are likely to be pulled over when police officers assume their expensive cars have been stolen, a cultural image that Injury Reserve plays on in the EP titled *Drive It Like It's Stolen*.[16] This is quite why "Jailbreak" is so powerfully charged, its recording teeming with dynamically expressive sounds, both anxious and ecstatic in its locomotion.

If we hear the song evoking the epitome of modern automobile technology and creatively maximizing its potential, then we are hearing the sounds of subversion in action on two levels. First, in terms of freedom and mobility in the African American context of post-emancipatory inequalities.[17] Second, as an intervention into technological limitations laced into the sounds of the track, amplified by the digitally enhanced voices, and made explicit by the lyrics. The politics of sound and identity make the racial-economic tensions encapsulated in the song's premise serious indeed. But it is also worth attending to the pleasurable dynamism of the track, leading one reviewer to dub it a "banger" (Hobbs 2019): start to finish, pedal-to-the-metal, exhilarating fun. The samples interlock in high-octane rhythms and the rap bars weave in and out of the beat with electrifying forward momentum. In the music video and live performances, the track is accompanied by a visual analogue: near-constant strobes of pulsing electric energy. "Jailbreak the Tesla" belongs to a long tradition of songs about cars but also updates rap braggadocio for the 21st-century computer hacker. The track demonstrates that the intricate sonic production of hip-hop music can breathe life into familiar tropes like car customization and create something breathtakingly innovative.

Notes

1 The Teriyaki Boyz track was used as the main theme of the popular 2006 movie *The Fast and The Furious: Tokyo Drift*. The iconic three-note loop and its metallic timbre have also been used in Lil Yachty and Tierra Whack's "T.D (feat. A$AP Rocky and Tyler, The Creator)," among other tracks. Where the "Tokyo Drift" melody deviates after a few repeats to follow a shifting chord sequence, Injury Reserve keeps the production focused, looping the B♭-C♭-E♭ climb throughout the track.

2 The score reduction simplifies the pitch and rhythmic content of the loop for purposes of depiction. Many of the notes in the loop could be considered slightly "sharp" or "flat," though they are undoubtedly pitched precisely, and in no sense "incorrect": see Walser (1995, 194–99) on the creative use of noise and pitch in hip-hop production. Note also that the first two notes of the mbira loop are often muted, and the A♭ in the lower line often repeats on the fourth and sixth notes of each six-note pattern (as can be seen in the spectrogram).

3 The video has accrued 1.4 million views as of January 2022, according to YouTube's count.

4 Tesla, Inc. is the best-known and best-selling all-electric vehicle brand worldwide, producing almost 1 million electric cars in 2021 alone (Kane 2022). The company is recognized for its "futuristic," technologically advanced premium cars with sleek designs, such as "falcon doors" which open vertically (Kain 2012). The Series X, the SUV model, which Injury Reserve reference, costs in the region of $80–110k and, as such, represents a significant symbol of status and affluence. Tesla is also known for controversies surrounding lawsuits, worker rights, and the public profile of CEO Elon Musk.

5 Such readings would undoubtedly be racialized in this context, especially considering the carceral connotations of the word "jailbreak."

6 Examples of Black creative technological intervention include the glass bottle (for slide guitar), turntable scratching, the aerosol can, pitch correction software, and the mobile phone, among many others (see, e.g., Weheliye 2005; Fouché 2006; Katz 2012). An in-depth case study is provided by Joel Dinerstein (2003, 5), who writes on big band swing as a stylization of the rhythms and aesthetics of the machine age, whereby "African American artists integrated the speed, drive, precision, and rhythmic flow of factory work and modern cities into a nationally (and internationally) unifying cultural form."

7 That this ideal is constructed does not deny that record producers often do emphasize the "stereo triangle" as a mixing environment, but it does help to avoid assumptions about the domestic or professional spaces available to the majority of listeners (Zagorski-Thomas 2014, 166–67). It also warrants a discussion of listener (dis)ability, especially how hard of hearing (non-culturally deaf) individuals and members of the Deaf community still find significant value in listening to music, in many cases through bass sounds that can be felt throughout the body (Holmes 2017, 171–73).

8 Our thanks to Will Moylan for this observation.

9 A recent example of car sounds in mainstream pop is Olivia Rodrigo's Grammy-nominated "drivers license." The song begins with an engine starting, keys jingling, a car door closing, and start-up chimes which blend into the keyboard timbre that accompanies the first verse. This produces a diegetic crossfade effect from the "foley" car starting to Rodrigo's musical performance of the song (and positioning the song's protagonist as the driver, a narrative position reinforced by the lyrics).

10 A similar approach to bringing the sound of the car directly into the beat production is taken by Charli XCX's "Vroom Vroom," produced by the late UK artist SOPHIE. The idiosyncratically metallic and elastic synthesized timbres for which the producer SOPHIE is celebrated are a significant sonic inspiration for "Jailbreak." Among other samples which may be borrowed, the sound effect we label "bubble" is taken from her sample pack ("SOPHIE_fx_03.wav"), which is commercially available through the sample library Splice.

11 While Tesla appears to be a desirable brand in contemporary hip-hop, it has not necessarily superseded other favorites such as Mercedes-Benz and Chevrolet, which still enjoy widespread lyrical mentions (McLeod 2020, 50–51).
12 For a more detailed engagement with Gilroy's expansive argument, see Williams (2014, 120–21), and for a critique which engages thoughtfully with gender, see Brown (2012).
13 Consumers increasingly appear to be engaging in such practices. In one article, reporter Aaron Gordon (2020) details a dispute between Tesla and used car buyers, where important software features like the driver assistance system appear to be removed by the company because it was not purchased directly. By way of response, users are deploying unofficial software patches to reinstall "features owners feel are rightfully theirs" (ibid.), in a similar manner to jailbreaking tools.
14 The casual use of "explicit language" is commonplace in hip-hop, by now a standard feature of the genre.
15 Injury Reserve announced that Groggs passed away in June 2020.
16 A well-known example performed in rap music is Jay-Z's "99 Problems," where Jay-Z recounts an encounter with police for "driving while Black." Jay-Z imitates the police officer's racially charged accusations by using a (white) Southern American accent, further emphasizing the political divide between the white police officer and the African American driver.
17 "Black (feat. A$AP Ferg)" by Buddy also captures the spirit of this gesture in the lyric, "400 years of oppression, I'm about to get me that black Tesla."

References

Baber, Brendan. 1997. 'The Incredible Shrinking Sound System—Interview with Composer and Record Producer Stewart Copeland.' *Interview*, September 1997. https://web.archive.org/web/20060619072644/http://findarticles.com/p/articles/mi_m1285/is_n9_v27/ai_20227116/.

Bijsterveld, Karin, Eefje Cleophas, Stefan Krebs, and Gijs Mom. 2013. *Sound and Safe: A History of Listening Behind the Wheel*. New York: Oxford University Press. https://doi.org/10.1093/acprof:oso/9780199925698.001.0001.

Brown, Adrienne. 2012. 'Drive Slow: Rehearing Hip Hop Automotivity.' *Journal of Popular Music Studies* 24 (3): 265–75. https://doi.org/10.1111/j.1533-1598.2012.01333.x.

Christgau, Robert. 2017. 'Expert Witness with Robert Christgau: Up, Down, and All Over the Place.' Noisey. February 3, 2017. https://www.vice.com/en/article/kbgg4x/expert-witness-with-robert-christgau-up-down-and-all-over-the-place.

Clarke, E. F. 2005. *Ways of Listening: An Ecological Approach to the Perception of Musical Meaning*. Oxford and New York: Oxford University Press.

Demers, Joanna. 2003. 'Sampling the 1970s in Hip-Hop.' *Popular Music* 22 (1): 41–56. https://doi.org/10.1017/S0261143003003039.

Dinerstein, Joel. 2003. *Swinging the Machine: Modernity, Technology, and African American Culture between the World Wars*. Amherst, MA and Boston, MA: University of Massachusetts Press.

Edgar, Amanda Nell. 2020. 'Listening to Music in Cars While Black: Popular Music, Automobility, and the Murder of Jordan Davis.' In *Popular Music and Automobiles*, edited by Mark Duffett and Beate Peter, 137–54. New York: Bloomsbury Academic. https://doi.org/10.5040/9781501352331.

Fouché, Rayvon. 2006. 'Say It Loud, I'm Black and I'm Proud: African Americans, American Artifactual Culture, and Black Vernacular Technological Creativity.' *American Quarterly* 58 (3): 639–61.

Gamble, Steven. 2021. *How Music Empowers: Listening to Modern Rap and Metal*. London: Routledge.

Gilroy, Paul. 2001. 'Driving While Black.' In *Car Cultures*, edited by Daniel Miller, 81–104. Oxford: Berg.

Gordon, Aaron. 2020. 'People Are Jailbreaking Used Teslas to Get the Features They Expect.' Vice. February 11, 2020. https://www.vice.com/en/article/y3mb3w/people-are-jailbreaking-used-teslas-to-get-the-features-they-expect.

Haupt, Adam. 2017. '"True to the Game": Straight Outta Compton's Affirmation of White Supremacist Capitalist Patriarchy.' *Black Camera* 8 (2): 208–25. https://doi.org/10.2979/blackcamera.8.2.12.

Hobbs, Thomas. 2019. 'Injury Reserve—"Injury Reserve" Review.' Highsnobiety. May 20, 2019. https://www.highsnobiety.com/p/injury-reserve-review/.

Holmes, Jessica A. 2017. 'Expert Listening beyond the Limits of Hearing: Music and Deafness.' *Journal of the American Musicological Society* 70 (1): 171–220. https://doi.org/10.1525/jams.2017.70.1.171.

Huber, David Miles, and Robert E. Runstein. 2017. *Modern Recording Techniques*. 9th ed. New York and London: Routledge.

Injury Reserve. 2016. *S on Ya Chest*. Mp3. Floss. https://open.spotify.com/track/1vup3vAmSmFLxSIzGOM38V.

———. 2019. *Jailbreak the Tesla (Feat. Aminé)*. Mp3. Injury Reserve. https://open.spotify.com/track/7w0YG5qn9w8LYA4tHpQe0o.

Kain, Erik. 2012. 'Tesla Goes Back to the Future with Model X Crossover SUV.' Forbes. February 11, 2012. https://www.forbes.com/sites/erikkain/2012/02/11/tesla-goes-back-to-the-future-with-model-x-crossover-suv/.

Kane, Mark. 2022. 'Tesla Production, Deliveries Graphed through Q4 2021: Huge Surge.' InsideEVs. January 2, 2022. https://insideevs.com/news/558255/tesla-production-deliveries-graphed-2021q4/.

Katz, Mark. 2012. *Groove Music: The Art and Culture of the Hip-Hop DJ*. New York: Oxford University Press.

Kendall, Lori. 2011. '"White and Nerdy": Computers, Race, and the Nerd Stereotype: White and Nerdy.' *The Journal of Popular Culture* 44 (3): 505–24. https://doi.org/10.1111/j.1540-5931.2011.00846.x.

King, Jason. 2001. 'Rap and Feng Shui: On Ass Politics, Cultural Studies, and the Timbaland Sound.' In *A Companion to Cultural Studies*, edited by Toby Miller, 430–54. Malden, MA: Blackwell.

Lacasse, Serge. 2000. 'Intertextuality and Hypertextuality in Recorded Popular Music.' In *The Musical Work: Reality or Invention?*, edited by M. Talbot, 35–58. Liverpool: Liverpool University Press.

Lee, Michael S. W., and Ian Soon. 2017. 'Taking a Bite out of Apple: Jailbreaking and the Confluence of Brand Loyalty, Consumer Resistance and the Co-Creation of Value.' *Journal of Product & Brand Management* 26 (4): 351–64. https://doi.org/10.1108/JPBM-11-2015-1045.

Magaudda, Paolo. 2012. 'How to Make a "Hackintosh." A Journey into the "Consumerization" of Hacking Practices and Culture.' *Journal of Peer Production*, no. 2 (January): 1–8.

McLeod, Ken. 2020. *Driving Identities: At the Intersection of Popular Music and Automotive Culture*. London and New York: Routledge.

Moore, Allan F. 2012. *Song Means: Analysing and Interpreting Recorded Popular Song*. Surrey: Ashgate.

Moylan, William. 2002. *The Art of Recording: Understanding and Crafting the Mix*. 1st ed. Waltham: Focal Press.

———. 2020. *Recording Analysis: How the Record Shapes the Song*. New York and London: Routledge.

Pearce, Sheldon. 2019. 'Injury Reserve: Injury Reserve.' Pitchfork. May 22, 2019. https://pitchfork.com/reviews/albums/injury-reserve-injury-reserve/.

Pettitt, Joe. 2003. *How to Design and Install High Performance Car Stereo*. North Branch: Car Tech Inc.

Prior, Nick. 2018. *Popular Music, Digital Technology and Society*. Los Angeles: SAGE.

Rose, T. 1994. *Black Noise: Rap Music and Black Culture in Contemporary America*. Connecticut: Wesleyan University Press.

Savage, Steve. 2011. *The Art of Digital Audio Recording: A Practical Guide for Home and Studio*. New York: Oxford University Press.

Seabrook, John. 2015. *The Song Machine: Inside the Hit Factory*. London: Vintage.

Seo, Sarah A. 2019. *Policing the Open Road: How Cars Transformed American Freedom*. Cambridge, MA: Harvard University Press.

Sheller, Mimi. 2004. 'Automotive Emotions: Feeling the Car.' *Theory, Culture & Society* 21 (4–5): 221–42. https://doi.org/10.1177/0263276404046068.

Vinti, Mike. 2019. 'Injury Reserve—Injury Reserve.' The Quietus. May 31, 2019. https://thequietus.com/articles/26561-injury-reserve-album-review.

Walser, R. 1995. 'Rhythm, Rhyme, and Rhetoric in the Music of Public Enemy.' *Ethnomusicology* 39 (2): 193–217. https://doi.org/10.2307/924425.

Weheliye, Alexander G. 2005. *Phonographies: Grooves in Sonic Afro-Modernity*. Durham and London: Duke University Press.

Williams, Justin A. 2014. '"Cars with the Boom": Music, Automobility, and Hip-Hop "Sub" Cultures.' In *The Oxford Handbook of Mobile Music Studies*, edited by S. Gopinath and J. Stanyek, 2: 109–45. Oxford: Oxford University Press.

Zagorski-Thomas, S. 2014. *The Musicology of Record Production*. Cambridge: Cambridge University Press.

SECTION II

Track Revelations and Aural Mirrors

3

HEARING THROUGH THE GRAPEVINE

Marvin Gaye, Norman Whitfield, and the Long Journey of Motown's Biggest Hit, "I Heard it Through the Grapevine" (1968)

Andrew Flory

Searching for a Single

In January 1967, Marvin Gaye was in a period of transition.[1] He had come to Detroit seven years before with friend and mentor Harvey Fuqua and signed to the upstart Motown Records. Gaye's goal was to work on the cabaret circuit. He spent several years recording standards and performing in nightclubs, hoping to become a Vegas-style entertainer (Flory 2019). But his work as a cabaret singer was starting to become incompatible with a teen-pop image. Motown had achieved big-time bookings for Gaye, with long runs at places like the Copacabana in New York and the Cocoanut Grove in Los Angeles. Reviews were tepid, however. After releasing four standards-oriented LPs in search of success in this market Motown was loathed to continue trying.

Duets had always been an important element of Gaye's record output. He worked for about a year with the company's star, Mary Wells, before she departed for the West Coast. Motown had recently issued a duet single called "It Takes Two" with a new singing partner, Kim Weston. But just as the record started to gain steam, Weston and her husband, Mickey Stevenson, announced their departure from Motown. Stevenson had been a major writing partner and catalyst for Gaye's pop hits, further calling into question the singer's future direction.

During Gaye's first half-decade at Motown, he had found his greatest commercial success with teen-oriented 45s. He worked with most of Motown's most prominent writers and producers—including Stevenson, Smokey Robinson, and Holland-Dozier-Holland—releasing more than 20 singles. His records achieved consistent but not overwhelming success. He had several tunes reach the top of the Black charts but had a hard time cracking the upper echelon of the pop listings. More recently, he had hit a dry spell in this area of his career. His last successful string of singles—"Ain't That Peculiar," "One More Heartache," and "Take

DOI: 10.4324/9781003089926-6

This Heart of Mine"—had been written and produced by Robinson, with the last released in May 1966. His last record, "Little Darling (I Need You)," had been lackluster and Motown was scrambling to find something to issue in his name.

Norman Whitfield

Two weeks of recording in early 1967 would become among the most significant of Gaye's career. He was concerned with completing material for a new ballads LP during these sessions, but Motown had two other objectives. One was to find a new line for Gaye's duet work. "It Takes Two" was currently rising the charts and Weston was no longer available. Through a complex series of moves, Motown executives brought together newly signed songwriters Nicholas Ashford and Valerie Simpson, producers Harvey Fuqua and Johnny Bristol, and a virtually unknown female vocalist named Tammi Terrell to create a new series of duet records during this time. During the first vocal session of this two-week stint, Gaye added his vocals to the duo's first single, "Ain't No Mountain High Enough."

The matter of a solo single was the other goal of these sessions. Among the handful of tracks that Gaye recorded for this purpose was "I Heard It Through the Grapevine," a song written by Norman Whitfield and Barrett Strong. The latter was a Motown performer who had sung one of the company's first big hits—"Money (That's What I Want)" in 1959—then left to release records for other R&B indies like Atco and Tollie. Strong had recently returned to Motown as a writer, signed to the firm's Jobete publishing arm. The first song he copyrighted in this capacity was "Grapevine," in mid-1966, which he brought to Norman Whitfield (Dahl 2001, 317).

A self-made musical visionary, Whitfield was a character among characters at Motown.[2] He came to the company with virtually no experience and learned the trade of record production and songwriting, eventually rising to the highest levels of success in the music business. His first session as producer and writer was a track recorded in late 1962 for Gaye called "Wherever I Lay My Hat (That's My Home)." He also shared co-writing credit on the early Gaye hit "Pride and Joy." Whitfield took the reins for Gladys Knight and the Pips after the group joined Motown in late 1965 and controlled much of the production for the Temptations after the mid-1960s, delivering them "Ain't Too Proud To Beg" and "Beauty Is Only Skin Deep." He was eccentric and a bit of a hustler. In the same sentence, Mickey Stevenson remembered him as a genius who "looked at the world in a unique way" as well as a "pain in the ass" who wouldn't sleep and had a hard time expressing himself outside of musical creation (Stevenson 2020).

The Composition

"Grapevine" is a complex song.[3] Its lyrics speak from the perspective of a lover confronting their romantic partner. The protagonist is about to be dumped and they know it, privy to this information via a fast-moving and highly effective gossip network. In one sense, it is unrealistic fantasy: a forestall. Anyone who has experienced abandonment would like to have known about it beforehand. Preparation for something like this is rare.

In another sense, the text of "Grapevine" depicts the moment of realization, the exact point when the lover comes to terms with the relationship's end. Throughout, there is a mythological element of the grapevine as a metaphor within the lyrics, a sense that there are larger forces in the community that are greater than any individual. The "grapevine" becomes a legendary topic along the lines of "the midnight hour" or "the crossroads." "It went all the way back to Confederate black soldiers," Whitfield once recalled in an interview. "They had a grapevine in order to pass on their words and experiences to each other" (Zollo 2000, 146).

Musically, the verse is based on a blues. It does not use a conventional 12-bar form but rather a song-oriented scheme that draws on blues elements such as non-functional dominant seven chords, a basic i/I–IV–V harmonic vocabulary, and pentatonic orientation in both the melody and harmony. It begins with a minor-key riff that becomes the touchpoint of the song. The riff is a dyad in parallel fifths that skips up a minor third and then works its way back home by descending stepwise. The starkness of the passage is evocative. It seems to connect to the mythology of the grapevine, its otherworldliness and exoticism. Each of the realizations of the song in its early years includes this riff in one way or another. In early versions, it includes a downward 3–2–1 melodic motion. (The second part of the riff was later sometimes interpreted a third higher [5–4–3]). This core idea was an essential element of the musical work.

The riff is also recast using triadic language, slightly modifying the intervallic movement and incorporating plagal motion common in gospel harmonization. There is a modal mixture in these instances: the tonic is minor, but the subdominant based on the major (with an added seventh). The verse proceeds with two consecutive eight-measure phrases based on this riff, followed by retrograde harmonic movement through V7 and IV7—a hint at blues form. An 8-measure pre-chorus then leads to a square, 16-bar chorus. The chorus uses the major version of the tonic chord, offering a modal slipperiness that further contributes to the otherness of the harmonic language.

Early Recordings

Whitfield's first attempt to record "I Heard It Through the Grapevine" was in early August 1966.[4] The track was assigned preliminarily to the Isley Brothers, an R&B group with several important chapters in its history. The group had signed to Motown in late 1965 and quickly scored a moderate hit with the HDH-produced "This Old Heart of Mine (Is Weak for You)." But there was no obvious follow-up single. Whitfield and Smokey Robinson were both charged with the task, cutting tracks for the Isleys on the same day of recording that summer.

One of the songs Whitfield put forth was "Grapevine." This first arrangement used a relatively fast pace (about 128 beats per minute) and was arranged with an instrumentation that was sparse by Motown standards. He clearly wanted a Wurlitzer electric piano as the featured instrumental voice. It offers a muddiness, a lack of distinction in its lower range that is central to the ambiguity of the tonal language of "Grapevine." The remainder of the band track features drums, bass, two electric guitars in unison, and conga drums. The drums and congas set the tone, playing a funky beat based on an eighth-note pattern in the hi-hat. The guitars emphasize hard backbeats, chunking on 2 and 4 throughout most of the track. The bass is busy and melodic—a Motown stylistic trademark.

The artist assigned to this early version was altered almost immediately to the Miracles. Given the proximity of Robinson to the tracking session, this may have even happened on the day the song was cut. The track was eventually completed during another session about two weeks later with Robinson in the lead and backing parts sung by the Miracles. Like so many recordings in the history of Motown, however, this one languished. The Miracles issued an HDH track, "(Come 'Round Here) I'm The One You Need," for their next single and this early version of "Grapevine" went to the vaults.

Creating the Iconic Tracks

Whitfield clearly understood the power of the recording studio. He revisited his own catalog to a great extent, exploiting material by cutting and re-cutting the same song with different artists. He often recommissioned and tweaked arrangements and explored different styles when doing this. The whole process was beneficial to him as a credited songwriter, but it also revealed the central element of his strategy: to make song material, arrangements, and sonic artifacts blend in a manner that is only possible with the modern recording studio. Robert Fink (2011) has written about this element of Whitfield's work with the Temptations several years later, noting a meta-narrative running through the group's early psychedelic soul work like "Cloud Nine" and "Runaway Child (Running Wild)." This sort of synergy was not easy to achieve, and "Grapevine" was one of many songs that took more than one attempt to hit the mark.

Whitfield initiated a second recording of "Grapevine" in early February 1967. Given his track record of hits with the Temptations, it is no surprise that he initially listed this group as the artist for this track. It is hard to know who was actually playing on the sessions. There is a union contract for the date, but it includes far more players than those on the recording itself.[5] The instrumentation was nearly identical to the first attempt at recording the song, but the band took a different turn with the feel of the performance. This started with a reduced tempo of about 115 beats per minute, which dramatically altered the sense of space in the track. Compounding this was a much different drum part, which kept an eighth-note emphasis but moved the main timekeeping voice from the hi-hat to the floor-tom. With a low-tom and kick-drum in alternating succession, the resulting beat pattern was murky and thuddy, offering another example of primacy that connects to the "grapevine" theme in the lyrics.

Accompanying the kit drummer was a melodic Motown bass and auxiliary percussionist, who alternated between tambourine and conga. After the presentation of the main riff in the introduction, the latter role, probably played by Eddie Brown, features a foreshadowing tambourine roll that takes on a serpentine quality—yet another connection to the lyrical theme. Robin Attas (2015) has cited this opening in her article about the role of introductions in pop music, noting that it uses a "process-based approach," which is related to Mark Spicer's (2004) idea of an "accumulative form." Instruments are added one voice at a time. Two guitars enter, playing chords under the introductory riff before settling into backbeat rhythmic stabs. The riff is still central to the arrangement in this version, but the opening keyboard instrumentation is augmented with a pulsating lower organ part. The two keyboard instruments, probably played by the same performer (Griffiths, Gittens, or Van Dyke), blend in a manner that creates a sound that is more than the sum of its parts.

Motown was a leader in recording technology, the evidence of which is embedded into this recording of "Grapevine." Since early 1965 the company had been tracking to bespoke one-inch eight-track machines, allowing for greater flexibility in mixing and far more specificity with sonic character during tracking. The initial tracking session for the 1967 "Grapevine" used five of these available slots, one each for bass, drums, percussion, keyboards, and guitars. The natural ambience of the 2648 West Grand Boulevard studio was such that generous room sounds were all that was needed for both percussion tracks. "The studio floor was a slab on the ground," with hardwood coverings that had recently been added during a remodel, remembers former technical engineer John Windt (Ohlsson and Windt 2021, 31:40). "There was no sound-deadening in there at all … there weren't any baffles. There was nothing to absorb any kind of sound in the room except the bottom end" (59:45).[6]

Guitars and bass were recorded through another homemade piece of gear, a large direct-injection unit that allowed for monitoring and included a single dial for each monitor level. Artificial ambience was added to two tracks, printed indelibly to tape: the guitars received a generous amount of chamber-generated reverb and the entire keyboard part was affected with a thick spring effect throughout the introduction, causing the unit to pulsate in a quarter-note rhythm. This adds to the spookiness of the sonic scene.

Following the common peripatetic path of song allocation at Motown, this new "Grapevine" track was reassigned almost immediately to Marvin Gaye, who added his vocals in early February 1967. His performance was stunning. Like many of Gaye's vocal parts during this period, the melody was at the very top range of his chest voice for nearly the entire song. Robinson had performed the song in the same key, but singing in his trademark, breathy *mezzo voce*. He also interpreted the melody slightly differently—lower in some spots. Robinson was going for a pop sound, while Gaye—compounded and aided by a tempo change and other instrumental alterations—evoked more of a blues. The loneliness and urge to refrain from crying in the lyrics are far more evident in his performance.

Some of this was aided by technology.[7] The dynamics of Gaye's vocals were altered through dynamic limiting, which is evident in a notable amplitude shelf on the final vocal track. This allowed him to sing loudly and passionately—probably through a Neumann U67—and not overpower the input. And the color of the limiting, most likely filtered through a Universal 175 or Fairchild 660 or 670, added to the grit of the performance. The effect was used while tracking, and it is clear that Gaye is reacting to the technology while singing. Electronics alone were not the driving force of the performance, however. By this point, Gaye was a seasoned vocalist with complete control over his instrument—a singer who had put in his 10,000 hours, even if reluctantly, with the ability to knock off a performance of a lifetime during an otherwise-mundane afternoon recording session.[8] "Grapevine" was one of the several-dozen vocal performances that Gaye cut during this short stint in the studio. (Four days later, the band track was cut for his next duet single with Tammi Terrell, "Your Precious Love.")

Development and Sweetening

Whitfield continued to develop the track. He quickly commissioned backing vocals, which were recorded several days later. Performed by the Andantes, Motown's go-to female support trio, these parts were markedly different from the ones created by the Miracles.[9] The Andantes used a far more dialogic tack, repeating lyrics using an echo technique and remaining slightly behind the pace of Gaye's vocal melody. Most striking is the group's precision. Using similar diction, phrasing, and microscopic details like vibrato onset, the sounds of the three singers blend in a manner that merges them into a single voice. Exactitude is compounded by striking textural changes in the tonal space. While maintaining a lockstep rhythmic profile, the three group members alternate between unison lines and thorny and dissonant chordal passages. They also introduce several important backing vocal figures that move into the spotlight of the arrangement. The most notable is the tag line after the first verse-chorus section, which comes after Gaye stops singing and the group's echo continues over the riff in an alternating pattern at the interval of a fifth ("I heard it through the grapevine/not much longer would you be my baby").

It is unclear when Motown arranger, Paul Riser, first entered the project.[10] He most likely arranged the lead sheets for the backing session. The same charts were probably used for both the Miracles and Gaye versions, with all of the differences between the basic backing tracks (tempo, melody drum pattern) stemming from the performative leeway allotted to producers and session musicians. Thus, the charts were most likely completed months before the Gaye recording. Riser's involvement in the backing vocal arrangement is unclear. There is no question, however, about his central role in the string and horn arrangement. An experienced and busy arranger, Riser was among Motown's best at marshaling the forces of orchestral instruments to support a soulful R&B single. Whitfield had not pursued an orchestral part for the first recording of

"Grapevine." Reaching this stage signaled that he was bullish on the prospects of this second version.

Riser's instrumental "sweetening" parts were recorded in early April. His chart is masterful. It is split between two main voices: one a trio of French horns, the other a group of violins, violas, and cellos. The arrangement starts slowly, entering at the end of the introduction with a dramatic octave leap in the horns. During the verse sections, Riser casts the strings with a slinky melody response to each of Gaye's vocal utterances. The pre-chorus material shifts its rhetorical approach, employing a thick, chordal texture that supports a rising line in the violins while backing vocals maintain the dialogue. All of the instrumental and vocal elements play together for the first time on the chorus when they collectively emphasize the rhythmic motive associated with the word "grape-vine." A striking flutter-tongue horn part at the end of this section offers one more sonic allusion to the primitive and mysterious grapevine theme in the lyrics. The climax of the entire song comes in the orchestral arrangement after the second verse. During a short instrumental bridge, a two-tone melody modified from the main riff—performed in the upper strings—takes the lead voice, accompanied by dramatic octave jumps in the horns.

Rejection

Even with new backing vocals and a completed string arrangement, this version did not proceed through Motown's Friday morning Quality Control meeting. It was rejected for release as a single and, like the Miracles version, seemed to fade quickly into the company's growing archive. For his Gaye's next solo single, Motown opted for an older HDH track called "Your Unchanging Love." It was a departure for the singer, who seemed to be moving more substantially into the earthy soul market. His solo singles were becoming less of a priority, however, after newfound success as a duettist with Terrell. "Ain't No Mountain High Enough," the duo's first single (whose vocals Gaye had recorded about a week before he sang "Grapevine") slowly ascended the charts during late spring and early summer.

Whitfield also moved on. In short order, he initiated a new set of tracks for "Grapevine," using a new arrangement for Gladys Knight and the Pips. During the months following, he continued to record the song, creating new versions with the Temptations, Bobby Taylor and the Vancouvers and a joint Temptations and Supremes project. He also revisited his first Miracles recording, recording a new vocal with Smokey Robinson, which was released on the group's *Special Occasion* LP. The standout of these tracks was the Pips version, however, which was released as a single in late September 1967. By Christmas, it has risen to the number two single in the country—just below "Daydream Believer" one week and "Hello Goodbye" the next. It was a bona fide hit, justifying all of the effort Whitfield had put into recording and promoting the song.

All the while, Whitfield still believed in the Gaye performance. The evidence for this is in the trail of mixes, a series of audio documents that reveal extended activity with the Gaye tracks during the 18 months after the musical performances were complete.[11] One early mix exists from the short period before the orchestral performances added: it was a draft, most likely created as a demonstration disc for Riser. The first proper attempts at assembling the track with all of its parts most likely occurred in late April 1967. Three mixes were created that day each experimenting more daringly with ambient sounds. In the first two, the drums are heavily affected, creating a low, tribal drone. The third mix was filtered through a sheen of tinny plate reverb. These were probably the recordings used to evaluate the single for release when it was first rejected. They remind us how many factors there were to keep at bay during the analog mixing process. The variances of tape speed, tube-based effect units, leveling, temperature, and humidity made each attempt into its own audio universe.

Persistence

The Motown tape logs tell us that Whitfield continued to revisit the multitrack master several more times during the coming year-and-a-half. In a set of two mixes from fall 1967, he eliminated most of the orchestral parts and backing vocals during the first verse; the strings did not enter until the onset of the second strophe and the backing vocals at the end of that verse ("you could have told me yourself ..."). Again, ambience was a major point of experimentation. So was length. Both of these mixes begin to fade during the third chorus, cutting off the iconic final riff section that includes the "honey, honey I know" lyrical coda. Whitfield later completed another version around February 1968, restoring the orchestral instruments and backing vocals to the first verse.

All told, "Grapevine" sat on the shelf for about a year-and-a-half.[12] This was a critical time in the history of pop recording, a period when the record business was in the liminal space between mono and stereo. Freeform FM radio was emerging as a significant movement and the LP was slowly becoming the primary format of the music industry. Even though Motown had always been interested in the latest recording technology, the company's success was firmly planted in older playback formats: AM radio and 45s. All of this became central to mixing Gaye's "Grapevine." Many of the song's early masters were run down to Scotch 201 quarter-inch tape in mono using a stereo machine, probably one of the many Studer C37 decks in the Motown mixing room. In these cases, both left and right tracks carried the same information. Comparing these tracks through phase cancellation reveals only small differences, encoded through imperfections in the media itself (tape speed, etc.) and stereo reverb encoded during the mixing process. This may have been the result of working with a mono desk, but in any case, it represents a moment when these technologies were in flux. It was old-school mixing in a stereo world.

It was not until preparations for the release of a Gaye solo LP during fall 1968 that "Grapevine" was scheduled for release. Mixes were completed during late summer for a collection that would later be titled *In the Groove*. The track was slated as album filler occupying a middle slot on side one. In order to release it on the LP, Whitfield needed to create a stereo mix. In this version, the instrumental tracks are panned dramatically: keyboard, orchestral parts, backing vocals, and bass on one side; guitars, drums, and percussion on the other. The vocals sit directly in the center. Like many two-channel mixes from the period, this rendition is not conceived in a balanced manner that most recordists consider foundational to a modern headphone-oriented audience. Accordingly, it creates several interesting moments of spacial dialogue between left and right channels: the serpentine tambourine and French horns occupy different sides of the mix in the introduction and throughout the track the driving rhythm section sits left while most of the melodic activity occurs on the right.

The LP *In the Groove* was issued in late September with "Chained" as the lead single. The Gladys Knight and the Pips version of "Grapevine" had been a huge hit already and ended its natural cycle atop the charts at the very beginning of the year. King Curtis had already popularized an instrumental version (of the Knight rendition) that summer. It seemed that the song had already lived a full life in the pop arena. Many listeners probably first encountered Gaye's "Grapevine" as a typical Motown album filler track—a new recording of a song made famous by someone else, which helped the company to capitalize on its publishing holdings. Advertisements for the album appeared in *Billboard*, and a short 50-word blurb with a clear memory of the Knight hit noted that the collection included a "first-rate reading" of the song (*Billboard* 1968a, 73).

Single Release

Motown eventually released Gaye's "Grapevine" as a single in November 1968, nearly two-and-a-half years after the first studio work on the song. Gordy (1994, 275) claims in his autobiography, "the DJs played it off the album so much that we had to release it." Common lore holds that the song's popularity originated on WVON in Chicago. The track began to appear on playlists all over the country in the middle of the month, in Chicago, St. Louis, Philadelphia, and locally in both Detroit and Windsor.[13] It seems that the company released it first to radio in select markets—drumming up interest—before it was issued and available in stores. By the time the disc was released at the end of November, it was already at the top of several local radio charts. The blurb in *Billboard* called it Gaye's "most exciting and commercial entry in a while" (1968b, 74). The latter was certainly prescient. In the weeks and months to come, it became the best-selling disc in Motown's decade-long history.

The single itself used a mix that betrayed the peripatetic life of the track. The most up-to-date rundown—the one included on *In the Groove*—was cast in stereo for the purposes of the LP environment. But Motown was still using mono for most of its single releases, in keeping with the company's popularity on AM radio. Rather than construct a new mono single version, the Motown brass issued a version of the LP stereo mix "folded" to mono. In most cases, combining these two channels would result in the loss of anything centered due to cancellation, but the phase of one stereo channel had been reversed in a manner that seemed to presage the track's eventual single release. The end result—a track that quickly took the world by storm—was far from a carefully constructed mono product. Rather, it was a clumsy superimposition of the left and right channels of an unbalanced early-stereo mix.

The single shot to the top of the *Billboard* singles charts in both the mainstream ("Hot 100") and Black ("Soul") markets. It became Gaye's first number-one pop hit and remained at the top of the record listings for the entire month of January 1969. Motown eventually renamed and rebranded *In The Groove* as *I Heard it through the Grapevine* later in the year, using the same master plates and different cover art. The success of "Grapevine" marked a coming-of-age for Gaye and Motown during the last year of the 1960s. Including stripped-down and gritty sonic representations, the performance and arrangement contained elements of older Motown while also pushing into the future of crossover Black pop. On the surface, it was honest, real, and unfabricated in a manner unlike many of the company's pop-oriented singles from years before.

The Seeds of the Vine

This is really only the beginning of the story. "I Heard it through the Grapevine" had a long and varied life as a song that far outlived the stretch in early 1969 when it was the most popular single in the country. In short order, there were dozens of cover versions recorded and released to a public. Motown acts released no less than half-a-dozen renditions before the end of the decade, while a huge variety of others throughout American popular music also tried their hand at the tune. Rock bands, soul singers, and jazz musicians alike recorded and performed the song, altering its place in modern society from a fleeting pop tune to a more broad instance of vernacular music known by listeners from all walks of life.

In January 1967, Marvin Gaye's "I Heard it through the Grapevine" was one of several dozen vocal performances recorded at Motown during another busy week of never-ending studio work. It was little more than an attempt at a new single. It grew to become something far larger. Today, it takes only the first several seconds for most listeners to recognize the track. The crack of a tambourine and a low and rumbling Wurlitzer passage that uses a limited and "primitive" musical vocabulary are all it takes. While seemingly simple and direct, the history of "Grapevine" is rife with intricacies that reveal the history of Gaye, Motown, and Black pop in general. It is the story of creativity at Motown, a central conduit for R&B during the 1960s, but also a tale that twists like rumors spreading through a tightly knit community. Like the relationship in the story, "Grapevine" had its own secrets along the way. It is this richness that helped to create one of the most recognizable recordings in the history of American popular music.

Notes

1 Biographical sources for Gaye include Ritz (1985); Gaye (2003); Turner (2000); and Dyson (2004).

2 There are few resources about the life of Whitfield. For more, see Zollo (2000); *Black Music* (1976, 14); Davis (2008); and Fink (2011).

3 "I Heard it Through the Grapevine" was certainly written much earlier, but the initial Library of Congress application for copyright was dated August 15, 1966, with Norman Whitfield and Barrett Strong as co-authors.

4 This study was aided by the use of proprietary archival material, the most important of which were the Motown session logs and a series of "tape file cards" used internally during the company's Detroit years to organize information about publishing, sessions, and tape archive locations. These materials are held privately by Universal Music Enterprises.

5 Motown has a long history of interaction with the Detroit American Federation of Musicians Local 5. At the time of writing, the branch still operates but with a fraction of the activity from the post-War period through the departure of Motown. The Local maintains has a paper archive and has generously provided the author with many contracts for Motown sessions. These "B4" contracts provide a list of tunes and musicians present at the session. The contract for the "Grapevine" tracking session, a morning session dated February 3, 1967, also includes two other Whitfield-produced (and co-written) tracks: "Ain't No Sun Since You've Been Gone" and "You Don't Love Me No More." It lists 17 musicians (with instrument in parens): Henry Cosby (leader, saxophone), Richard Allen (drums), William Benjamin (drums), Eddie Brown (hand percussion), George Bohannon (trombone), James Gittens (keyboards), Johnny Griffith (keyboards), James Jamerson (bass), Floyd Jones (trumpet), Joseph Messina (guitar), Eugene Moore (saxophone), John Trudell (trumpet), Daniel Turner (saxophone), Earl Van Dyke (keyboards), Don White (trombone), Robert White (guitar), and Eddie Willis (guitar). Taking out the horn players, who clearly did not play on the "Grapevine" track, leaves two drummers, a percussionist, a bassist, three keyboardists, and three guitarists. This makes it difficult to determine which drummer, guitarists, and keyboardist actually played on the "Grapevine" track.

6 Bass traps were installed "all four walls in the ceiling up to about three feet," adds Windt (59:10).

7 A great source for detailed information about the technical aspects of recording at Motown can be found in Ohlsson and Windt (2021), an online lecture and conversation delivered to the Audio Engineering Society Pacific Northwest Section.

8 Stories abound of how difficult it was to get him into the studio. Dozier (2019, 110–14) offers an insightful anecdote about recording Gaye from the perspective of a producer.

9 For more information on the Andantes, see Wright, Demps, Barrow-Tate, and Hicks (2007).

10 There are few scholarly or print resources about Paul Riser. Among the most visible of the Motown arranger corps, he has delivered numerous lectures and led workshops about his work. Notable instances of this include Riser (2007, 2011).

11 Motown mixes can only be dated approximately. Dates are sometimes included on the larger reel, which contain compilations of mixes (or "duplicate masters") loosely grouped in order of accession. More often, looking at the other mixes on a reel is the best way to give a good approximation of the general time period that the mix was completed.

12 He released three singles as a solo artist ("Your Unchanging Love," "You," and "Chained") during this time, in addition to all of the memorable records that he recorded with Tammi Terrell.

13 A good source for radio surveys is the online crowd-sourced Airheads Radio Survey Archive (ARSA), http://www.las-solanas.com/arsa/index.php. Using the surveys listed in the archive, though not complete, gives a good picture of the national emergence of Gaye's "Grapevine."

References

Attas, Robin. 2015. "Form as Process: The Buildup Introduction in Popular Music." *Music Theory Spectrum* 27: 279–96.

Billboard. 1968a. "Album Reviews." September 21, 1968.

Billboard. 1968b. "Spotlight Singles." November 16, 1968.

Black Music. 1976. "Norman Whitfield." 14–15, 52–53.

Dahl, Bill. 2001. *Motown: The Golden Years*. Iola, WI: Krause.

Davis, Sharon. 2008. "Norman Whitfield." In *Chiwaggin': The Classic Soul Interviews*. New Romney, UK: Bank House.

Dozier, Lamont. 2019. *How Sweet It Is: A Songwriter's Reflections on Music, Motown, and the Mystery of Muse*. BMG.

Dyson, Michael Eric. 2004. *Mercy, Mercy Me: The Art, Loves and Demons of Marvin Gaye*. New York: Basic Civitas.

Fink, Robert. 2011. "Goal-Directed Soul? Analyzing Rhythmic Teleology in African American Popular Music." *Journal of the American Musicological Society* 64 (Spring): 179–238.

Flory, Andrew. 2019. "The Ballads of Marvin Gaye." *Journal of the American Musicological Society* 72 (no. 2): 313–362.

Gaye, Frankie. 2003. *Marvin Gaye, My Brother*. San Francisco: Backbeat.

Gordy, Berry. 1994. *To Be Loved: The Music, the Magic, the Memories of Motown*. New York: Warner.

Ohlsson, Bob and John Windt. 2021. "Motown: In the Studio Where It Happened!" Online lecture delivered to the Audio Engineering Society Pacific Northwest Section. February 23, 2021. https://www.aes-media.org/sections/pnw/pnwrecaps/2021/feb2021/index.htm.

Riser, Paul. 2007. "Paul Riser Master Class." Lecture delivered at American Society of Music Arrangers and Composers at Los Angeles Valley College. June 2, 2007. https://asmacblog.org/2007/06/02/paul-riser-master-class-saturday-june-2-2007/.

Riser, Paul. 2011. "Funk Brothers Gonna Work It Out." Lecture and Interview with Jeff Mao delivered to Red Bull Music Academy. https://www.redbullmusicacademy.com/lectures/paul-riser-funk-brothers-gonna-work-it-out.

Ritz, David. 1985. *Divided Soul: The Life of Marvin Gaye*. New York: Da Capo.

Spicer, Mark. 2004. "(Ac)cumulative Form in Pop-Rock Music." *Twentieth-Century Music* 1: 29–64.

Stevenson, William "Mickey." 2020. Telephone interview with author. July 25, 2020.

Turner, Steve. 2000. *Trouble Man: The Life and Death of Marvin Gaye*. New York: Ecco.

Vickie Wright, Louvain Demps, Marlene Barrow-Tate and Jackie Hicks. 2007. *Motown from the Background*. New Romney, UK: Bank House.

Zollo, Paul. 2000. "The Motown Memories of Norman Whitfield: Legendary Writer of 'War,' 'I Heard It through the Grapevine,' and 'Just My Imagination' Looks Back." In *Calling Out Around the World: A Motown Reader*, edited by Kingsley Abbott, 144–48. London: Helter Skelter.

4

"LET THE MUSIC PLAY" (1975)

Hearing the Disco Mainstream in Barry White

John Howland

In a December 1975 advertisement for *Billboard*'s 1976 "First International Disco Forum," the magazine asked: "marketing disco product—how is it different?" (*Billboard* 1974a, 33). It was a prime moment for this business question. The advertisement parallels the release of the appropriately titled Barry White single, "Let the Music Play" ("Play"). While this essay will focus on the analysis themes of this book, through an examination of "Play," it will also illustrate how a context-minded aural analysis of chart-oriented mainstream pop can build enhanced understanding of a recording. A history-informed approach can particularly enrich analyses of cross-market, dance-oriented tracks as their social meaning and cultural aura are innately entwined with historical context, listener experience, and creative and marketing intention. With this enhanced framework, recording analysis can reveal how production, arrangement, mix, musical detail, lyrics, and performance fulfill the social and cultural musical expectations and manifest experiences of the listener/dancer. My analysis elicits understandings both of a hit record and how releasing multiple mixes shape distinct listener experiences and meanings.

Despite White's influence on 1970s mainstream disco, this single is the first White track to glorify discotheque culture. I suspect this was meant to reinforce the White disco brand. Top trade magazines reviewed "Play" through a similar lens. *Billboard* notes: "White is back with his patented … style, a fine disco arrangement and his usual catchy melody …. Nothing really different from past efforts, but why mess with success?" (*Billboard* 1975, 58). *Cash Box* observed: "Barry has an … unmistakeable style that has netted … thousands of radio hours and a mountain of record sales. With 'Let the Music Play' he sticks to his guns" (*Cash Box* 1975, 28). As explored in my 2021 book, *Hearing Luxe Pop* (*HeLP*), White's music received considerable backlash as guilty-pleasure, formula-driven "black Muzak" (Howland 2021). This context is sensed in rock-critic Robert Christgau's dually

DOI: 10.4324/9781003089926-7

acidic-appreciative reviews of White. In 1977, Christgau notes that White's "two previous albums"—including the 1976 *Let the Music Play* (*Play*)—"slipped precipitously (last one didn't even make top hundred)." Reviewing the March 1975 *Just Another Way to Say I Love You*, from which "Play" was an outtake, Christgau contends that "with product as uniform as White's, subtle differentiations take on … significance—I'd swear he's a shade more turgid … [T]his [album] failed to crack top ten. Statistics never lie. C+." Nonetheless, White's 1975 *Greatest Hits* album—released one month before "Play"—was awarded A—despite how "White manipulates studio technology with as much originality as … [1950s producer] Mitch Miller: the strings-versus-rhythm dynamics, as well as his much-maligned baritone, resonate with physical authority" (Christgau 2021).

By late 1975, White was enjoying industry accolades and back catalogue sales, but as Christgau suggests, White might have been chasing the mainstream market in sales, radio play, and the discotheque rather than leading. As a dance-oriented, mid-1970s commercial hit, "Play" received multiple release mixes. In this historical context, a mainstream disco hit was intended to be experienced in multiple mixes and varied listening/technology contexts—the radio, discotheque, car stereo, home stereo, etc. And while disco tracks—like mainstream, radio-targeted commercial pop—were routinely derided for adherence to convention, such formulaic qualities were essential for their discotheque proponents.

Sound-recording analyses within historical musicology (my field) rarely require analyses that are comprehensive and are typically framed through the interpretation of primary sources, style issues, and genre histories. Further, historical-musicology analyses must be in dialogue with larger research goals. As such, there should be an argumentative reason for why formal musical considerations might be central to a historical study—such as considerations of how large-scale structure is relevant to understanding studio practice or recording development or markets; how defining idiosyncratic or formulaic production elements are employed; how aspirational or artful elements of arranging (such as orchestral sweetening) are manifest; or how multiple mixes/tracks might compare in relation to their targeted listeners; among other possibilities. Several such topics frame this present study. Commonly, with longer analytical examples in historical musicology studies, integrated narrative framing demands both big-picture summaries and well-chosen considerations of limited passages that reveal insight into larger points. The analysis below retains this two-part approach.

While maintaining dialogue with William Moylan's *Recording Analysis*, this chapter articulates how one might *historically* hear White's "Play." "Hearing" here means attuning analytical listening to relevant historical, musical, technological, and cultural discourses. This essay pairs analytical intentional listening with a historical musicologist's understanding that a recording is an object whose meaning and experience are understood to be entwined with contextual history, whether experienced, imagined, or informed by subsequent media. I balance Moylan's argument for the unique status of a recording with its role as a *commercial* object whose intention and appreciation is entwined with the marketplace of pop. This essay suggests ways in which historical primary sources enhance evaluation through insights into the aesthetic-appreciation field that listeners/dancers brought to their experience of, and engagements with, a recording like "Play."

This chapter has a further concern for hearing the recombinatorial mechanisms of mainstream pop (see Leydon 2013). Such skills of "hit-making" and market success are central to hearing White's music. Riikka Hiltunen's research on the pop hit-making process posits "foresightfulness" as a companion to creativity. This notion includes predictive intuition, interpretations of taste trends, and balancing continuity (genre, style, etc.) with individuality (University of Helsinki 2021). White was undeniably skilled in foresightfulness. From this vantage, an analysis of "Play" must hear how historical popular-music products achieve hit status through targeting of the marketplace. To paraphrase Albin Zak, both pop production practice and audience experience are filtered through literacies in "a universe" of other recordings (Zak 2013, 43). Every track is thus unique in its positioning within a constellation of other recordings. Lastly, as a historical musicologist, record collector, and recordist, my relational filters (Moylan 2020, 60) to these recordings are a mix of professionalism, intimate *and* objective historical relations, and personal experiences with the physical recordings studied.

This chapter begins with observational considerations via Moylan's recording domains (music, lyrics, and recording), details about functions (primary, supportive, ornamental, and contextual), and perspective levels (Moylan 2020, 44, 45). My evaluations reflect on how these musical observations might be enhanced through insights gleaned from historical context. While the music's referential constellation is only narrowly articulated, I consider four recordings:

- The "single mix": The hit commercial 45-rpm single (released November 1975), a 3:25 mix primed for radio, the youth market, and lower-fidelity playback for home or dance/DJ situations (White 1975a).
- The "instrumental mix": The single is backed with a 4:15 instrumental mix for listening or dancing/DJ'ing entertainment (ibid.).
- The "extended mix": The album, *Let the Music Play* (released January 1976), includes a 6:08 title track (White 1976a).
- Lastly, I will also consider the original multitrack master.[1]

20th Century Records further released a 12-inch, 6:16, "For Disco DJs Only" mix (early 1976; this non-commercial release is the basis for the extended mix), and a 7-inch single with a 4:30 "full length disco" mix (late 1975; this is likely a vocal mix of the instrumental's arrangement).

Listening to "Let the Music Play"

While White's longtime engineer, Frank Kejmar, was central to the album's studio work, the *Play*/"Play" studio sessions are notable among White's 1973–1976 releases for the absence of his longtime co-arranger, studio conductor, and orchestral facilitator, Gene Page. *Play* instead offers "thanks to Mr. Don Peake for Orchestration" while crediting production, "arrange[ment] and conduct[ing]" to White (see the credits on White 1976a). As evidenced in the multitrack, White's track development likely first involved foundational rhythm-section work followed by subsequent part overdubbing, with the orchestral "sweetening" session(s) typically happening late in the process.[2] This aligns with his pre-1975 studio practice (see Howland 2021). The following lays out analytical observations progressively from formal considerations, to how such elements enhance the vocals and lyrical narrative, to how these core elements are enhanced through mix and arrangement choices in relation to materials present on the 24-track master.

Form

The "Play" mixes and master retain the structure—but not the mix—of the single as a central frame, as seen in Table 4.1. With this in mind, I first discuss the single mix, and then offer accounts of the instrumental and extended mixes. Table 4.1 outlines sectional material, timings, measure count, select texture

TABLE 4.1 Structural outline of the formal divisions in the mixes to "Let the Music Play": (A) the LP extended mix; (B) the instrumental 45-rpm mix; and (C) the commercial-single 45-rpm mix. Basic harmonic materials and primary textures are further noted

(A) Hit Single* [opening 26 mm. cut]** ***(B) Instrumental, 45-rpm, side B, mm.1–26												
m.1 / 0:00				m.9 / 0:20				m.17 / 0:39				
Intro A1 [8 mm]				Intro A2 [8]				Intro A3 [8+2]				
|	|	|	|	|	|	|	|	|	|	|	|	|
(C) "Disco" mix, LP, mm.1–56												
m.1 / 0:00				m.9 / 0:19				m.17 / 0:38				
Intro A1 [8 mm]				Intro A2 [8]				Intro A3 [8+2]				
|	|	|	|	|	|	|	|	|	|	|	|	|
PRIMARY TEXTURES: *Street drama and protagonist soliloquy*: cut-time; "street" audio; spoken R&B intro; high-register violin pads, slow cello-section countermelodies, majestic piano responses, straight guitar comping, sparse bass, eighth-note hi-hat, and subdued drums.												
KEY: C major												F9
INTRO A PROGRESSIONS Cmaj7 | ·/· | Em7 | ·/· | Dm7 |·/· | Dm11 | G7												

(*Continued*)

TABLE 4.1 "Table 4.1 Structural outline of the mixes to 'Let the Music Play'" (*Continued*)

<table>
<tr><td colspan="4"><u>(A) Hit Single, 45-rpm, side A, mm.1–30</u></td></tr>
<tr><td>m.1 / 0:00</td><td>m.9 / 0:20</td><td>m.17 / 0:40</td><td>m.27 / 1:03</td></tr>
<tr><td>Intro B [8 mm]</td><td>Chorus 1 [8]</td><td>Chorus 2 (extended) [8+2]</td><td>Verse 1 [10]</td></tr>
<tr><td>| | | |</td><td>| | | |</td><td>| | | |</td><td></td></tr>
<tr><td colspan="4"></td></tr>
<tr><td colspan="4"><u>(B) Instrumental, 45-rpm, side B, mm.27–56</u></td></tr>
<tr><td>m.27 / 1:02</td><td>m.35 / 1:22</td><td>m.43 / 1:42</td><td>m.53 / 2:06</td></tr>
<tr><td>Intro B [8]</td><td>Chorus 1 [8]</td><td>Chorus 2 (extended) [8+2]</td><td>Verse 1 [10]</td></tr>
<tr><td>| | | |</td><td>| | | |</td><td>| | | | |</td><td></td></tr>
<tr><td colspan="4"></td></tr>
<tr><td colspan="4"><u>(C) "Disco" mix, LP, mm.1–56</u></td></tr>
<tr><td>m.27 / 1:02</td><td>m.35 / 1:21</td><td>m.43 / 1:40</td><td>m.53 / 2:05</td></tr>
<tr><td>Intro B [8]</td><td>Chorus 1 [8]</td><td>Chorus 2 (extended) [8+2]</td><td>Verse 1 [10]</td></tr>
<tr><td>| | | |</td><td>| | | |</td><td>| | | | |</td><td></td></tr>
<tr><td>Enters disco: 4/4; spoken; disco groove; reduced texture, high strings, funky guitar, hi-hat/kick</td><td colspan="2">Lost in the dance-floor groove: four-beat cowbell; guitars, hi-hat, kick, cowbell, and bass move to the front of the sound stage; fuller sound; violins duet with sung vocal. (Instrumental mix: strings as voice/lead; piano, harp, celli, horns, vibraphone further back)</td><td>cowbell emphasis; guitar lead bubbles to surface</td></tr>
<tr><td colspan="4">Key: B-flat major</td></tr>
<tr><td>INTRO B, 4 mm twice:
B♭Δ9|Gm11|Cm6|F9</td><td colspan="3">CHORUS PROGRESSIONS
E♭maj7–E♭6 | ·/· | Dm7 | Gm7 | Cm7 | ·/· | B♭Δ7 | Dm7–G7</td></tr>
</table>

<table>
<tr><td colspan="4"><u>(A) Hit Single, 45-rpm, side A, mm.31–64</u></td></tr>
<tr><td></td><td>m.37 / 1:27</td><td>m.45 / 1:46</td><td>m.55 / 2:11</td></tr>
<tr><td>[Verse 1 cont.]</td><td>Chorus 3 [8]</td><td>Chorus 4 (extended) [8+2]</td><td>Verse 2 [10]</td></tr>
<tr><td>| | |</td><td>| | | |</td><td>| | | | |</td><td>| | | | | | | | | |</td></tr>
<tr><td colspan="4"></td></tr>
<tr><td colspan="4"><u>(B) Instrumental, 45-rpm, side B, mm.57–90</u></td></tr>
<tr><td></td><td>m.63 / 2:30</td><td>m.71 / 2:50</td><td>m.81 / 3:15</td></tr>
<tr><td>[Verse 1 cont.]</td><td>Chorus 3 [8]</td><td>Chorus 4 (extended) [8+2]</td><td>Verse 2 [10]</td></tr>
<tr><td>| | |</td><td>| | | |</td><td>| | | | |</td><td>| | | | | | | | | |</td></tr>
</table>

A + B + C (*Continued*)

TABLE 4.1 "Table 4.1 Structural outline of the mixes to 'Let the Music Play'" (*Continued*)

<table>
<tr><td colspan="4">(C) "Disco" mix, LP, mm.57–90</td></tr>
<tr><td></td><td>m.63 / 2:29</td><td>m.71 / 2:48</td><td>m.81 / 3:12</td></tr>
<tr><td>[Verse 1 cont.]</td><td>Chorus 3 [8]</td><td>Chorus 4 (extended) [8+2]</td><td>Verse 2 [10]</td></tr>
<tr><td>| | |</td><td>| | | |</td><td>| | | | |</td><td>| | | | | | | | | |</td></tr>
<tr><td></td><td colspan="2">(see earlier)</td><td>(see earlier)</td></tr>
<tr><td colspan="4">VERSE PROGRESSIONS
4-times 2 mm [Cm7 | B♭/F–D/F♯–Gm] + 2 mm [Cm7 | F11–F]</td></tr>
</table>

<table>
<tr><td colspan="3">(A) Hit Single, 45-rpm, side A, mm.65–86</td></tr>
<tr><td>m. 65 / 2:35</td><td>m.73 / 2:55</td><td>m.83</td></tr>
<tr><td>Chorus 5 [8]</td><td>Chorus 6 (extended) [8+2]</td><td>Chorus 7 (fade)</td></tr>
<tr><td>| | | |</td><td>| | | | |</td><td></td></tr>
<tr><td colspan="3"></td></tr>
<tr><td colspan="3">(B) Instrumental, 45-rpm, side B, mm.91–105</td></tr>
<tr><td>m.91 / 3:38</td><td>m.99 / 3:58</td><td></td></tr>
<tr><td>Chorus 5 [8]</td><td>Chorus 6 (fade)</td><td></td></tr>
<tr><td>| | | |</td><td>| | |</td><td></td></tr>
<tr><td colspan="3"></td></tr>
<tr><td colspan="3">(C) "Disco" mix, LP, mm.91–112</td></tr>
<tr><td>m.91 / 3:36</td><td>m.99 / 3:55</td><td>m.105 / 4:10</td></tr>
<tr><td>Chorus 5 [8]</td><td>Ch. 6 (short) [6]</td><td>Bridge (= Intro B) [8]</td></tr>
<tr><td>| | | |</td><td>| | |</td><td>| | | |</td></tr>
<tr><td colspan="2">(see earlier)</td><td>Intro B textures of breakdown/buildup</td></tr>
</table>

<table>
<tr><td colspan="2">(C) "Disco" mix, LP, mm.113–130</td></tr>
<tr><td>m.113 / 4:30</td><td>m.121 / 4:49</td></tr>
<tr><td>Chorus 7 [8]</td><td>Chorus 8 (extended) [8+2]</td></tr>
<tr><td>| | | |</td><td>| | | | |</td></tr>
<tr><td colspan="2">heavier cowbell and harmonizing vocal on these choruses</td></tr>
</table>

<table>
<tr><td colspan="3">(C) "Disco" mix, LP, mm.131–154</td></tr>
<tr><td>m.131 / 5:13</td><td>m.139 / 5:33</td><td>m.149 / 5:57</td></tr>
<tr><td>Chorus 9 (instr.) [8]</td><td>Chorus 10 (instr.) [8+2]</td><td>Chorus 11 (fade)</td></tr>
<tr><td>| | | |</td><td>| | | | |</td><td>| | |</td></tr>
<tr><td colspan="3">strings act as the lead vocal; bass moves forward</td></tr>
</table>

details, and harmonic progression content. The intro-verse-chorus divisions are of mostly conventional. The choruses have an eight-bar harmonic progression (E♭maj7–E♭6 | ·/· | Dm7 | Gm7 | Cm7 | ·/· | B♭maj7 | Dm7–G7) but are presented in two-chorus pairings. The second statement adds a two-bar extension to heighten the lyrics' "on and on" moment. The progressions of the ten-bar verses are built from four repetitions of a two-bar vamp (Cm7 | B♭/F–D/F♯–Gm) and an appended two-bar variation (Cm7 | F11–F). There are two distinct subsections employed in the introductions of the three mixes. The single only uses the second introductory subsection, which is labeled "Intro B" in Table 4.1. Intro A (discussed below) appears on the instrumental and extended mixes. White's characteristic bedroom-baritone spoken rap is featured in the single and extended-mix introductions.[3] As compared to Chorus 1, Intro B foregrounds a less-energetic, reduced texture of strings, funk-lite guitar, and hi-hat. Intro B is built on a repeated four-bar progression (B♭maj9 | Gm11 | Cm6 | F9). This subsection starts the single directly with a propulsive, disco groove foregrounding hi-hat above a four-on-the-floor bass drum pattern (striking one–two–three–four–*and*). (The hi-hat groove involves a half-open, 8th-note strike on each pulse beat, with off-beat double-16th strikes in-between on a closed hi-hat.)

The single arrangement follows Intro B with two statements of these two-choruses-plus-verse sectional couplings, an additional chorus pair, and a fade-out on Chorus 7. The late arrival of Verse 1 is a dance-music strategy that foregrounds the primacy of the chorus in a disco track. Here, the Chorus 1 arrival is further enhanced through the first full-toms drum fill at the end of Intro B, and Chorus 1's percussion expansion with a four-beat cowbell pulse (midway back in the sound stage) and a fuller instrumentation overall.

As was common in mainstream disco, "Play" aimed for success across discotheques, Top-40 radio, and private consumption. This market-segment sequence is tiered, with success at each level (hopefully) building popularity at the next, leading to mass-market sales. Such multi-market targeting was achieved via the multiple mix releases seen with "Play." The extended mix was both central to initial discotheque success and the title track for the LP, the most profitable consumer medium.

As seen in Table 4.1, the single's arrangement forms the foundation of the three mixes. The extended and instrumental mixes firstly expand this framing via Intro A, which involves three statements—labeled Intro A1, A2, and A3–of an eight-bar harmonic progression (Cmaj7 | ·/· | Em7 | ·/· | Dm7 |·/· | Dm11 | G7). Underneath soaring strings, a lushly held F9 in the two-bar extension of Intro A3 (an 8 + 2 bar construction) allows a C to Bb modulation. This shift is paired with an increase in rhythmic intensity from a repetitive, cut-time eighth-note hi-hat pattern in Intro A, to the 4/4 eighth-plus-two-sixteenth-note pulse-level pattern of Intro B. Both Intro A and subsections feature White's spoken-rap R&B intro narrative. While the song maintains a dance-floor-friendly 100 BPM, between Intro A and Intro B, there is an upshift in intensity felt both via a cut-time to 4/4 pulse shift and an enrichment of instrumentation and rhythmic energy between the subsections. In instrumentation, Intro A involves long-held high-register violin pads, slow cello-section countermelodies, majestic piano responses, straight-four-beat guitar comping, sparse bass, and subdued (cut-time) kick plus hi-hat drums. Following a downward-cascading string run at the Intro A3 turnaround, Intro B activates each of these instrumentation areas: there are fuller, middle-high violins (an octave below Intro A) set on an active melodic riff, as well as rich, low cello support and a newly groove-locked rhythm section (rhythmically funky guitar and piano comping, activated bass, and more active drums). This enhanced Intro B material aligns with the protagonist's narrative arrival in the discotheque. Thus, there is a tiered-energy design mapped onto the formal sections, from the "street" interior soliloquy of Intro A to Intro B arrival in the disco to the lush personal drama of the dance-floor themed Chorus 1.

The instrumental and extended mixes also involve alternate formal endings. The instrumental mix concludes with a Chorus 6 fade-out. Bar 6–7 of the chorus' harmonic progression involves a Cm7 to B♭maj7 shift. In the extended mix's shortened Chorus 6, this seventh-bar move to B♭maj7 is spliced with Intro B's B♭maj9 opening to provide the Bridge section, which is a reworking/reprise of Intro B. Where Intro B previously built up to Chorus 1, here its reduced texture has a dual disco-style breakdown/buildup and Bridge function, albeit with new lyrics and vocals (again spoken). This is precisely timed for the standard 1975 disco-mix breakdown. This Bridge then leads to two paired-chorus statements, first with vocals (7 and 8), then instrumental (9 and 10), followed by a Chorus 11 fade-out.

Vocals, Lyrics, and Narrative

Listening to "Play" also involves an engagement with the lyrics, which create both the imagined cinematic drama and persona of the song. Contrary to the reduced content of disco lyrics in contemporary tracks like Donna Summer's "Love to Love You, Baby" (on the chart at the same time), "Play" merges an early R&B story-narrative and contemporary disco-themed lyrics. The former maps onto White's spoken/rap introduction, a White convention that extends a long R&B intro-narrative tradition. With White, these are often bass-baritone bedroom seductions professing/explaining love, but "Play" is the rare White breakup song. "Play" also adds audio that suggests real-world ambient interactions. Intro A involves an ambient space akin to an apartment-block lobby—one hears car noises; shuffling shoes; interactions involving a man ("hey, man what's happenin'?!?") and several ladies (inaudible words, giggling); whistles; and a door opening. Intro B involves background party screaming. White's lyrics interact with these scenes. There's a street party going on and the protagonist is not interested. The interior monologue tells us that the protagonist's woman is "at home," and he has "been out here walking and talking for about four hours," reflecting on romantic "hurt." He interacts with the other vocals. After midnight, he decides to drown his troubles in the immersive discotheque. The end of Intro A tells us, "everybody's going in, lemme step on up," and Intro B then frames the entry: "One ticket please. Lord have mercy, everybody's here."

The sung chorus' vocals, melody, and phrasing are akin to White's best-known hits, though here he glorifies the immersive disco experience as opposed to love: "Let the music play, I just wanna dance the night away," etc. This is phrased to three short descending arch-like phrases before White shifts upwards to his highest register on "Ooh, ooh, ooh, ooh-wee!" Over Chorus 2's last five bars, he sings of losing "misery" through "movin', kickin', groovin'," and wanting this to go "on and on and on." With the latter, all the mixes employ a fader to cinematically pull White's passionate vocals back down into the mix. The verses employ a talk-sing performance with interjections ("hah, hah, hah!"). The lyrics here—set to rising, assertive mid-register lines—reflect on "what's killing me inside." In the song's design, the lovelorn dancing protagonist shifts repeatedly between losing himself in the groove and momentary "reality" reflections.

From 24-Track to Mixing

Each of the "Play" releases targets a different audience engagement via a mix restaging of the foundational 24-track master recording. The master is outlined in Table 4.2, which describes the contents and audio qualities of each track. Access to a multitrack offers unprecedented insights into the specifics of a recording, its arrangement, its steps of construction, the production, and performative decisions of the development and the mix. Many of the points made throughout this essay have been clarified, refined, and even identified through the study of this multitrack. Historically multitrack masters have been inaccessible to scholars. Research access to this multitrack was granted by Universal Music Group, but there is now a multitude of leaked historical multitracks. With an orchestral-pop track like this, an equally valuable (but difficult to obtain) primary document is an extant score arrangement, which was unavailable here. In its absence, the multitrack offers the best source for accurate transcription and assessment. What material is left in a mix and how it is adjusted in the final edit is balanced with information about what was left out, altered, or otherwise adjusted to achieve a unique mix's audio image. The multitrack offers insights unavailable to, or potentially overlooked by, a study of the commercial mix alone. In the analytical pursuit of articulating the creative process and the contents and design of the recordings studied, the whole is most completely understood through the added study of the constituent multitrack details, something that is difficult with dense commercial audio recordings.

Beyond the details of Table 4.2, there are additional master-multitrack audio comments to be made here. First, the three electric-guitar tracks offer complementary parts and timbres, and there is a composite (submix) track with a distinctly varied mix of the four guitar parts. The "disco-horns OD" ("OD" is "overdub") track includes a late (and brief) horn-section enhancement as well as

TABLE 4.2 Layout and audio contents of the 24-track digital master recording to "Let the Music Play." Examined at the New York offices of Universal Music Group, May 2015

Track no.	*Track name*[4]	*Audio contents*
1	Barry1	Lead track of White's vocal. Performed to full backing tracks (evident via headphone audio bleed).
2	Barry2	Harmonizing overdub track of White's vocal. Performed to full backing tracks (evident via headphone audio bleed, which includes track 1).
3	Bass	The electric bass is recorded flat via a small, miked amplifier.
4	Guit1	"Guit1" (a semi-hollowbody?) has a slightly gritty tone, performs full chords in mid-low register and single lines on low strings.
5	Guit2	"Guit2" is a solid-body Stratocaster-type guitar, which performs a rhythm/scratch[5] part in mid/high registers, emphasizing 2–3-note chord stabs, bright timbre, and light spring reverb.
6	Guit3	"Guit3" has the same tonal qualities as Guit2, playing both glissando phrasings and secondary rhythm parts, alternating high and mid registers.
7	Guits-composite	This composite mix involves a varied mix of the four guitar parts.
8	Hi Hat-Snare	Tightly miked on the hi-hat, but the track captures other percussion bleed (e.g., tom, tambourine, etc.). This is isolated for enhancement in the dance mix(es).
9	Kik	Tightly miked on the kick drum, but the track captures other percussion bleed (e.g., tom, tambourine, etc.). This is isolated for enhancement in the dance mix(es).
10 11	PianoL PianoR	Piano "L" and "R" offer two mic positions on White's keyboard.
12 13	celli1 celli2	Two differently miked celli tracks from the sweetening session.
14	cowbell	This cowbell track is an overdub that enhances the choruses and verses.
15	disco-horns OD	The "disco-horns OD [overdub]" track includes a brief horn-section enhancement late in the recording as well as 16 seconds of crowd noises for the introduction.
16	guit4	The guitarist for Guit2/Guit3 likely performed "Guit4," a more active track (a 16th-note scratching groove) with a non-reverb solid-body rhythm guitar.
17	harp	The harp track has string-section bleed that suggests it was performed live in an isolation booth.
18	horns	The French horns are muted both via hand muting and a nasal-sounding straight mute. The track has string-section bleed that suggests it was performed live in an isolation booth. The track also has notable ambient noise from the musicians.
19	toms	This audio includes both toms and cymbal accentuations from the drum-kit tracking.

(*Continued*)

TABLE 4.2 "Layout and audio contents of the master recordings of 'Let the Music Play'" (*Continued*)

Track no.	*Track name*[4]	*Audio contents*
20	traffic-violin OD	This audio includes the core pre-recorded street-drama audio. (I did not document the late-track "violin" presence.)
21	vibes	The vibraphone part involves four-part mid-range chords on beats one and three.
22	violins1	
23	violins2	The three violin-section tracks capture three mic positions.
24	violins3	

16 seconds of introduction crowd noises. Track 20, "traffic-violin OD," includes the core street-drama audio. This sounds to be pre-recorded in a building lobby on a portable recorder. Lastly, the initial live rhythm-section tracks (with only Guit2; the other guitar tracks are overdubs) form the foundation of the recording. As is evident in the string-bleed heard in the background of the two White vocal tracks, the string-sweetening session was overdubbed *before* the vocal session. It is unknown whether there was a scratch (temporary) vocal that was erased.

The five guitar tracks are central to the groove, which is fundamental for understanding the mixes. A well-known commercial single might be described as a static museum-like presentation of a track's art. With the multitrack, the museumgoer can step inside that static audio image and examine (and experience) its unique components from multiple dimensions and perspectives. By simultaneously studying disco-remix releases, one sees that the art is indeed not a static identity. The "Play" multitrack offers a mutable, multivalent script that can be manifest into a multiplicity differentiated realizations within dance-floor culture both then and now. Indeed, a 2012 reissue of the *Play* album offers three additional remixes alongside the single, instrumental (here 5:48), and extended mixes: the "M+M Throwback Mix" (9:07); the "Funkstar's Club Deluxe Mix" (5:50); and an "Alternate Version" (4:34) (White 2012). The CD thus offers two mixes for dance idioms and sonic palettes of the 2010s and four original-era realizations. All are unique, but collectively they are "the track."

As is especially evident in the instrumental, the strings frequently embody a vocal-melody role throughout the three mixes studied here. This is distinct from the harmony-pad, stab, and decorative-fill roles of strings on a disco hit like "Love to Love You, Baby." While White's strings employ similar textures, the violins (and sometimes cellos) frequently *duet* in melodic, vocally phrased unison octaves above White, as heard in Chorus 1. Excitement is generated by this juxtaposition of sweet high strings and White's bass-baritone, which adds phrasing and melody liberties. The violins are in a larger ambient space further back (and above) both the groove elements and White. For the extended and single mixes, White's voice is the protagonist and melody, but in the instrumental-mix and the extended-mix's instrumental sections, the strings are the voice (lead) part. In these instrumental mixes, groove elements gain importance and the guitar-composite, hi-hat, kick, cowbell, and bass move to the front, without much reverb/ambience. Their foregrounding lies between primary and supportive levels depending on the listening mode—the more lost-in-the-groove one is, the more central the groove is, hence their frontal role in the disco mix. The gospel-ish, middle-low-register rumblings of White's piano add supportive harmony and groove accompaniment, as well as ornamental material. The harp, celli, horns, and vibraphone add ornamental color further back.

The commercial-mix audio studied here is from original analogue media. My 45-rpm single sounds poor in comparison to the LP. While some fidelity loss comes from the "love" (frequent playing) of the original owner, the disc's material might play a role. Because of their higher rpm, 45s can potentially offer better fidelity than LPs, but commonly, cheaper material (polystyrene) was used. This seems to be the case here. The single's mid-range-dominant mix is also targeted to the technologies of its intended market (teens, narrow-band AM radio, DJs of all levels, etc.), and its sound works well on mid-/lower-range playback systems and AM and car radios. Overall, White and the lush strings dominate. In the introduction, the hi-hat (other kit components are barely audible) and the street audio are brought up. At the Intro B turnaround and Chorus 1, the snare, toms, and bass gain prominence, while the hi-hat lies underneath. The kick is heard, but midrange snare drives the mix. Harp interjections are faint, and horns, vibes, and other details are buried in the background. This balance also colors the verses, although guitar interjections bubble to the surface. The piano and composite guitar (not always heard) are mid-level. The fade-out emphasizes strings and piano glissandi.

While the instrumental retains qualities of the single, various materials are reframed. Intro A brings two guitars to prominence, with the piano at midlevel, support from balanced drums, and street audio further back for Intro A1. In Intro A2, the cellos rise to midlevel, the piano repeatedly leaps to the foreground, and snippets of street audio pop out. With Intro A3, the ethereal violins add single-line harmony support. The Intro A subsections have more dramatic buildup. At the arrival of Intro B, with violins as vocals, the groove components leap to the foreground. The composite guitar(s) are brought up further, with more cowbell and piano heard. Throughout, the harp glissandi rush forward above the kick and snare accentuations. The instrumental mix offers grooving, Black Muzak (see Howland 2021). The "rocket" string effects (discussed below) also burst forth in a way not experienced in the single. At the verses, the cowbell and three-part guitar lines rise to groove prominence.

The extended mix reveals different priorities and higher fidelity, both likely to benefit top hi-fi and discotheque sound systems. With Intro A, the same dramatic design is evident (as in the instrumental) but with White foregrounded, the earlier-mentioned instrumental parts back slightly, street audio raised, and a car horn to open the drama. All elements are crisp, and the kick dominates the low end. As Intro B arrives, the groove mix is reordered with low-end kick and sizzling hi-hit dominating, snare further back, and midlevel tom accentuations. The bass is lower for Intro B but rises at Chorus 1. The composite guitar retains prominence, and White's backing vocals have a clear presence. At Chorus 1, White's vocals are slightly under the drum groove, a facet that privileges the discotheque experience. At Verse 1, the cowbell moves forward, and groove elements are in a dry foreground space in front of the reverbed strings. While White's voice also has more ambient reverb than on the single from the Bridge, the vocal shifts to a dry presence (symbolizing internal thought?) for the new spoken dialogue ("I'm not out here looking for a one-night affair," etc.). At bar 5 of the Bridge, the break-style buildup qualities are manifest in part through a flurry of piano glissandi and the full Intro B drum-groove tracks dropping in. At Chorus 7, there is enhanced harmony vocal and cowbell. From Chorus 9, the orchestra shines.

Figures 4.1 and 4.3 present mm.7–8 of Chorus 1 and mm.1–2 of Verse 1. These illustrate the multitrack audio available. Figure 4.2 represents the three

FIGURE 4.1 Mm.7–8 in Chorus 1 of "Let the Music Play." Transcription based on multitrack material (with conflations of both piano and string track divisions, and silent tracks omitted). This corresponds to mm.23–24 (0:51 ff.) in the commercial-single 45-rpm mix; mm.49–50 (1:54 ff.) in the instrumental 45-rpm mix; and mm.49–50 (1:52 ff.) in the LP extended mix.

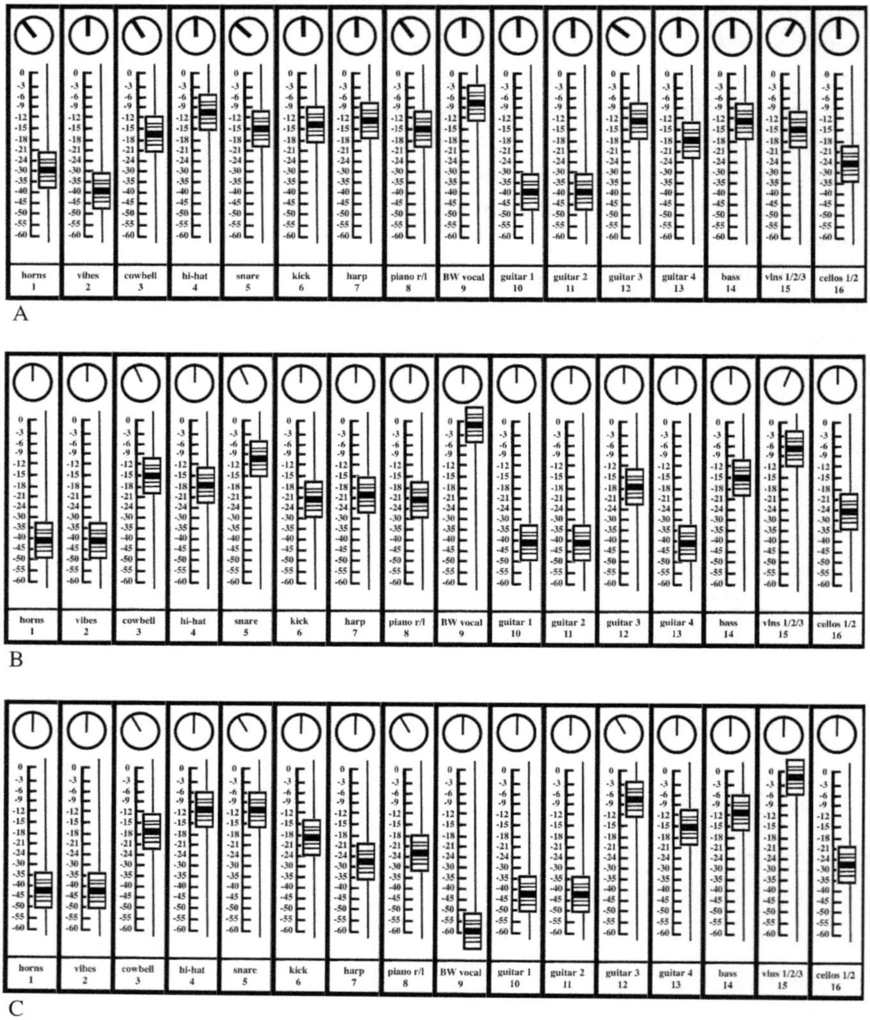

FIGURE 4.2 Perceived mix levels of track audio in the mixes to "Let the Music Play," mm.7–8 Chorus 2 (see Figure 4.1): (A) the LP extended mix; (B) the instrumental 45-rpm mix; and (C) the commercial-single 45-rpm mix. These are symbolic and relational, not actual mix levels. Below –35 db on the graphic indicates that track audio is inaudible within the mix and may or may not be present.

perceived mix balances heard for the Chorus 1 material in Figure 4.1. Lastly, the use of stereo-field separation is modest and employed for clarity in optimal private listening. This includes the slight right-field strings in the single and extended mixes (because White is center). The cowbell and snare have slightly more left-field presence in the mixes.

FIGURE 4.3 Mm.1–2 in Verse 1 of "Let the Music Play." Transcription includes all multitrack material. This corresponds to mm.23–24 (1:03 ff.) in the commercial-single 45-rpm mix; mm.49–50 (2:06 ff.) in the instrumental 45-rpm mix; and mm.49–50 (2:06 ff.) in the LP extended mix.

Groove components have varied levels of primacy in the mixes. The drumset and cowbell are central. Despite its machine-like, disco 16th-note hi-hat pattern (played half-open), "Play" avoids a straight four-on-the-floor kick. In Figure 4.1, the kick pattern involves a 16th-note pickup before the one and three strong beats (the one and three have snare reinforcement). This pickup note adds groove propulsion. Figure 4.3 shows a variation that drops beat three. The cowbell reinforces the four-beat disco pulse. Depending on the section and mix, bass and guitar have near groove equality to the percussion. While the guitar parts are deftly coordinated, the mixes generally favor one source at a time, although occasional secondary guitars can be heard. In Figure 4.1, in the extended mix, for instance, Guit3 dominates while Guit4 is faintly in the right field (Figure 4.2A). Figures 4.1 and 4.3 reveal a harmony-foundation role for the bass, with emphases on locking into the kick groove. The piano functions at a low-level supportive or ornamental level.

While the strings employ textures and gestures similar to earlier White hits (see Howland 2021), Figure 4.1 illustrates a new massed high-strings effect that is quite successful. Underneath White's "on and on" fade-out, the divided strings stagger scalar glissandi that rush from low to high register, and back again (a distant cousin to the late 1700s Mannheim "rocket" string effect). Figure 4.3 illustrates the vocal-style string textures. By contrast, the horns and vibes tracks are somewhat like vestigial limbs—while important to earlier White hits, here they are lost in the mix background.

"Play[ful]" Hearing: Evaluation and Conclusions

White talks in 1974 about his listening "homework" in following production practices in popular music (Robinson 1974). He references his studies of the sweet-soul of Philadelphia International and Motown records and how he adapted those models to the "Barry White sound" of 1973–1974. My observations on the "Play" mixes though reveal a moment when White shifts from a stylistic leader to an artist reacting to the market. This perception is seen in period reception.

Historical sources suggest framings for understanding White's relation to mainstream disco circa 1976. There is, for instance, an unprecedented 52-page White "Tribute" in a 1976 issue of *Record World* ("RW"). This arrives two weeks after the *Play* album. As *RW* notes, "the sound of the discos has become synonymous" with White (*Record World* 1976b). That said, while he personally emphasizes his business acumen, when asked about discos as a market for promotion early in his career, White feigns a period (1973) lack of awareness of discotheques. And when later asked, "do you … gear your records to that market?," he responds by outlining trend developments but saying, "No, I still don't … think about discos" (*Record World* 1976a, 8, 24). White was routinely framed as a leading disco act. Christgau's assessment of White's sinking chart successes is accurate though. As such, the analysis observations above suggest adjustments to the White sound for the changing disco market. "Play" is the most successful of White's 1975–1976 singles and marks his last *Billboard* top-10 hit (peaking at no. 9).

The transition from late 1974 to early 1976 was when industry recognized the potential of discos to make hit songs and started to accommodate the disco DJ's needs for extended disco-mix releases. By 1976, the latter was just starting to be disseminated narrowly in "DJ only" 12- and 7-inch extended-mix singles or album-side cuts. In 1974, one year before "Play," *Billboard* twice gives front-page coverage to extended disco mixes (see, e.g., *Billboard* 1974b, 1, 10). Mention is made of several four-plus-minute seven-inch mixes by pioneering disco-producer/DJ Tom Moulton (Shapiro 2005, 43–45). As a columnist, Moulton mentions White in his discussion of both the extended-mix album releases that were popular among DJs and the 7-inch practice of backing the "commercial" 3-minute 45-rpm single with a longer instrumental of the same tune. This practice gave DJs a nearly seven-minute long mix if two copies were played back-to-back (Moulton 1974b, 19). In December, *Billboard*'s disco-industry survey argues for the proven strengths of disco DJs in making sales hits, both through the affluence of disco customers and by breaking songs that later enter Top-40 radio rotation (Melanson 1974, 1).

While *Billboard* notes in December 1974 that White's label, 20th Century, had $2 million in monthly sales from hits first broken in discotheques, 20th Century was not among the pioneering labels providing extended edits to DJs. The same month, Moulton's column mentions the discotheque success of the extended mix of Al Downing's "I'll Be Holding On" (Moulton 1974a, 36). Moulton neglects to note that this was the first commercial 12″ extended single (Shapiro 2005, 45). After adjusting to these multiple-mix release strategies, 20th Century again reported $2 million sales in December 1975, and "the biggest scores were with … 'Let the Music Play' … and White's 'Greatest Hits' album" (*Billboard* 1976, 10). Again, "Play" reveals that White and his engineer were adapting their mix, production, and arrangement to this market.

The above overview of "Play" documents White's negotiations with what disco-historian Peter Shapiro calls the mid-1970s "art of the mix and the development of the twelve-inch single." Shapiro expounds on Moulton's influential early-1970s mix-tape edits (Shapiro 2005, 41–42), and notes Moulton's lesson that Fire Island's gay dance clubs—where White gained fame with "Love's Theme"—required mixtapes that capitalized on familiarity before taking audiences "somewhere they'd never been before" (ibid., 42). Moulton's mixtapes from 1973 led him to seeking master multitracks that might extend his remixes of singles. He also introduced the prototype disco break (see Mason 2020, Shapiro 2005, 43–44). By early 1975, Moulton had innovated the extended disco-mix 12-inch single (Shapiro 2005, 45), first via one-off acetate lathe cuts and then via limited vinyl pressings. Moulton later remarked to Shapiro that hearing the fidelity of the 12-inch acetate made "forty-fives sound like such shit. The levels were so much [better] … the dynamics, the bass, the everything, … [A]ll of a sudden [it's like] you're playing the master tape." Shapiro adds that 12-inch singles enhanced the discotheque primacy of the bass, allowed for grooves with greater volume, and allowed "seemingly infinite possibilities of restructuring songs" (ibid.). These concerns highlight the increasing desirability of the high-end, powerful sound systems then appearing in discotheque culture. The late-1975 "Play" mixes and media reveal a middle ground among these mainstream-disco trends.

White had been doing his homework concerning the discotheque market. All of White's 45-rpm vocal singles dating from 1973 forward included instrumental mixes on their flip sides. By 1975, the additional "Play" 7- and 12-inch disco-mix ("DJs Only!!") releases map onto this chronology of early maxi singles, as they are later called. "Play" marks the first White single after his 1975 platinum *Greatest Hits* LP which was drawn from two years of four hit albums and multiple singles (White 1975d). White's albums are far more eclectic than his media criticism suggests, but *Greatest Hits*—released at the first peak of mainstream disco—collects all his top singles, which not coincidentally were discotheque hits that defined the White sound. 20th Century also released a "For Disco DJs Only!!" 12-inch offering extended mixes of seven of the ten album tracks, with two being nearly eight minutes long (White 1975c). These releases canonized the White (dance-floor) sound in the month before "Play" was released. That said, with "Play," White was adjusting his mix, arrangement, and instrumentation parameters to better align his established sound with disco-production trends and hopefully continue his run of Hot 100 and disco market success. The extended mix's groove-enhancement is part of this attempt, as is the reduced orchestral sweetening. Late 1975 reveals some industry movement away from the fuller disco sweetening that White influenced—strings *plus* horns, brass, woodwinds, harpsichord, etc.—in favor of just strings. The reduced mix sweetening in "Play" suggests he had an ear to these trends (though his multitracks and live performances can include such enhancements). Nonetheless, as a producer, he self-consciously retains the integrity of the familiar framework of the single across these mixes. I suspect this strategy kept people on the dance floor just as they expected the single's fade-out. In sum, as stated in the lyrics, White's interest in success on the dance-floor ("I'm gonna … see what's going on in this discotheque") was central to his market aims ("yeah, right here, … where I'm gonna stay"). Let the music play.

Notes

1 Thanks to Harry Weinger, Universal Music Group, for generously allowing research access.
2 The "Play" multitrack includes rhythm-section tracks that only include microphone audio bleed from those rhythm-section instruments; the strings are isolated recordings (presumably recorded separately); and the horn track includes string bleed (likely from headphones, suggesting an isolation booth or a playback overdub at that session or later).
3 See Howland (2021, chapter 8) on White's voice. This use of "rap" here is in accordance with period references to spoken sections within contemporary soul.
4 These are the exact track names and ordering on the Pro Tools digital master file for "Play" at Universal Music. It is unknown how this ordering or track-title information aligns with the original master tapes and documentation.
5 "Scratch" guitar refers to a largely percussive part of the funk/soul rhythm-guitar techniques of "chanking." The latter style emerged in the 1960s, centrally via James Brown's guitarist Jimmy Nolan. Chanking—which conflates "choking" and "yanking"—partly references the "choking" sound heard from not fully fretting a struck/picked guitar note or chord voicing. This can create a percussive "scratching" instead of clear notes when performed with rhythmic vitality and an appropriately percussive right-hand strumming or picking technique. This comping (accompaniment) style also includes "stabs"—briefly fretted dyad and chord voicings (and individual notes) that are struck sharply before loosening the fretting hand (thereby choking the note/chord). The technique can further include slight string-bending ("yanking") to sharpen the pitch of a note briefly within this rhythmic context. Thus, in practice, such a scratch guitar parts includes a rich spectrum of percussive accompanimental sounds from noise to clear pitches and harmonies.

References

Billboard. 1974a. "First International Disco Forum." *Billboard*, December 6, 1974.
Billboard. 1974b. "Labels Mix Records for Club Scene." *Billboard*. November 2, 1974.
Billboard. 1975. "Top Single Picks: Barry White—Let the Music Play." *Billboard*, December 20, 1975.

Billboard. 1976. "$2 Million Dec. Sales for 20th Century." *Billboard*, January 17, 1976.

Cash Box. 1975. "Cashbox/Singles Reviews/Picks of the Week/Barry White." *Cash Box*, December 20, 1975.

Christgau, Robert. 2021. "Consumer Guide '70s, 'W': Barry White." Accessed July 7, 2021. https://www.robertchristgau.com/get_chap.php?k=W&bk=70.

Howland, John. 2021. *Hearing Luxe Pop: Glorification, Glamour, and the Middlebrow in American Popular Music*. Berkeley, CA: University of California Press.

Leydon, Rebecca. 2013. "Recombinant Style Topics: The Past and Future of Sampling." In *Sounding Out Pop*, edited by Mark Spicer and John Covach, 193–213. Ann Arbor, MI: University of Michigan Press.

Mason, Andrew. 2020. "Beat Doctor: Tom Moulton's Extended Disco Remix Forever Changed Recorded Music." *Wax Poetics*, December 11, 2020. Accessed October 12, 2022. https://www.waxpoetics.com/article/tom-moulton-beat-doctor.

Melanson, Jim. 1974. "Discos Demand Five-Minute Singles." *Billboard*, December 14, 1974.

Moulton, Tom. 1974a. "Disco Action," *Billboard*, December 14, 1974.

———. 1974b. "Discotheque Club Dialogue." *Billboard*, November 2, 1974.

Moylan, William. 2020. *Recording Analysis: How the Record Shapes the Song*. New York: Routledge.

Record World. 1976a. "Dialogue: The Maestro Tells It Like It Is." *Record World*, January 17, 1976.

Record World. 1976b. "Record World Salutes Barry White." *Record World*, January 17, 1976.

Robinson, Richard. 1974. "Barry White: Nobody in the World Sounds Like Me." *Hit Parader*, August 1974.

Shapiro, Peter. 2005. *Turn the Beat Around: The Secret History of Disco*. London: Faber and Faber.

University of Helsinki. 2021. "Creating Hit Songs Also Means Sounding Out Possible Futures." Accessed July 7, 2021. https://www.helsinki.fi/en/news/culture/creating-hit-songs-also-means-sounding-out-possible-futures?fbclid=IwAR1PW4RuxX8nky7Dv-LNSp8NMk3p778fuftCR754R1rg_LAlm-Xb89hbits.

White, Barry. 1975a. *"Let the Music Play,"* backed with *"Let the Music Play (Instrumental)."* 20th Century Records TC-2265, 45-rpm.

———. 1975b. *"Let the Music Play" (For Disco DJ's Only!!/The Full-Length Disco Version of …)*. 20th Century Records TCD53, 45-rpm.

———. 1975c. *Barry White's Greatest Hits: For Disco DJ's Only!! (Special Mix for Disco Play)*. 20th Century TCD-50, LP.

———. 1975d. *Barry White's Greatest Hits*. 20th Century T-493, LP.

———. 1976a. "Let the Music Play," on *Let the Music Play*. 20th Century Records (distributed by Pye, UK) BT502, LP.

———. 1976b. *"Let the Music Play" (For Disco DJ's Only!!)*. 20th Century Records TCD53, 12-inch single.

———. 2012. *Let the Music Play, Expanded Edition*. Mercury, B0016463-02, CD.

Zak, Albin. 2013. "No-Fi: Crafting a Language of Music in 1950s Pop." In *The Art of Record Production: An Introductory Reader for a New Academic Field*, edited by Simon Frith and Simon Zagorski-Thomas, 43–55. London: Ashgate.

5

REACHING FOR STARDOM

Live and Studio Sound in Prince's "Purple Rain" (1984)

Steve Waksman

What makes a particular chord progression work? If the history of rock and pop tells us anything, it is that harmonic complexity is hardly the key to musical interest. Few things can be more satisfying than a basic three-chord rave-up. And yet, some progressions stand apart, giving the songs that they breathe life into a powerful sense of singularity. Prince's "Purple Rain" is one such song. The opening chord, a B♭ add9, hangs in the air for a full bar, establishing an atmosphere of delicacy and ambiguity that infuses the remainder of the progression, which follows in sequence: Gm7sus4, Fadd9, and E♭ add9.[1] Outlining the root chords of the progression suggests nothing out of the ordinary: I–vi–V–IV. Yet the voicings of the chords, individually and in combination, lend the song a sort of otherworldly quality. The harmonic openness of the chord shapes may be characterized as "jazzy," and yet "Purple Rain" in its affect is not a jazz song. Like Prince himself, "Purple Rain" exists in a space between genres, drawing in the listener by suggesting that there is more to the song than meets the ear.

Other qualities further enhance the impact of this chord sequence, qualities that may be said to distinguish "Purple Rain" not only as a song but as a *song recording*. Timbrally, the rhythm guitar bears no trace of distortion but is hardly "clean." A pronounced chorus effect colors the sound. In the language of guitar effects, chorus refers to a slight delay of the original signal that is then repeated in an additive fashion, generating a kind of shimmering wash that can be heard especially prominently when applied to chords. Chorus was a strongly favored effect in much 1980s pop and rock and was especially common in the new wave genre then current. Andy Summers of the Police, for instance, relied heavily on the effect for his signature sound. Foregrounding the chorus effect draws attention to the genre hybridity of Prince and his co-guitarist, Wendy Melvoin—who is credited with devising the song's harmonic voicings.[2] From early in his career, and especially with the release of his breakthrough 1980 album, *Dirty Mind*,

DOI: 10.4324/9781003089926-8

Prince had been characterized as an artist who fused funk and new wave. "Purple Rain" carried that association forward but also recast it, transplanting characteristic new wave sounds into an extended rock ballad more reminiscent of 1970s epics such as "Stairway to Heaven" and "Freebird."

The "Purple Rain" guitar chords are also notably reverberant. Here, we move into an area that concerns more than timbre. Reverberation brings us into the acoustic space of the recording. As Peter Doyle observed about early rock 'n' roll recordings, "Echo and reverberation made it seem as though the music was coming from a somewhere … and this 'somewhere' was often semiotically highly volatile" (Doyle 2005, 5). So it is with "Purple Rain." Although reverberation has the suggestion of distance, the reverb on the opening guitar chords of the song is relatively contained—enough to broaden the sound field, but not so much as to create a space in which the listener might feel subsumed. Indeed, heard unaccompanied throughout the entire first run of the chord progression, the guitar sounds proximate despite the reverberance. To use the terms of William Moylan, the opening chords of "Purple Rain" exist in a space between the zone of "personal" listening, marked by a "kinesthetic sense of closeness," and that of "social" listening, with more sense of separation (Moylan 2020, 315). That vacillation between proximity and distance comes even more into focus upon the entrance of the drums toward the very end of the progression, where a highly reverberant snare drum sound makes it clear just how much more depth to the sound field there is beneath the zone occupied by the guitar.

Another sort of spatial ambiguity also runs throughout "Purple Rain." The foundation of the recording was made during a live performance at the First Avenue nightclub in Minneapolis on August 3, 1983. That concert was pivotal not only in the making of the song "Purple Rain" but in the trajectory of Prince's career. He introduced his new band, the Revolution, to the public at the show and played for the first time in concert several of the songs that would comprise the finished *Purple Rain* album and accompanying film, including "Let's Go Crazy," "Computer Blue," "I Would Die 4 U," and "Baby I'm a Star," along with the title track (Tudahl 2018, 119–24). Prince's road manager, Alan Leeds, arranged for the full concert to be recorded using a remote truck provided by the Record Plant, with the express hope that several of the resulting tracks might be used for the album in progress (Daley 2009). In the end, material from three of the "live" tracks was included on the finished album: "I Would Die 4 U," "Baby I'm a Star," and "Purple Rain." Yet the final versions of these songs were very far from straight live recordings. "Purple Rain" in particular was significantly edited and altered in the recording studio, with Prince re-recording and overdubbing multiple parts himself and pressing for the addition of a string section, heard most prominently in the song's coda.

Live recordings have a history all their own, albeit one that has been scantly documented to date. "Purple Rain," though, represents a more complex recorded artifact, a hybrid live/studio creation that retains some key markers of liveness but does not purport to be a "live" recording as such.[3] In the remainder of this chapter, I will explore how this fundamental tension between live and studio sound shapes the version of "Purple Rain" that closes the album of the same name. My analysis will revolve around three primary features of the song: crowd noise, Prince's lead guitar, and the string section accompaniment. The traces of crowd noise audible in the recording may appear as the most explicit index of the performance's liveness but also occupy an ambiguous space because they are often buried in the mix or coexist in a complex relationship with other song elements more identifiably tied to the recording studio. I would suggest that the crowd noise, rather than signifying liveness as such, mediates between the live qualities of the lead guitar and the studio crafting of the strings. The interplay between these three elements gives "Purple Rain" much of its dramatic arc and combines to create a sonic space that is defined by a deliberate blurring of boundaries.

Sounds of Liveness

In his influential study of the phenomenon of "liveness," Philip Auslander demonstrates how forms of electronic and digital media have encroached upon the sphere of live performance, rendering obsolete any effort to draw a strict line between the "live" and the "recorded" (Auslander 1999, 7). Less acknowledged in Auslander's framework are those instances where recorded media seek to emulate qualities of liveness, but this tendency exists as a crucial inverse to the incursion of recorded media into live performance practices. On radio and then television, the laugh track and studio audience have historically served this function. In the realm of film, forms of direct cinema or *cinema verité* have sought to counter the polished surface of classic Hollywood film in both fictional and documentary forms, with the concert documentary itself emerging as a primary medium for such experimentation.[4] With regard to sound recordings, the "live" record emerged as a distinct phenomenon in the 1940s, not coincidentally contiguous with the moment when recording technology began to allow for processes of overdubbing and recording of multiple tracks that significantly transformed the relationship between music as it was played "live" in the studio and music as it sounded on the completed recording.

The first commercially released live recordings are generally acknowledged to have been made from the groundbreaking Jazz at the Philharmonic concert series begun by Los Angeles promoter and record producer Norman Granz in 1944. Granz arranged for his first Jazz at the Philharmonic concert to be recorded through an arrangement with the Armed Forces Radio Services and continued to record the concert series moving forward (Hershorn 2011, 64). Seeking an outlet to release the recordings, Granz was rebuffed by all the major record labels he approached and eventually reached an agreement with the small independent label Disc, headed by an upstart record executive named Moe Asch, then at the start of a career that would turn him into a major figure in the history of American folk recordings. The first Jazz at the Philharmonic recordings came out in October 1945 and would be followed by several more volumes. Granz's vision of live recording valorized their "documentary" value. Songs were allowed to run for their full performance time (often ten minutes or more), mistakes were not edited out, and crucially, the sounds of the audience were left clearly audible, expressing surprise and enthusiasm whenever one of the all-star soloists (Coleman Hawkins, Illinois Jacquet, Nat King Cole) played a particularly stirring passage.

Live albums have not always been issued in such a relatively raw state as those produced by Granz, but the value assigned to the sound of the audience as an index of the "realness" of the events caught on record remained in place for years to come. During the rock era of the 1970s and 1980s, the live album gained new currency alongside the standardization of arena concerts as a primary mode of live music production. On arena concert albums from these years, the crowd is heard as a voluminous mass, mainly audible in the space between the tracks as a way to mark the continuity between one song and the next. The sheer presence of the audience on record was what mattered. Blues-rock heavyweights Foghat went so far as to include a note on their 1977 live album: "Applause: Audience."[5] More than a simple index of documentary realness, the crowd noises on live albums of the rock era also certified the scale and scope of a given band's popularity, just as the live album itself typically worked as a sort of career marker, representing that moment in an artist's trajectory where they had achieved a certain breakthrough status after a series of studio releases.

In Prince's own career arc, the *Purple Rain* project occupied an analogous space, coming as it did after a prolific run of five albums in five years, from his 1978 debut, *For You*, to his 1982 double album, *1999*. As a multimedia event—a planned, near simultaneous release of a soundtrack album and film, then to be followed by an extensive national concert tour—*Purple Rain* begs for comparison with the Beatles' *A Hard Days Night*, released in 1965. Like its forerunner, *Purple Rain,* the film is a lightly and loosely fictionalized version of the artist's own march toward success, albeit with more emotionally difficult content at its core. Moreover, in both films, live performance figures significantly in the shape of the narrative. In *A Hard Days Night*, the film climaxes with a live performance that occurs, importantly, before a television studio audience, making explicit the extent to which the Beatles' celebrity relied on a complex web of media relationships that nonetheless had the intensive emotional attachment of the band's live, screaming, and crying fans at its center. *Purple Rain* uses live performance more as a central narrative device throughout, from Prince's first encounter with his romantic interest, Apollonia, to the certification of his stardom with the movie-ending (but not album closing) "Baby I'm a Star." Shot at the same First Avenue club where the basic tracks for "Purple Rain" were recorded, the performance scenes in the film exist—much like the corresponding songs on the soundtrack recording—in a space between live performance and re-enactment, with an audience of stand-ins and the band clearly playing along with the prerecorded tracks.[6] Much as they might draw attention to the sheer excitement of Prince as an onstage performer, they also foreground the artifice of the live performance as such.

The crowd noise heard on the recorded version of "Purple Rain" is most legible when considering the way in which the album, as soundtrack, correlates with the film's narrative. Although "Purple Rain" does not conclude the musical segment of the film, it does serve as an emotional climax. When Prince introduces the song to the First Avenue audience, his remarks tie together two of the film's main narrative strands. Dedicating the song to his father, Prince's character "the Kid" evokes a troubled paternal relationship that underlies his development, including a pattern of domestic abuse that he both reproduces and works to overcome. Meanwhile, crediting the song to the "girls in the band," guitarist Wendy Melvoin and keyboardist Lisa Coleman, the Kid demonstrates a willingness toward collaboration and acknowledgment of other contributions to his work that his character—self-possessed to the point of narcissism—previously lacked. The resulting song transmits a level of emotional openness that reaches the club's audience in a way that the Kid and his band never before achieved and sets the stage for his rise to stardom.

On record, the audience first becomes clearly audible in "Purple Rain" at the start of the third verse. Until this point, just past the 2:30 mark, there is little in the song that would suggest anything uniquely "live" about the recording. The trigger for the entrance of the audience is the dramatic shift in Prince's vocal delivery as the third verse begins. Having sung the first verses and choruses in a fairly subdued style that stayed comfortably in the middle range of his voice, here Prince issues a more visceral utterance that is infused by a sort of sonic excess drawn from gospel or soul—his singing verging on a shriek—as he sings the line, "Honey, I know I know I know times are changing." In the pause that follows his delivery of the line, we hear the crowd response, with one high-pitched yell followed by a more collective expression of assent and then a single voice screaming "OWOO." This outburst stands out in the overall scheme of the song. The audience will not be so clearly audible again until the closing moments. Nonetheless, the presence of the crowd has now been implanted into the song, making the listener aware that "Purple Rain" occupies a peculiar sonic space confined by neither stage nor studio.

Guitar Leads

Prefiguring the eruption of crowd noise into "Purple Rain," the texture of the song becomes notably denser in the second chorus. Albin Zak defines texture as "that quality of composite sound images created by the interaction of diverse elements" that may include "timbre, relative amplitudes, rhythm, ambience, frequency range, chord voicings, spatial configuration, and so forth" (Zak 2001, 85). Zak further observes that the texture of a song may vary significantly between the verse and chorus sections. Of such variation, he notes: "while such changes of texture often simply mirror song structure, they can also provide the track with a narrative layer that is more through-composed in conception, spanning the song's usually sectional and strophic form with an overarching gesture" (ibid., 93). The textural change that surfaces during the second chorus of "Purple Rain" has this sort of narrative significance described by Zak. Once the new level of textural density arises, it carries into the remainder of the song, and the sound of the audience becomes part of that larger mix of elements that signal a shift from the more introspective and personal space that the song inhabits for its first two verses toward more outwardly directed expression that, to an extent, exchanges intimacy for publicity.

The most explicit shift that happens with the second chorus comes in the vocals. Whereas Prince sings the first two verses and the first chorus unaccompanied, with the start of the second chorus, the voices multiply, adding a rich layer of harmonization to the consecutive iterations of the refrain, "Purple rain, purple rain." A more subtle change enters the song immediately prior to this surge of vocal harmony. As Prince sings the last line of the second verse, a power chord sounds on his guitar. Prior to this point in the song, the guitar timbre has been free of distortion, preserving the tone established by the opening chords. This surge of distortion thickens the texture of the song considerably, although the sound of the guitar remains relatively low in the mix. Prince's distorted guitar sound does not overtake the song, but from this moment forward, his guitar playing is heard far more prominently and strays farther from the regularity of the chord progression that continues to provide a solid rhythmic and harmonic foundation. Another power chord marks the transition to the second chorus, undergirding the vocal harmonies; and as the chorus proceeds, his lead guitar fills come more to the foreground, adding a new element of counterpoint to the sung melody.

Within the narrative of *Purple Rain*, the film, the song "Purple Rain" marks the moment when Prince's character, the Kid, reaches his audience in a new and more emotionally impactful way, which establishes the terms of his emergence as a star. The rising presence of Prince's lead guitar within the song can be heard as an index of this narrative thrust within the song and the film as a whole. Prince's stardom was founded upon his multi-talented musical persona—a gifted singer, a talented and resourceful producer, a unique songwriter, and a versatile multi-instrumentalist. Amidst this range of talents, Prince's brilliance as a guitarist was often overlooked or downplayed during his early career for reasons to be examined further below. Yet *Purple Rain*—as film, soundtrack album, and song—makes that feature of Prince's artistry central to his appeal. Consider the narrative arc of the album, which opens with "Let's Go Crazy" and its climactic unaccompanied guitar solo, where Prince merges Jimi Hendrix-inflected timbre and phrasing with the speed and precision of 1980s "shred" technique; and closes with "Purple Rain" and its extended, emotion laden guitar solo where the exuberance of "Let's Go Crazy" transforms into something more somber, the speedy runs of the former giving way to sustained cries of the latter. These two bookend tracks suggest the dualism of Prince's persona: hedonistic and contemplative, extroverted and introspective, virtuosic, and expressive, all qualities that he captures and conveys through his guitar.

In this light, it matters that the success that Prince achieved—both fictively through the film narrative and in reality through the massive commercial success enjoyed by both the film and the album—registered as "crossover" success. The film makes this explicit through the representation of the First Avenue audience, within which white and black clubgoers mix freely and are variably singled out in close-up shots that grant momentary individuality to members of the crowd. David Brackett has shown that the term "crossover" gained increasing currency in discussions of the music industry beginning in the early 1970s, referring to instances when "minority" genres such as country or soul music achieved more widespread mainstream success (Brackett 2016, 281–82). By the early 1980s, the consequences of racially based crossover—and specifically of black-identified genres crossing over toward white acceptance—drew heightened scrutiny as major radio formats such as AOR and Top 40 became more racially homogeneous, a process that became magnified by the advent of MTV in 1981, which severely limited its play of black artists in its first years of programming. While the music industry appeared to withdraw toward a more racially polarized separation of genres and audiences, a series of African American artists exposed the permeability of these categories throughout the decade, with the monumental success of Michael Jackson's 1982 *Thriller* album being followed by Prince's *Purple Rain*, Run-D.M.C.'s *Raising Hell*, Whitney Houston's *Whitney*, and a host of others.

Examining the musical contours of 1980s crossover, Brackett observes a dichotomy that existed between two musical instruments whose timbres coursed through the era's music: synthesizers and guitars. The synthesizer was, in effect, a unifying device of 1980s pop, cutting across genres and often being integrated with a general aura of "funkiness" that pervaded the music of white artists such as Hall and Oates or Culture Club and black artists including Michael Jackson and Prince. Electric guitars, on the other hand, retained a strong association with rock which, by the 1980s, was very much white-identified music; and so, in Brackett's terms, "guitars in a black recording could signify a higher crossover differential," meaning a more direct appeal to white audiences who may otherwise be inclined to tune out music by black artists (ibid., 311). In this, Prince's "Purple Rain" or "Let's Go Crazy" were most aligned with other crossover works such as Michael Jackson's "Beat It" (featuring a guitar solo by Eddie Van Halen) or Run-D.M.C.'s "Walk This Way" (a collaboration with Aerosmith vocalist Steven Tyler and guitarist Joe Perry). Unlike those other crossover moves, though, Prince does not rely on collaboration with a white artist to lay claim to the symbolic associations that inhered in the electric guitar. His own virtuosity on the instrument personifies the crossover tendencies that course through the music on *Purple Rain*.[7]

Given how much this was the case, why did Prince's guitar playing not become more central to his emerging stardom? Cultural historian and critic Jack Hamilton, writing just after Prince's 2016 death, suggested that the problem was not that his abilities on the instrument went unrecognized but that they were treated as a curiosity, one of many facets of Prince's persona that did not fit the expectations of black pop stardom. By Hamilton's account, the source of this problem was twofold: that since the 1960s, the electric guitar had been largely reimagined as a "rock" instrument, and that rock had been reimagined to be a predominantly "white" musical form (Hamilton 2016). Studying similar fault lines in connection with the cultural advocacy group the Black Rock Coalition, cultural anthropologist Maureen Mahon asserts, "black rock was a critical intervention into what was for black musicians an impossible situation ... [and] their particular situation was part of a larger and long-standing problem that went well beyond the realm of music: the refusal, disinclination, and inability to deal with the breadth and complexity of contemporary African American people and culture" (Mahon 2004, 8). If Prince's standing as a lead guitarist made him into an agent of crossover, on the one hand, it also made him into the embodiment of the socio-musical category crisis concerning race that had overtaken rock music by the 1980s, on the other.

Adding another layer to these associations in "Purple Rain" is the way that the electric guitar—and specifically, the distorted timbre that marks Prince's lead guitar playing in the song—serves as a signifier of the song's liveness. It is no idle coincidence that this shift in the song's sonic character foreshadows and accompanies the entry of crowd noise into the recording. The third verse, where the audience's whoops and screams arise, is also the first where Prince's lead guitar fills are clearly audible. In the chorus that follows, during a pause that precedes the final line, he plays an ascending guitar melody that culminates in the sound of striking the instrument's muted strings several times (this occurs between 3:32 and 3:40 of the recording), generating an aggressive, percussive sound as though he cannot quite bring himself to stop. Sure enough, the succeeding vocal line gives way to the guitar solo that fills the next ninety seconds of the song, a solo that *Rolling Stone* called "arguably the greatest power-ballad guitar solo in history" ("100 Greatest Guitarists" 2015).

Although it is difficult to say with certainty which portions of the album version of "Purple Rain" were recorded live and which were added later during studio sessions, available evidence suggests that Prince's solo was taken directly from the live recording made at First Avenue during that pivotal August 3, 1983, performance. Video of the original performance from that concert has circulated for years on a range of websites (YouTube, Vimeo, Dailymotion), with added commentary that purports to explain where the live recording can be heard in the "official" album version and where it has been edited.[8] The video performance is substantially longer than the version that appeared on *Purple Rain* and several parts of the live rendition would be edited out completely, including a long extended introductory passage during which Wendy Melvoin plays the song's opening chords repeatedly while waiting for Prince to join the band, and an entire second verse that appears to have been excised from all subsequent performances. The solo, too, has been edited down, lasting a full minute longer in the live video. However, the shape of the solo—the opening double-stop bend; the succession of sustained, bent high-pitched notes interspersed with rapid runs; and the three-note melodic refrain that repeats throughout the last portion of the solo, culminating in an outro riff that propels the song toward its conclusion—are all here and retained in the version that closes the album. While Prince may have overdubbed certain sections later to correct what he heard as "mistakes," the guitar solo he plays in "Purple Rain" is one of the primary features that signals the song's origins in a live club performance.

Strings Attached

By contrast with Prince's lead guitar, the strings heard in the album version of "Purple Rain" are nowhere to be found in the 1983 First Avenue performance. Of the song's defining components, they stand as the clearest marker of the extent to which "Purple Rain" was reimagined in the recording studio. Additional work on the song commenced on August 15, just twelve days after the live recording had been made (Tudahl 2018, 135). Prince worked on the overdubs for "Purple Rain" with recording engineer Peggy McCreary at Sunset Sound studio in Los Angeles, which served as his favorite recording location throughout this phase of his career. During one of the marathon sessions for which Prince was known, he decided to add strings to "Purple Rain," first using an Oberheim keyboard to provide a guide for the arrangement and then quickly inviting a trio of string players to the studio—Novi Novog (violin and viola), Suzie Katayama (cello), and David Coleman (cello), the brother of Prince's keyboardist Lisa Coleman. Prince designated Lisa to supervise the recording of the parts, about which her brother David recalled:

> I walked into the control booth and [Prince] said, "Hi David. Do something!" He got up from the console and I sat in his seat and they played me the track and that was it. I went in and joined Suzie and Novi and we started from there … Lisa would come in and she'd do some things on the piano so we'd learn our parts by ear … So we'd go through the song and punch in our string parts.
>
> *Ibid., 137*

Work on the string parts continued for an additional session three days later, on August 18, at which point the recording of the strings for the song was complete. It would take several further sessions for the mix of the song to be finalized, however, and it was in the mixing stage that Prince and McCreary would have made the key decisions about just how the strings would be heard in relation to the song's other elements.

Timbrally speaking, strings are often positioned as the antithesis to the effect of a distorted electric guitar or a wailing saxophone. Against the raw or coarse qualities of the latter, strings may bring an air of smoothness and refinement to a recording. This connotation is compounded by the association of orchestral string instruments with the "high" side of the boundary between high and low, or classical and popular music. However, as John Howland has argued in his analysis of the category of what he calls "luxe pop," the assumption that violins or cellos inherently signify the "classical" in pop music ignores the long tradition of string-laden pop that, at this point constitutes a tradition of its own (Howland 2018, 193). Within that longer history lay several distinct styles and practices, from symphonic jazz to symphonic soul, and chamber pop to disco.

Prince's decision to integrate strings into the sonic textures of "Purple Rain" evokes multiple styles within this longer history. Regarding his persona as a black male pop artist, it connects him most directly to the symphonic soul of 1970s figures such as Isaac Hayes and Barry White. As Mitchell Morris and Alice Echols have shown, symphonic or "soft" soul was linked to the refashioning of black masculinity, presenting a version of manhood that projected vulnerability over aggression and equally valorized romance and sexuality (Morris 2013; Echols 2010). Soft soul further linked blackness with upward mobility, favoring trappings of middle-class comfort over the sort of street or ghetto sensibilities that have often underpinned prevailing notions of black realness or authenticity. Myriad commentators have observed that Prince's own gender persona persistently went against the grain of conventional trappings of masculinity in ways that also suggested a challenge to other kinds of limits. In the words of Francesca Royster,

> His performances reveal an eroticism attached to multiple objects: lovers … but also God, his guitar, James Brown, and himself. Prince forms "strange relationships" with his listeners, with his instruments, and with his own performing body … His gender and sexual fluidity might also be linked to his musical style – a repertoire of mastery over a range of styles and past performers.
>
> *Royster 2013, 2*

In effect, the dual prominence of the strings and the electric guitar in "Purple Rain" can be said to stand for the fluidity with which Prince crosses boundaries of both gender and genre in the song. The strings, moreover, indicate another path toward crossover, countering the rock-based referentiality of the guitar with the lush sounds of upwardly mobile, adult-oriented pop.

At the same time, the sound of "Purple Rain" hardly qualifies as symphonic. With just three string instruments added to the mix, the arrangement is more akin to chamber music—what has in more recent parlance come to be known as "chamber pop" and has roots in the melodic inventiveness of the Beatles and other rock artists of the 1960s and 1970s. In the larger scope of Prince's own work, the strings on "Purple Rain" prefigure the more concerted turn toward psychedelia that would take hold in his follow-up project, *Around the World in a Day*, which featured strings prominently on such songs as "Raspberry Beret" and "Paisley Park."

On "Purple Rain," the strings announce themselves in the first chorus, where they enrich the texture and thicken the harmony. The doubling of the cellos means that the pitch of the string contributions is concentrated on the low end. Moving into the second verse, the strings rest until the midway point, at which point they swell into a slowly ascending melody that accompanies the line, "Baby, I could never steal you from another" (heard from 1:43–1:48). Significantly, this immediately prefigures the moment when the timbre of Prince's guitar becomes notably more distorted. From this point forward, the strings and distorted electric guitar engage in a sort of counterpoint that considerably alters the texture of the song, staging a sort of reconciliation of the "live" and "studio" recorded elements along the way.

From the moment that Prince embarks on his "Purple Rain" guitar solo, the strings can be heard as part of the foundation underpinning his lead guitar lines. Their role throughout the first, most dynamic part of the solo is to play a melodic embellishment that can be heard most clearly at the start of each succession of the chord progression. The strings mesh with the surrounding timbres of the electric bass, drums, rhythm guitar, and keyboards to produce the sonic bed on top of which the guitar lines move fluidly. A key transitional moment comes about a minute into Prince's solo. The guitar reaches a climactic point, with Prince bending into the root note of B♭ high on the neck of his E string, pushing it up a full step to C multiple times and releasing it. He then plays a swiftly picked series of triplet figures that leads him to yet another dramatic note bend two frets further up the neck, the resulting D being the highest note he will play during the entire solo. Here again, he leans into the note a second time to further accentuate its importance to the solo's trajectory and then eases back. During the ensuing pause in the lead guitar's energy, the strings come to the foreground, playing an eight-note descending sequence that follows a roughly pentatonic logic and connects two iterations of the song's root note of B♭ an octave apart. The motion of the strings here has no counterpart in the live version of Prince's solo, but its addition on record propels the solo into its final passages, where Prince repeatedly plays a three-note melody once again centered around the song's root, reestablishing a base of melodic stability on which the song will build to its end.

As much as the timbral interplay between electric guitar and strings shapes the musical trajectory of "Purple Rain," it is the relationship between the strings and the song's undercurrent of crowd noise that defines its conclusion. After seven minutes, "Purple Rain" reaches a point that seems to signal an ending, but after some final lead guitar flourishes from Prince, the instruments all recede except for the two keyboards. Lisa Coleman maintains a droning synthesizer chord over which her bandmate Matt Fink plays a cyclical piano melody. Out of the surrounding quiet, the strings arise to a position of prominence they have not had at any earlier part of the song. They embark on a harmonic sequence that moves fitfully, with each pair of notes swelling first upward and then back down as though a series of waves crashing against the shore. At every eighth note in the progression, an extended period of sustain gives the sense that the song could end imminently, but the swells continue for a minute and a half, each cycle involving some subtle reharmonization of the core elements while the droning chord of the synthesizer remains unchanging. The overall effect is both restful and unsettling, the intimations of dissonance in the chord voicings suggesting that the song's subject is not yet fully at peace but is continuing toward that goal.

While the strings perform this final movement, the piano melody becomes more faint and eventually disappears entirely. As it fades, traces of the audience begin to be audible. A whistle from an audience member can be heard (7:21), and other crowd sounds enter the aural frame, but at this stage, they are mixed so low that one can easily assume they are audio hallucinations. When the song passes the eight-minute mark, these intimations of crowd noise start to be more unmistakable. By the time the strings reach the end of their closing passage, the crowd noise has achieved a position of near parity in the mix, and the audience cheers unaccompanied for the final seconds of the track before fading out. At this point, any boundary that might be drawn between "live" and "studio" recorded elements has collapsed entirely. The audience cheers for a part of the performance they were not present to hear, and the recording fabricates a space of collective fantasy in which, one might say, self-realization is achieved through mass acceptance—which is, in effect, the fantasy of pop stardom writ large.

"Baby I'm a Star"

The three songs on *Purple Rain* that feature elements of the live recording made at First Avenue—"I Would Die 4 U," "Baby I'm a Star," and "Purple Rain"—close both the film and the album, but as noted previously, they appear in a different order in each. In the film, "Purple Rain" immediately precedes the other two songs, while on the album, it immediately follows them. Both works, film and album, culminate with the suggestion that Prince's stardom has either already been achieved or is inevitably on the horizon; yet the flipped order of the songs makes a difference in terms of how this stardom fulfills the demands of narrative closure. Ending the film with "Baby I'm a Star" seeks to ensure that Prince's stardom is the last thing that viewers take away with them. That his assertion of stardom follows from the statement of emotional vulnerability represented by "Purple Rain" signals that stardom succeeds what the film takes to be the Kid's most direct statement of his true or authentic self, and further suggests that his stardom essentially resolves the conflicts that had previously shaped his character's development up until that point. On the album, by contrast, stardom does not resolve Prince's emotional burdens but unearths them, opening the space in which they can find their deepest expression.

Prince's stardom involved the crossing of multiple boundaries or thresholds. Marked as a "crossover" artist, he challenged and confounded some of the ways in which racial categories were coded onto prevailing popular music genres in the 1980s. Apart from the racialized dimensions of his appeal, he blended genres sonically and stylistically, a tendency that was much in evidence in the rich textures and diverse timbres of "Purple Rain." Amidst these various acts of boundary crossing, the divide between "live" and "studio" sound may appear benign. However, I would argue that this breach of sonic spatial logic is fundamental to the dramatic force of "Purple Rain" as a song unto itself and as a key piece of the larger narrative that surrounded this moment in Prince's career. If, as William Moylan has claimed, recorded tracks "present environments in unnatural ways; they change space and our relationships to spaces," then Prince and recording engineer Peggy McCreary's creative blending of live recording and studio production in "Purple Rain" stands as a compelling instance of spatial transformation (Moylan 2020, 333). The fusion of live and studio sound that comprises the final track creates an otherworldly space in which Prince's vision of musical and social transformation might be realized and, as the final sound of the cheering audience bears out, affirmed.

Notes

1 All music analysis in this essay is based on the versions of the songs found on Prince and the Revolution (1984), unless otherwise stated.
2 Recalled Melvoin about the process of composing "Purple Rain," "I started playing this chord … And then we tried to change the chord, and Prince's chords at the time were much more simple then. We kept all those suspended chords in there and then I put the 9 in there" (Tudahl 2018, 106).
3 There are numerous examples of "live" albums that are in fact hybrid live/studio creations, as well as a smaller number of studio sessions that seek to create the air of a "live" recording. The 1975 Kiss album *Alive!* is a prominent example of the former, known for including multiple studio overdubs; and Cannonball Adderley's 1967 jazz release *Mercy, Mercy, Mercy!* is emblematic of the latter, billed as "Live at 'The Club'" but actually recorded in a Los Angeles studio. "Purple Rain," the song, and the additional album tracks that were similarly produced from the live recordings at First Avenue, constitute a different sort of recorded artifact because nothing on the album makes a claim about the music's "liveness." The audience sounds and other trappings of live performance heard on the album exist in a kind of unmarked space and so play upon the boundary between "live" and "studio" recorded work in a more subtle and suggestive way than the more common examples of live recordings that may feature studio effects.
4 On the cinematic form of concert documentaries, see James (2016).
5 Liner notes for Foghat (1977).
6 All details from the film version of *Purple Rain* are drawn from Magnoli (1984).
7 This assertion begs for some immediate qualification, given that Prince's band, the Revolution, was a racially integrated ensemble and that his co-guitarist in the band, Wendy Melvoin, was a white woman. Fully accounting for Melvoin's role in shaping the sound of Prince's music would require a separate essay. Here, I will note that as integral as she was, Prince was typically the one to occupy the role of lead guitarist in the band, and the solos on "Let's Go Crazy" and "Purple Rain" are his.
8 I have consulted several of these sites to ensure consistency, and they all repost a version of the video performance with added supertitles that explain the details of the song relative to the album version. Credit for the information in the supertitles is never clearly granted, although one source says that they were added by someone who was "a former member of Prince's team." See "Prince—Purple Rain First Performance," Vimeo, https://vimeo.com/44801620. Accessed November 9, 2021.

References

"100 Greatest Guitarists." 2015. *Rolling Stone*, December 18, 2015. https://www.rollingstone.com/music/music-lists/100-greatest-guitarists-153675.

Auslander, Philip. 1999. *Liveness: Performance in a Mediatized Culture*. London: Routledge.

Brackett, David. 2016. *Categorizing Sound: Genre and Twentieth-Century Popular Music*. Oakland, CA: University of California Press.

Daley, Dan. 2009. "Prince and the Revolution." *Mix* 33, no. 1 (January). ProQuest.

Doyle, Peter. 2005. *Echo & Reverb: Fabricating Space in Popular Music Recording, 1900–1960*. Middletown, CT: Wesleyan University Press.

Echols, Alice. 2010. *Hot Stuff: Disco and the Remaking of American Culture*. New York: W.W. Norton & Company.

Foghat. 1977. *Live*. New York: Bearsville.

Hamilton, Jack. 2016. "Why Prince May Have Been the Greatest Guitarist Since Hendrix (and Why That Shouldn't Seem Like a Surprise)." *Slate*, April 28, 2016. https://slate.com/culture/2016/04/why-prince-was-the-greatest-guitarist-since-jimi-hendrix.html.

Hershorn, Tad. 2011. *Norman Granz: The Man Who Used Jazz for Justice*. Berkeley, CA: University of California Press.

Howland, John. 2018. "Hearing Luxe Pop: Six Degrees of Symphonic Soul." In *The Relentless Pursuit of Tone: Timbre and Popular Music*, edited by Robert Fink, Melinda Latour, and Zachary Wallmark, 185–211. New York: Oxford University Press.

James, David. 2016. *Rock 'n' Film: Cinema's Dance with Popular Music*. New York: Oxford University Press.

Magnoli, Albert, dir. 1984. *Purple Rain*. Burbank, CA: Warner Bros.

Mahon, Maureen. 2004. *Right to Rock: The Black Rock Coalition and the Cultural Politics of Race*. Durham: Duke University Press.

Morris, Mitchell. 2013. *The Persistence of Sentiment: Display and Feeling in Popular Music of the 1970s*. Berkeley, CA: University of California Press.

Moylan, William. 2020. *Recording Analysis: How the Record Shapes the Song*. New York: Routledge.

Prince and the Revolution. 1984. *Purple Rain*. Burbank, CA: Warner Bros. Records.

Royster, Francesca. 2013. *Sounding Like a No-No: Queer Sounds & Eccentric Acts in the Post-Soul Era*. Ann Arbor, MI: University of Michigan Press.

Tudahl, Duane. 2018. *Prince and the Purple Rain Era Studio Sessions, 1983 and 1984*, expanded edition. Lanham, MD: Rowman & Littlefield.

Zak, Albin. 2001. *The Poetics of Rock: Cutting Tracks, Making Records*. Berkeley, CA: University of California Press.

SECTION III

Layers Make the Record

6

"IN A SKY FULL OF PEOPLE"

Spatial and Cinematic Staging in Seal's "Crazy" (1990)

Mike Alleyne

This chapter is supported by a webpage.

Introduction

This chapter primarily engages aspects of music production studies and recording analysis while also employing elements of cultural studies and film studies to deconstruct and survey the sonic subject matter. Records produced by Trevor Horn are meticulously and purposefully constructed (at times perfectionistic),[1] with each episode within each work calculated to achieve memorable impressions of both commercial and aesthetic value. In theory, this is the ultimate goal of all record producers, however for many observers, Horn's incisive aural realizations epitomize the attainment of that creative objective. Essentially, the analysis explores the spatial design and hyperrealistic contexts of Seal's "Crazy," with the aim of illuminating Horn's sound staging as fundamentally cinematic in both execution and aural effect. This critical exposition also directly employs the perspectives of the recording's primary figures. As Horn himself has observed, "a record is a process in the same way that a film's a process" (Allinson and Levine 2008), while Seal has tellingly stated that "To me, the sound of a guitar or the tone of a voice is as cinematic as volumes of lyrics" (Flick 1991, 26).

A key role of the record producer is to achieve aural clarity and focus while facilitating the listener's engagement through the manipulation of sonic spaces. Simon Zagorski-Thomas observes that "The notion of staging refers to the treatment of sound in ways that add meaningful context for the listener to a performance or a perceived musical 'event'" (2014, 78). The strength of "Crazy" resides not only in the considerable dynamics of the vocal and instrumental performances but also in the conscious utilization of spatiality as a fluid multidimensional medium through which the record's aesthetic and artistic goals

DOI: 10.4324/9781003089926-10

are articulated, replete with cinematic narrative presentation. This thoroughly staged record exists in its own sonic universe and cannot be faithfully reproduced live, even by the same personnel, partly because of the simultaneously occurring multiple overdubbed performances populating the aural spaces, but also due to the number of contrasting spatial environments embedded in the record. Zagorski-Thomas identifies this idea of contradictory aural spaces coexisting in a recorded work and being perceived by listeners as "a single interpretation" and as a "conceptual blending" (2014, 80–81). This interpretive effect embodies a central paradox of modern recording wherein the vocal and/or instrumental voices are multiplied and amplified, becoming hyperreal in relation to real-time human aural cognition and performance. Fundamentally, "Crazy" is a collective series of performances consciously constructed outside of standard linear temporal considerations. Popular music records, in the forms that we have come to know them, were at one point entirely surrealistic propositions, and many of the track elements that recording analysis sometimes identifies as surreal are now commonly accepted by casual listeners as "normal" content.[2]

Moylan comments that "In recording analysis, we examine the finished artifact, what it *is*, not how it got there" (2020, 95). However, in certain cases, the processes leading to the creation of the finished artifact may have a direct and inextricable critical bearing on its manifest form, and this proposition applies to key aspects of Seal's "Crazy."

Within academic critiques, Timothy Warner's focused analysis of Horn's production work ends around 1985, five years before his first recording with Seal emerges.[3] Warner identifies key features of Horn's production, noting that "The manipulation of space as a significant musical parameter is a characteristic of much of Horn's work" (2003,103). This spatial emphasis works in conjunction with the kinetic effects of Horn's production and its aural continuity, all components that manifest themselves clearly on "Crazy."

As the introduction's first quote suggests, process is paramount in Horn's creative philosophy, and his reference to film underlines the inherently cinematic nature of his work and the importance of creating audiovisual spectacle within the imaginations of listeners. In fact, the unfolding soundscape of Horn's work on "Crazy" represents a counterpart to the multiple perspectives and dimensions of a film, inclusive of dramatic shifts in focus, tone, and intensity. One author has even described Horn's work with Seal in terms of "widescreen epics," thereby reemphasizing the cinematic character of the material (Batey 2010). Tagg (2012, 551) refers to "episodic markers" in film music, and although his discussion did not appear to have direct application to pop hit singles, the description applies well to Trevor Horn's work and the sequencing (pun intended) of its narrative formulations and aural punctuations. Indeed, the inclusion of the Horn-produced track "Tinseltown" on *Seal IV* (2003) more than a decade after the release of "Crazy" emphasizes the resonance of filmic conceptualization as a defining aesthetic feature in his work, whether nuanced or overt. The cinematic parallels discussed in this chapter are neither rigid reflections of film theory nor of a specific category of filmmaking but instead aim to underscore broad interrelationships evident on "Crazy" that extend from Horn's sonic design. Jim Morrison observes that the production process of collating sonically isolated recorded fragments to create an illusion of simultaneous performance not reproducible live "is the aural analog of the process of shooting a film" (2009, 87). This is the precise aspect of audiovisuality being emphasized in the forthcoming interpretation of "Crazy." While there are other parallels with visual realms, the dynamic movement and power to affect and project mood that is inherent in cinema is most appropriate for consideration of Horn's work on the track. "Crazy" is as much a spatial composition as it is a musical one (while also bearing relationships to filmic construction), and the combinations, isolations, and juxtapositions that occur within it are intrinsically vital to its aural identity. Horn's overt manipulation of the sound stage leads to interrelated cinematic scenes that reinforce the song's aesthetic and emotional energies.

In interpreting recorded aural hyperreality, Moylan recognizes that "the listener accepts the virtual reality as part of the context and expression of the track – part of what makes it unique. This happens no matter the level of realism of the spatial properties" (2020, 287). Despite the listener's willing assimilation of collated and often seemingly incongruous sonic environments as manifested on "Crazy," part of this chapter's mission is to reexamine the record's aural paradoxes and ways in which they create collective coherence.

Locating "Crazy"

During his production peak in the 1980s, Horn commented that "The information in a single has to be arranged in the right bursts. Everything else is secondary. The drum sounds, the tricks, the lyrics, everything, they're all less important than the arrangement of the *feel*. Records with a bad feel to them are never hits" (Freff 1984, 94). My analysis focuses on the album version of Seal's "Crazy" that achieved its British chart debut as an edited single in December 1990, eventually reaching number 2 in early 1991 before arriving in the top ten in America in September of that year.[4] The eponymous debut album (first released in May 1991) later achieved platinum sales status in the United States, with "Crazy" as the primary vehicle driving its success and laying a foundation for Seal's career.[5] At about one and a half minutes longer than the single, the album version is the definitive representation of the record, written by Seal and Guy Sigsworth (the song's demo producer) but shaped and molded by Trevor Horn's production vision.[6] Horn's auteur imprint and strategic aesthetic decisions inclusive of sound stage placements are intrinsic to my analysis, which is primarily concerned with the record's holistic sonic environment, the peculiarities of its various components, and the collective cohesion of elements that create definitive aural and emotional experiences. Remarkably, Horn was initially unconvinced of the song's potential and was reluctant to feature it, relinquishing only after Seal's contractual insistence that it be released as a single. Horn commented that "We just had to work on it until it was a hit" (Cromer 1998, 53).[7] It should also be noted that recording engineers have continually played vital roles in Horn's productions (with three credited on the single mix of "Crazy") and that components of the creative agency often attributed to Horn as the producer may also arise from other personnel in the studio contributing technological and creative ingenuity.

"Crazy" and its accompanying album capture an energy not since repeated on record in Seal's career, warranting further investigation of its unique characteristics. Notably, unlike most other artists with whom he has worked, Horn has produced five Seal albums. The sophomore *Seal* (1994, including the Grammy-winning hit, "Kiss from a Rose") was a more contemplative, introspective affair that was much less declaratory and incendiary than its predecessor. The third album *Human Being* (1998) was shrouded in a sinister ontological gloom (signaled by the spectral visual intensity of its album cover),[8] perhaps unwittingly reflected in that record's murky, subterranean mix—and, more specifically, a mix that was arguably untypical of Horn's often emphatic sonic presence normally prioritizing lucidity amidst density. This was, in Seal's own words, "a very dark album"[9] articulating forms of psychic claustrophobia, whereas Seal's debut collection including "Crazy" embodied the effervescence and altruistic philosophy characteristic of escape from such conditions, supported by energetic optimism and a pronounced aural brightness, clarity, and vitality.[10] Horn also produced the long-delayed *Seal IV* (2003) following Warner Bros.' regrettable abandonment of the preceding *Togetherland* project (2001).[11] *Togetherland* suggested a triumphant return to the sonic and emotional energies of the debut album (as typified by the compelling animation of "Let It Ride," the dynamism of "Love Is Better," and the nakedly emotional ballad "All I Wanted to Say"). Despite *Seal IV*'s best moments, it remains a world apart from the soundscape of the record it replaced, leaving guitarist Gus Isidore, a participant on both projects and the debut album, to lament *Togetherland*'s demise (Leroy 2007, 10–11).[12] After a commercially unsuccessful hiatus from Horn's creative supervision (on the EDM-driven *System* in 2007 produced by Stuart Price, and *Seal 6: Commitment* in 2010 helmed by David Foster), *Seal* 7 (2015) reunited them with poignant reminders of their combined effectiveness. Despite the five-album collaboration, the very first project arguably remains the most memorable, with "Crazy" as a crucial artistic benchmark for both Horn and Seal.[13]

One might theorize that by the time that his work began with Seal, Horn's stylistic application of production technology had matured. The overt technological exhibitionism displayed in some earlier productions is less acutely pronounced in its rendering on Seal's recordings. Horn stated that previously "I was more interested in doing perverse things with sound" when the appropriate technology arrived in the early 1980s, as opposed to following industry norms: "I had to find a way of making my records sound extreme to some degree because that was the only way to get people's attention" (Cunningham 1998, 301; also see Buskin 1999, 285). Horn's propensity for creating sonic spectacles is typified by his work in several contexts. There is the product of his own group, The Buggles, best known for the prophetic single "Video Killed the Radio Star" (1979) replete with historical staging (in both the record and music video). That record also simulates the sound of the human voice on vintage radio broadcasts to "capture" the aural imprint of the past on record (Buskin 2011).[14] Further, in concept and in title, that record valorizes the role of the *visual* presence, establishing and reinforcing the echoes of this theme throughout his work (Warner 2003, 8; Zagorski-Thomas 2014, 88, 90, 91). The three months Horn spent crafting that single (Zak 2001, 136; Massey 2009, 183) also created a template for the type of detailed soundscape and sound stage formation that would characterize "Crazy." Horn has confirmed that "we spent two months doing 'Crazy' – and then about two months on the whole album. 'Crazy' wasn't an easy record to make, because we were aiming high" (Lindvall 2014).

Degrees of earlier sonic spectacle supervised by Horn appear on albums such as ABC's *The Lexicon of Love* (1982, with its elegantly calculated retro-styled cover seemingly situated on a stage, further inviting dramatic and cinematic interpretations), Frankie Goes to Hollywood's *Welcome to the Pleasuredome* (1984) and *Slave to the Rhythm* by Grace Jones (1985). The latter album even subverted normative mixing approaches by occasionally shifting the entire sonic content to one sphere of the stereo sound stage to the total disorienting exclusion of other potential spaces and, at other points, rapidly shifting sounds across the stereo spectrum to create an artificial shimmering effect.[15] Horn's emergence as a successful producer was coincident with (though certainly not entirely reliant upon) the arrival of MIDI and associated digital technologies, facilitating his dramatic spatial reconfiguration processes. Whereas *Who's Afraid of The Art of Noise,* the debut album by The Art of Noise (1984, not solely produced by Horn) includes numerous sound source features in its rampant application of sampling, its composite presence as a collection of musical tracks differs recognizably from the comparatively more conventional musicality of Seal's debut, despite the layers of technology also embedded in that record.[16]

It is even possible that Seal's multi-genre fluidity contributed to a sense of production freedom for Horn working with the then unknown artist. The absence of fixity was highlighted in Dave Laing's (1992) article headline for *Billboard* magazine: "Brits Around The World: Seal? Dance? Rock? Black Rock? Pop," while incidentally acknowledging the stereotypical genre quarantines within which black artists, in particular, are placed.[17] The many issues and complex histories surrounding Britain, Blackness, and cultural space within the music industry are beyond this chapter's scope, but the emergence of Seal also coincided with an era in British pop characterized by degrees of escape from those enclosures, reaching into international space. This was especially signified by the 1984–1989 American chart successes of singer/songwriter Billy Ocean, the less well-known Ephraim Lewis,[18] whose only album *Skin* yielded a marginal US chart single in 1992 and the massive 1989 commercial impact of the Soul II Soul collective.[19]

As an additional and crucial contextual point, the tumultuous political events of 1989, including the fall of the Cold War's symbolic Berlin Wall in Germany and the Tiananmen Square Massacre in China, contributed to the lyrical and instrumental urgency embedded in "Crazy," inevitably shaped by the cataclysmic time in which it was made. Seal confirmed this global sociopolitical interrelationship in the 2005 concert video *One Night to Remember* recorded in Germany as he introduced "Crazy," also citing "new discoveries in fractals," pointing to a lyrical reference in the song's opening verse.

Time, Space, and Texture in the Mix

An engagement in the realm of cinema simultaneously demands an audiovisual acceptance of and detachment from the text in order to momentarily accept "the transparent illusion of unmediated representation, made less remote by our capacity to 'believe' in the realism of what we are hearing" and seeing (Alleyne 2009, 19). More specifically referring to recorded sound, Moylan notes, "Worldly limitations are ignored, and listeners are willing to experience these illusions of distance and size, and accept them as the unique reality of the record" (2020, 291). Effectively, "Crazy" represents Trevor Horn's ability to establish a cohesive aural reality from multiple illusions to evoke sensory and emotional responses from the listener that are directly linked to the musical concept.

In discussing key features of his production style, Horn emphasizes that his typical work "has an interesting arrangement. It shifts perspective at some point so that it draws you in and takes you on a little journey" (Warner 2003, 152). This production summary is also readily applicable to "Crazy," and notably, the journey's continuity on the record totally and boldly excludes crash cymbals as one of the standard pop episodic markers.[20] It is among a handful of rhythm-driven major hit singles to adopt this approach, as well as the absence of a heroic guitar solo that might typically adorn a song of this dynamic character. This underscores Horn's willingness to discard standard production strategies to create the desired soundscape and mood, quite literally staging the sound for dramatic effect by exploiting vacancy over presence.

Given the impossibility of fully translating key aspects of the aural experience to words and paper in this analysis, my analytical approach draws on an image from the first verse of the song's lyrics, "through a fractal on that breaking wall," peering inside fragments of the production and artistic processes that result in this memorable work. The recording soundscape of "Crazy" successfully intensifies the urgency of the message, conveying meaning(s) in a way markedly different from Seal's other recorded electric or acoustic versions of the song, an idea to which I will return in the conclusion. Warner (2003, 90) compares key aspects of the introduction of "Crazy" to previous Horn productions, particularly "Relax" by Frankie Goes to Hollywood. Both records employ "a mysterious, beatless introduction, often tonally ambiguous in character … and draws attention to the strength and immediacy of the backbeat." On "Crazy," this approach utilizes a hypnotic miasma of multiple reverb-soaked synthesized ostinato textures, creating and maintaining a sense of aural dislocation even after the first vocals are half-spoken/half-sung. Figure 6.1 maps the multiple layers of sonic sources and effects, showing their placement within the sound stage.

Hyperrealism asserts itself from the very beginning with two atmospheric overlapping rhythmic keyboard parts, the second of which appears after four

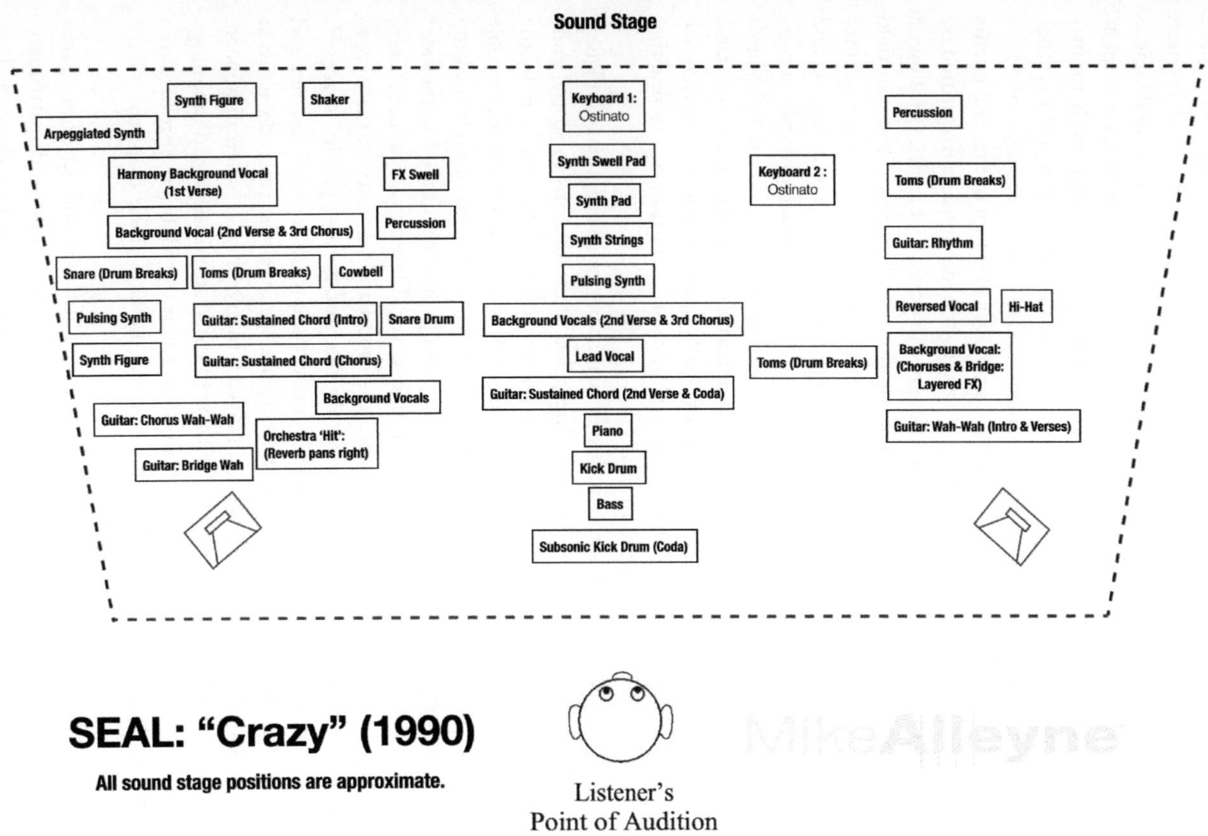

FIGURE 6.1 The sound stage of "Crazy."

seconds, underpinned by a sweeping synth pad chord (all provided by session player Richard Cottle). Moylan notes that in such recorded moments, "we find spatial qualities and relationships that do not align with life experiences, but that are readily accepted by the listener" (2020, 339). Three simultaneous keyboard parts followed by an emphatic fourth keyboard-triggered sampled orchestral "hit" invite listeners to embrace the recorded performance as a staged event, suspending temporal logic as well as performance realism. Quite similar to an amorphous dream sequence opening a film, the purposeful disorientation of the prologue's ethereal synth timbres accentuates the hyperreal effect. The crucial functions of the multilayered synthesizer textures establish tonal and timbral character and sustain the record's mood.

The sole orchestral "hit" on the record (a vastly overused tool in that era) occurs on the first beat of bar 9 with a significantly exaggerated emphasis magnified by its initial location on the left of the sound stage; the downbeat attack is followed by a sonic vapor trail of dense reverb that rapidly travels across the stereo field through the center and toward the right. This rhythmic effect works in conjunction with an arpeggiated filtered synthesizer to provide concrete punctuation and form, shifting attention away from a loosely floating liquid space toward a recognizably metric rhythmic pulse. The reverbs and delays applied to the keyboards create a series of otherworldly electronic spatial textures alluding to altered states of consciousness, verified by the content of Seal's opening vocal lines that are ushered in by a menacing yet controlled electronic surge of uncertain source that never recurs with quite the same intensity.

The song lyrics, delivered in the guises of various personae ranging from kaleidoscopic omniscience to the first person, are filled with allusions to spatial confinement and liberation, both physical and psychological, and the tone is set by the protagonist's reflection on a man in spiritual crisis who "talks about the people going under." In conjunction with the lyrical perspectives, the vocal multitracking exploits time and space to expand the narrative scope and reinforce the theme.[21] As the narrative of "Crazy" unfolds, overdubbed vocal layers, overlapping lines, and a completely reversed voice underscore the mood and intensity of the hyperreal soundscape, immersing the listener within the protagonist's psychic spaces.[22]

A wah-wah processed guitar phrase, panned right with a more distinct and isolated sound, prepares us for the next cinematic scene (as the intro develops at 0:20). The guitar, performed by session player Chrissy Shefts,[23] then disappears until the opening bar of the first verse with a chorused sustained chord unambiguously announcing a transition. Shefts' rhythm guitar overdubs are spread across the stereo field, recalling Horn's production on the title track from the Grace Jones album *Slave to the Rhythm* (1985).[24]

Throughout the song, lyrical juxtapositions of interior and exterior spaces find a variety of counterparts in the record's sonic treatments and instrumentation. In the first verse, we hear of a man metaphorically unlocking a door "While those around him criticize and sleep," counterposing the spatial possibilities of revelation with the implied claustrophobia of collective and individual unconsciousness. The corporeal presence of the bass in contrast to the ethereal arpeggiated synthesizer complements these lyrics, simultaneously invoking the coexistence of materiality and dreamscape.

The drums and bass (the latter performed by current Living Colour member Doug Wimbish[25]) only begin in bar 13, until which point the listener floats in a sphere of suspended animation. The establishment of the rhythmic template provides the equivalent of an identifiable camera angle through which the depth of the spatial contrasts can be perceived and more readily located, establishing a tangibly discernable narrative thread.

The first chorus heightens the record's structural focus as the elements' spatial relationships change. Instrumentally, the left-panned choppily intermittent wah-wah guitar is balanced by a right-panned heavily chorused rhythm guitar, both of which are anchored in the middle of the chorus by the return of the more centrally located sustained guitar chord. Collectively, in conjunction with the ambient synth chords and accentuated bass punctuation, the track now gains momentum, as the multiplicity of layers function within distinct sonic and harmonic environments.

Apart from a harmony in the second half of verse 1, Seal's singular vocal commands the sonic space until the end of the first chorus. As the chorus closes, multiple overlapping background utterances of the word "crazy" rapidly alternate between locations on the left and right sectors of the sound stage, overlaid at the start of verse 2 by the lead voice track also singing the title. This rapid-fire sequence spreads across the stereo space and temporarily creates an aural mosaic from the vocal stack, presenting one of Horn's dramatic strategies for repeatedly building and diffusing tension. The multiplicity of voices here represents the protagonist's internal voices or the "people walking through my head" at the start of verse 2. The multitracked ad-libbed background vocals that follow—positioned around and behind the lead vocal and delivered with considerable intensity—advance the protagonist's sense of interior conflict and disarray. In the second chorus, overdubbed harmony vocals accentuate the repetition of the main chorus line, effectively altering the chorus's effect. The additional expressive punctuation of "crazy," sung with overdubbed vocals that are treated with delay, reverb, and right-panning, deepens the sense of tension. As I emphasized earlier, no faithful rendering of the recorded artifact is possible in a live setting due to the replication of overdubbed parts, and the complex mix choreography further separates the studio from the stage while paradoxically utilizing aspects of the latter environment in creating the recording's dynamics.

The soundscape increases in complexity with the inclusion of reversed (and consequently almost atonal) vocals panned to the right (bars 75–77 [2:53–2:59]), creating a sound effect that reinforces the mood and color, and propels the record forward after its first four-bar drum break, an idea developed in the demo stage by producer Tim Simenon (Horn 2022, 3452). That unaccompanied reverb-soaked drum segment consumes the entire stereo spectrum with the timbrally-higher toms and the artificially loud opening hi-hat cymbal, panned right, while the lower toms are panned left and anchored by the central kick drum. An eight-bar contemplative segment follows (edited from the single) to achieve greater dramatic gravity through the absence of bass, as the drums provide a framework for the airy synth and solitary reverb-enshrouded vocal line.

Interestingly, the middle eight (bars 87–94 [3:22]) was based on a wordless demo vocal melody recorded by Seal that Horn felt needed actual lyrics (Allinson and Levine 2008). This section ultimately enhances the staging of the record by effectively creating another unique scene in its cinematic dramatization. As the drums halt, new space is created: the left-panned enclosed host environment of the wah-wah guitar is thrust acutely to the foreground, underpinned by sustained bass and elevated by string synth, piano, and an arpeggiated percussive synth. These elements in the first four bars of the middle eight are next accompanied by a right-panned chorused rhythm guitar that accentuates the intensity of the scene while preparing the listener for the reintroduction of the drums and a change in bass rhythm. The newly expanded spatiality at the start of the middle eight provides the ideal soundscape as Seal sings committedly about "a sky full of people" in which only some want to fly. As the title of this essay suggests, the record itself is brought into its magnificent existence by population of its sonic stratosphere through multiple overdubs, instrumentally laid bare in these transitional moments (as Seal is himself learning to "fly" as a solo artist). Moreover, the "sky full of people" is given a degree of visual presence in the "Crazy" music video (1990) in which seven embodiments of Seal are simultaneously seen in various surreal scenes, particularly in the second verse and chorus (see Figure 6.2).

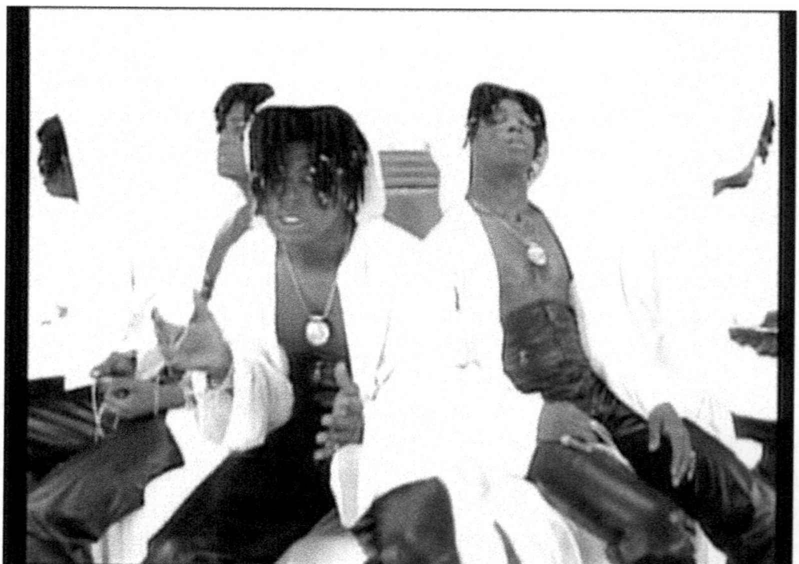

FIGURE 6.2 Seal in the official music video for "Crazy," with seven versions of himself (two of which are partially obscured at the back by the white background echoing the debut album cover setting). In this case, it's a screen full of people reflecting the sky full of people embedded in the recording. ©1990 ZTT/Warner Bros./Sire.

However, the recorded performance multiplicities are only partially reflected in the video since at least ten performers would be required to accurately represent the full contents of the record in the manner in which they are rendered. Nonetheless, the aura of multiplicity aurally projected is therefore given a tangible cinematic counterpart, derived at least in part from the expansive, ambitious nature of Trevor Horn's production.

The closing two bars of the middle eight include an overdubbed Seal ad-libbed background vocal, building anticipation toward the eight-bar vamp.[26] The record's free manipulation of space appropriately parallels the protagonist's dichotomous strategy of lyrically counterposing interior and external psychological states, achieving liberation by adopting the craziness of dreaming a better future into existence to ensure individual and collective ontological survival. Perhaps Horn's production approach, inhabiting the record with strategic overdubs and fluid spatiality, typifies not only the ideological freedom implied in the lyrics and their vocal expression but also the creative freedom offered by technology that facilitates the hyperreal representation of aural ideas.

Reflecting on the unconventional nature of the song's recording, Horn recalls that "we did most of it in this ballroom .... We just played it for days. I had a hand mic for Seal and we just kept working on the track ... and eventually after about a week we sort of had it," meaning that the heavily compressed principal vocals had been captured in a rented Los Angeles house (Allinson and Levine 2008), though it remains unclear whether this was the setting for just some or all of the vocal recording. In the absence of further tracking information, we can only speculate on the degree to which the unconventional space and place affected the vitality of the record's primary vocal performance.

The effectiveness of "Crazy" as a record lies in the integrated holistic aural presence of its elements. However, the vocals—lead and background—act as primary vehicles for emotional transmission of the song's themes and psychologically interior explorations. The constantly shifting matrix of their presence at different stages in the song occasionally requires a momentary muting of the instrumental text in order to more clearly reveal the purposes to which they have been applied within the sound stage. Moreover, Seal's narrative omniscience allows a focal shape-shifting between time, place, situation, tone, and personae, inclusive of amorphous philosophical, spiritual, and other contemplative angles. The "sky full of people" (bars 87–94 [3:21]) actually becomes a spatial sphere largely occupied by the singer himself and guitarist Chrissy Shefts, as well as Trevor Horn, through the series of interactive overdubs carefully applied throughout the record. The space created in this segment recalls Warner's observation[27] (in particular relation to Yes's "Owner of a Lonely Heart") that

> Trevor Horn's productions are often characterized by this device of returning for a few moments to absolute simplicity which provides an excellent foil to the otherwise highly complex textures. And these contrasts are given greater emphasis through technological manipulation.
>
> *2003, 71*

The introspective vocal utterance of "ohhh" leading into the final chorus extends the deeply contemplative tone with which the song (instrumentally and vocally) begins. Moreover, in this context, we can revisit the contention that the sounds of the sung words and not merely the words themselves express the embedded meanings in pop songs (Frith 1981,164; Warner 2003, 72; Moore, 2016, 189). Even a single word or phrase uttered once in a song can convey meaning that not only underlines thematic and emotional intent but also enhances overall narrative coherence and synergy. I suggest that the chorus conceptually functions as a cinematic flashback (in the term's filmic, theatrical, and production senses). The manner of the vocal expression literally and inexorably brings the song toward its logical conclusion and vitally transports the listener along this journey. After a four-bar drum break repeating the rhythms and dark percussive timbres following verse 2, another dramatic dissipation of intensity occurs across eight bars with both sustained and chorused overdubbed guitars gradually interwoven with a rising arpeggiated synth, heavily accented light percussive accompaniment, and the absence of either kick or snare drums reinforcing the pronounced sense of aural vacancy. This crucial section is notably absent from the edited single version.

Amidst the coda's many climactic textural layers, a new subsonic kick drum appears, providing an emphatic sense of conclusion to the record's philosophical discourse. The fade-out—itself a sign of modern hyperreal audience conditioning—features delayed vocals, shaker, chorused rhythm guitar, and an arpeggiated synth. This coda is also absent from the single edit, stripping that version of one of the recording's most dramatic moments.

Cinematic Drama and the Aural Shadow of "Crazy"

Horn's own observations about record production and the textual evidence of "Crazy" clearly demonstrate the cinematic intent informing its staging. The record is effectively a self-contained psychodrama in which the protagonist's narrative compression of time and space is solidly reflected in the sonic parameters of Horn's production. Furthermore, it exemplifies Moylan's concept of "invented space" (2020, 11) with multiple vocal and instrumental overdubs existing in contrasting sonic environments, the distortion of temporality, and artful manipulation through aural illusion. The production strategy and conceptual focus of the record are intrinsically unified. The observation that "film may be considered a fabricated dream which can therefore operate outside of normal boundaries" (Alleyne 2009, 20) is readily applicable here as the production of "Crazy" draws directly from the cinematic sphere. It is intriguing that the group Cinema was the nascent version of the Yes line-up that recorded the hit *90125* album (1983) that helped propel Trevor Horn's pop production conquests, eventually leading to work with Seal.[28] The soon-abandoned group name accidentally implied much about Horn's cinematic production strategies that would become his trademarks across the Eighties and beyond.[29] It is also noteworthy that the title of the

Yes album on which he debuted as a performing member in 1980 was *Drama*, further underlining the recurrent allusions to staging.

In a highly revealing 2004 interview, Seal discussed his views on producers and the production process, and he noted that Horn

> takes care and pays attention to my vocal track, so that when a person is listening to the record, they have all these things, all this depth going on around it, but they never lose sight of what they're buying the record for, which is the vocal. Very difficult to do.
>
> *Peel 2004*

Earlier I quoted Horn's remarks about feel as a critical, defining factor, and the attention to vocals to which Seal refers contributes directly to this idea. Ultimately, the dimensional depth, the episodic drama, the textural complexities, and the elaborate staging that occur on "Crazy" are executed in service of the vocal presence and the messages it carries—or at least its sonic totality, tonally and timbrally enhanced by a dynamic instrumental text and a sound stage in constant flux to establish holistic feel. This is an elusive intangible that is nonetheless materially experienced. Horn's treatment of Seal's vocal demanded considerable technical manipulation of the track around him, as the producer recalled in 1994:

> If you ever listen to 'Crazy,' Seal is absolutely bone dry. There isn't an effect on him. He's also flat, there's no EQ on him, and he's incredibly loud, while the whole track is compressed around him.
>
> *Buskin 1999, 283*

Although Seal is clearly cognizant of the production attention to his vocals both on "Crazy" and his debut album as a whole, he also acknowledges that audiences only connect with fragments of lyrical phrases and their relationships to song melody (Flick 1991, 26).

Significantly, each subsequent recorded live full band version of "Crazy" has suffered from absences of the sonic textures and substance evident on the richly layered spatially strategic debut album recording. It is an emphatic example of how a well-produced and performed record creates a unique aura and that even in an era featuring unprecedented sound reproduction technology options, live interpretations by the original artist only constitute shadows and alternative impressions of the textual portrait that the original recording represents.[30] Perhaps paradoxically, the recorded acoustic versions of "Crazy"[31] more effectively convey the song's essence than the live full band counterparts, pointing toward its compositional origins and the creative foundation from which it was initially built.[32]

The resonance of the imagery in "Crazy" and the distinctiveness of the 1991 album version are forcefully reemphasized by Seal's own interpolation in his commercially successful self-produced 1996 cover version of The Steve Miller Band's hit single "Fly Like an Eagle" (1976).[33] At 3:17, as the song moves toward its conclusion, he once again sings the line "In a sky full of people" just as he did (though in a lower key) in the middle eight sections of "Crazy."[34] The articulation of that phrase represents an apogee on both records, but the stratospheric elevation in "Fly Like an Eagle" is facilitated by the collective sonic power of "Crazy," a connection made more perceptible by the reiterated, powerfully projected lyric. Ultimately, the record's dreamscapes repeatedly enhance our experiential realities as a direct result of Horn's cinematic production process on a record that withstands critical scrutiny even decades later.

Notes

1 Producer, session drummer, and author Richard James Burgess who worked with Horn in the late 1970s and early 1980s, recalls Horn's attention to detail, often in search of mechanistic consistency in performances, seeking virtually identical instances of drum tone and timbre in played events ("Work with Trevor Horn," email, 2020).

2 Les Paul's sound-on-sound recordings with then wife Mary Ford, as a solo act on—for example—the 1948 instrumental "Lover," and later multitrack tape recordings manipulated time, space, and performances in ways that were so uncommon for the time that the industry failed to fully embrace the creative potential of overdubbing (Cunningham 1998, 25–26). Phil Spector's bombastic early 1960s "Wall of Sound," featuring duplicated instruments and overdubbed lead vocals, was purposely hyper-real, especially with the sonic layers (inclusive of reverb and echo) being compressed into mono rather than stereo. The Beatles' harnessing of the recording studio as a creative instrument in both mono and stereo extended the boundaries of aural normalcy as they freed themselves and their audiences from confining notions of live reproduction. However, the dub form of reggae represents the ultimate manifestation of spatial exploration and reconfiguration in popular music, initially characterized in the 1970s by the visionary mixes of producers/engineers King Tubby and Lee Perry (Alleyne 2012, 73–75). They and other Jamaican (and Afro-Caribbean) figures effectively introduced remixing as both an aesthetic and economic vehicle, and the contribution of dub pioneers in employing recording technology has been favorably compared to the expansive studio adventures of The Beatles (Stanfield 2020, 24).

3 Warner (2003) does occasionally reference Horn's work with Seal, but those recordings are not the primary subject matter and no in-depth analysis of Seal's work occurs.

4 In March 2022 (Mix Staff 2022; Rogerson 2022), news emerged that the album's original producer Trevor Horn had completed remixing Seal's solo debut in the cutting-edge Dolby Atmos format for future streaming access. Other accounts (Sinclair 2022) noted that a now rare "pre-mix" edition of the record was released and recalled in 1991 containing different mixes, edits, and recordings of tracks (including a shorter version of "Crazy"), soon replaced by the album version (bearing the same catalog number) discussed in this chapter with which many listeners are now familiar. As another example of alternate mixes of the debut album's material, its version of "Future Love Paradise" is only 4:20 in length, whereas the single/EP version extends to 5:35 with a much later fade point. A 2005 surround sound version of the album (CD & DVD-Audio) was released only in America and is out of print. Perhaps the most notable observation made in the stories about the 2022 release is that Trevor Horn surprisingly could not locate all of the original multitrack tapes and that some elements were transplanted from 5.1 mixes and live recordings. Ultimately, this raises the possibility that one or more altered versions of "Crazy" may be issued among

the Dolby Atmos mixes, incorporating previously unheard or recontextualized track content.

5 Seal first drew attention as the featured vocalist on producer Adamski's version of "Killer" (MCA) (for which Seal wrote the lyrics) that debuted on the British singles chart in April 1990—about eight months prior to Seal's first solo chart presence—eventually peaking at number 1. Another version was recorded for Seal's debut album, which garnered Brit Awards in 1992 for Best British Male Solo Artist, Best British Album, and Best British Video (Rees and Crampton 1999, 886), making Seal the first artist to win all of these awards in the same year at this event. In addition, as a result of his work on Seal's album, Trevor Horn won the Best British Producer award in 1992. "Killer" (first charting on November 16, 1991) was the album's fourth British single, preceded by "Crazy," "Future Love Paradise" (on the *Future Love* EP), and "The Beginning." "Violet" debuted in Britain as the fifth and final hit released in February 1992, but in America, "Crazy" was the only single from the album to achieve hit status, with "Killer" only reaching number 100 in March 1992 in a disappointing two-week run. This occurred almost a full year after the record's release in the United Kingdom on the superb *Killer ... On The Loose* EP that peaked at number 12 there. Besides remixes of "Killer," that EP also included an explosive live-in-the-studio cover version of "Hey Joe," made most famous by the Jimi Hendrix Experience, one of Seal's many musical influences. He would later cover the group's "The Wind Cries Mary" as a non-album bonus track on the "Newborn Friend" CD single (1994).

6 Despite the emergence of vastly varied official remixed versions of "Crazy," none of the commercially supplementary reinterpretations incorporated the same degree of spatial breadth or adventure as Horn's album version production. For example, the William Orbit Mix on the CD single and the promotional Urban Remix/Radio Mix by Wolf & Epic for Peace Productions (both 1990) are centered around dance grooves featuring wholesale replacement of instrumental components, yet neither remix emulates the sonic staging strategies and the stereo dimensionalities of the source material. Two later official remixes appearing on the "Killer 2005" CD Maxi-Single feature faster tempos and transplanted drums and percussion that substantially diminish any sense of spatiality. The compositional credits for "Crazy" are inconsistent, with Seal sometimes listed as the sole author (as on 1990 copies of the single) or as the co-author with Guy Sigsworth (as included in the *Best: 1991–2004* compilation).

7 Since Seal was signed directly to Horn's ZTT label (later acquired by Universal Music in 1998), this may have provided additional motivation to achieve commercial success with the debut album project.

8 The starkly nightmarish cover photo of the unclothed Seal taken by Jean-Baptiste Mondino loosely parallels his naked depiction of Prince a decade earlier on the cover of the *Lovesexy* album. Both records also represented significant musical departures and reinterpretations of identity for the artists. The guitar, keyboard, and vocal duo of Wendy & Lisa, members of Prince's mid-1980s band The Revolution, also appear on Seal's debut album (with their background vocals especially evident on the song "Whirlpool"). In addition, they are credited on his *Human Being* album.

9 Zahlaway (2001) interviewed Seal prior to the anticipated release of the subsequently aborted *Togetherland* album, and the character and commercial failure of *Human Being* occupied much of the narrative.

10 In *The Greatest Music Never Sold* (Leroy 2007, 12), author Dan Leroy notes that "Horn and Seal came to blows during the sessions, when Seal objected to Horn's meticulous rearranging of his vocal tracks ...." Despite subsequent reconciliation, the discordant undertones arguably became embedded within the sonic fabric of *Human Being*.

11 *Togetherland* started as a self-produced enterprise (Leroy 2007, 31) and was later co-produced by British film composer Henry Jackman. (Jackman's work with Art of Noise circa 1999 establishes another link to Trevor Horn, as does his subsequent soundtrack work with Buggles collaborator Hans Zimmer.) Avid Seal fans have

constantly lamented Warner Bros. Records' decision not to release the record, particularly as some entire songs (such as "Eliz," "All I Wanted to Say," and "Love Is Better") were briefly accessible on Jackman's website and were subsequently circulated on YouTube.

12 Leroy (2007, 10–11). Isidore is described as one of several people familiar with the work who rate *Togetherland* as a more definitive, representative, and better album than *Seal IV.*

13 This overview of Seal's recorded catalog excludes his cover song albums *Soul* (2008), *Soul Live* (2009), *Soul 2* (2011), and *Standards* (2017). Trevor Horn's involvement with these releases was limited to partial production on *Soul 2*, shared with David Foster, who produced Seal's previous *Soul* albums.

14 It may be merely coincidental that the now-famous film music composer Hans Zimmer appeared in the video clip for "Video Killed the Radio Star" and also appeared with the group on the BBC's *Top of the Pops* performing the follow-up single "Living in the Plastic Age" (1980), but the cinematic parallel is intriguing.

15 As Warner (2003,133) notes, Jones's "Don't Cry—It's Only the Rhythm" is "explicit in its use of unnatural effect, drawing the listener's attention to the artificial nature of the recorded product." However, my unpublished interview with the album's main recording engineer, Stephen Lipson, also revealed that the described aural effects were simply the results of extreme initial experimentation with a Synclavier sampling keyboard software upgrade (Stephen Lipson, interview by author, November 17, 2021).

16 Formed in 1983, The Art of Noise was initially a studio-centered technologically propelled collective including Horn, engineer Gary Langan, programmer J. J. Jeczalik, journalist Paul Morley, and Oscar-winning composer/arranger Anne Dudley. All of them had worked on ABC's Horn-produced *The Lexicon of Love* album released in the preceding year. Since the mid-1980s, the group members have repeatedly splintered and reformed in various configurations and under different group names. Anne Dudley played a prominent role on the first three Seal albums and *Seal 7*, providing dramatic and emphatic string arrangements.

17 The article was published on February 15, 1992 in the wake of the debut album's transatlantic success, but its central point remained relevant years later. In a review of the third Seal album *Human Being* (Williamson 1998, 110), it was recognized that "Seal draws widely on white rock influences such as Joni Mitchell and Crosby, Stills & Nash, which doesn't endear him to those who expect black artists to stay in the R&B ghetto." Seal's stylistic eclecticism also highlights the paradoxes surrounding Black artists with white producers and the often accompanying blurring of sonic identity.

18 The complexities surrounding the character of his recordings (released and unreleased), Lewis's brief career and his untimely death in 1994 are chronicled in my unpublished 2007 IASPM conference paper "'This Floating World': Ephraim Lewis & The Discourse of Desire." In numerous album reviews and articles discussing the Black British scene, Lewis was frequently compared to Seal. The introductory solo music videos by both artists were shot by the same production company (Big TV) within a year of each other.

19 Seal's *Future Club EP (The Nellee Hooper Remix)* (1991) featured the sonic reinterpretations of "Future Love Paradise" by the Soul II Soul producer/remixer/songwriter, potentially widening the dance, Black and pop audiences for the recording.

20 The nine songs on Seal's debut album either completely exclude crash cymbals or embed them deeply within the mix's texture. This strategy was also true of Horn's work with The Buggles.

21 In discussing the multitracked vocals on another Horn production, the US chart-topping single "Owner of a Lonely Heart" by Yes (1983), Warner notes that "the listener is being 'spoken to' by several voices, which clearly belong to the same singer, simultaneously, and this extremely artificial situation has the internalized, ethereal

quality of thoughts or dreams: something which is both magical and outside reality" creating "an overall sound that dislocates both time and space" (2003, 72).

22 Undoubtedly, the later recording of "Kiss from a Rose" represents the peak of Seal's recorded vocal multiplicities, though notably, its demo (still unheard by Horn until preparations began for the second album) preceded the first album and may have partially informed aspects of Seal's performances on his debut.

23 It is especially notable that this otherwise male-centered and constructed production features the multiple guitar parts of session player Chrissy Shefts, thus inviting consideration of the significant role of female musicianship in creating the record. She played Horn's twelve-string Rickenbacker guitar (Horn 2022, 3492).

24 This parallel is especially evident in the 12-inch remixed version of the title track.

25 The album credits are not track-specific and the absence of detailed track listing has the unfortunate effect of anonymizing the session musician contributions. However, my contact with Wimbish confirmed his participation in "Crazy." The bassist has also performed live gigs with Seal, circa 1991/1992.

26 Two-thirds of the 1990 CD single's third track "Crazy (Acoustic Version/Instrumental Version)" consists of this eight-bar vamp being deconstructed and looped, expanding its textural and rhythmic potential.

27 Also see Inglis 2016 that includes a video deconstruction of the Yes track and its constituent elements by Trevor Horn.

28 The Yes *90125* album was also Horn's last completely analog production, by his recollection (Massey 2009, 186).

29 This does not ignore Horn's earlier work with, for example, British pop vocal group Dollar, but it aims to emphasize later growth in his conceptual execution.

30 This statement, of course, assumes that the idea of live performance is taken literally and that no pre-recorded multi-instrumental elements not emanating from the stage in real-time are included.

31 Here I am particularly referencing the 1991 CD single of "Crazy" including an acoustic rendering among its multiple remixes and representations, the 1990 acoustic Church Sessions captured on video, the 1992 live performance on *Acústico MTV* recorded in Brazil, and the Seal compilation, *Best 1991–2004* that includes "Crazy" on a disc of specially recorded acoustic interpretations of key songs from his catalog.

32 Despite energetic performance, the band-assisted versions on *Live at The Point* (1992), *Live in Paris* (2005), and on *One Night to Remember* (2005) all display spatial deficiencies in comparison to the studio recording of "Crazy." While each of the cited live versions is intended to only re-present the song without verbatim reiteration, a significant sonic gulf remains. Contrastingly, the stripped-down acoustic versions all embody the song's essential energy more effectively by not attempting to recapture the soundscape of an unrepeatable studio construction. The "Crazy" eight-remix CD single (1990) includes an acoustic interpretation that continues into the second verse, another was captured on video in 1990 with acoustic guitars and drums, and a re-recorded 2005 rendition on the CD *Best: 1991–2004* featuring piano and acoustic guitar reasserted the splendor of the minimalist approach freed from the sonic constraints of Horn's typical studio strategies. Ironically, the 2005 version of "Crazy" (and indeed the entire disc of Seal acoustic tracks) was produced by Trevor Horn, suggesting that the producer was fully conscious of the song's primary characteristics and had purposefully cinematically magnified their scope and sonic undercurrents for Seal's debut album. When interviewed for the short-lived Seal fan club newsletter *Out On The Void* (1:3, August 1992), the singer confirmed that "I write from the acoustic guitar."

33 Seal's recording of "Fly Like an Eagle" roughly coincided with the termination of his management relationship with John Wadlow with whom his successful recording career began as "Crazy" became an international hit. The explicit sense of freedom reflected in the lyrics of Miller's "Fly Like an Eagle" and their interrelationship to themes of liberation in "Crazy" may, therefore, also have found an additional tangible manifestation mirroring the spatial expanses of the latter song's recording. Perhaps

ironically, subsequent legal conflicts with Seal's erstwhile manager reverberated up to 2007 (Anderson 2007), with a longevity almost parallel to the debut single.

34 The fully reprised lyric is "In a sky full of people / Only some want to fly / Isn't that crazy?," ensuring the inclusion of sonic and historical context by directly referencing the title of his first hit. "Fly Like an Eagle" appeared on the 1996 *Space Jam* soundtrack album.

References

Alleyne, Mike. 2009. "Sounds Reel: Tracking the Cultural History of Film Sound Technology." In *Sound and Music in Film and Visual Media: An Overview*, 15–41. New York: Continuum.

Alleyne, Mike. 2012. *The Encyclopedia of Reggae: The Golden Age of Roots Reggae*. New York: Sterling.

Allinson, Richard, and Steve Levine. 2008. "'Trevor Horn.' The Record Producers." Radio show. BBC Radio 2.

Anderson, Charlie. 2007. "Ongoing Management Commission gets Seal of Approval: Wadlow v Samuel." *Music Law Updates*, April 2007. http://www.musiclawupdates.com/?p=2570

Batey, Angus. 2010. "Seal: The Reluctant Pop Star." *The Guardian*, September 16, 2010. https://www.theguardian.com/music/2010/sep/16/seal-interview-pop

Burgess, Richard James. 2020. "Work with Trevor Horn." Email.

Buskin, Richard. 1999. *Inside Tracks: A First-Hand History of Popular Music from the World's Greatest Record Producers and Engineers*. New York: Avon.

Buskin, Richard. 2011. "The Buggles 'Video Killed the Radio Star.'" www.Soundonsound.Com. Sound On Sound. December 2011. https://www.soundonsound.com/people/buggles-video-killed-radio-star

"Chrissy Shefts—Guitarist, Vocalist, Songwriter." 2012. Www.Chrissyshefts.Com. 2012. http://www.chrissyshefts.com

Cromer, Ben. 1998. "Studio Action: Pro File." *Billboard (Archive: 1963–2000)*, March 21, 53–54. https://ezproxy.mtsu.edu/login?url=https://search.proquest.com/docview/1506059073?accountid=4886

Cunningham, Mark. 1998. *Good Vibrations: A History of Record Production*. London: Sanctuary.

Freff. 1984. "Trevor Horn." *Musician (Archive: 1982–1999)*, March 01, 74–76, 78, 94. https://ezproxy.mtsu.edu/login?url=https://search.proquest.com/docview/964129450?accountid=4886

Flick, Larry. 1991. "Seal Slides from Club to Pop Success." *Billboard (Archive: 1963–2000)*, July 27, 26–29. https://ezproxy.mtsu.edu/login?url=https://www.proquest.com/magazines/talent-seal-slides-club-pop-success/docview/1505902298/se-2?accountid=4886

Frith, Simon. 1981. *Sound Effects: Youth, Leisure, and the Politics of Rock 'n' Roll*. New York: Pantheon.

Horn, Trevor. Adventures in Modern Recording. London: Nine Eight Books, 2022. Kindle.

Inglis, Sam. 2016. "Trevor Horn: The Art of Record Production." Soundonsound.Com. 2016. https://www.soundonsound.com/people/trevor-horn-art-record-production

Laing, Dave. 1992. "Brits Around the World: Seal? Dance? Rock? Black Rock? Pop." *Billboard (Archive: 1963–2000)*, February 15. https://ezproxy.mtsu.edu/login?url=https://search.proquest.com/docview/1505965806?accountid=4886

Leroy, Dan. 2007. *The Greatest Music Never Sold*. New York: Backbeat.

Lindvall, Helienne. 2014. "Trevor Horn: 'I Had Delusions the Label Would Be a Hub of Creativity.'" *The Guardian,* February 13, 2014, sec. Media. https://www.theguardian.com/media/media-blog/2014/feb/13/trevor-horn-video-killed-the-radio-star-ztt

Lipson, Stephen. 2021. Interview by Mike Alleyne. Online. November 17, 2021.

Massey, Howard. 2009. *Behind the Glass, Volume II: Top Record Producers Tell How They Craft The Hits.* Milwaukee, WI: Backbeat.

Mix Staff. 2022. "Trevor Horn Remixes Seal's Debut in Dolby Atmos." Mix Online, March 8, 2022. https://www.mixonline.com/recording/trevor-horn-remixes-seals-debut-in-dolby-atmos

Moore, Allan F. 2016. *Song Means: Analysing and Interpreting Recorded Popular Song.* New York: Routledge.

Morrison, Jim. 2009. "Acoustic, Visual, and Aural Space: The Quest for Virtual Reality in Musical Reproduction." *Explorations in Media Ecology* 8 (2): 81–98.

Moylan, William. 2020. *Recording Analysis: How the Record Shapes the Song.* New York: Routledge.

Peel, Ian. 2004. "Sounding Off: The Producer's Role." www.Soundonsound.Com. Sound On Sound. August 2004. https://www.soundonsound.com/people/sounding-producers-role

Rees, Dafydd, and Luke Crampton. 1999. *Q Rock Stars Encyclopedia.* London: Dorling Kindersley.

Rogerson, Ben. 2022. "Trevor Horn on Mixing in Dolby Atmos for the First Time: 'It Was an Interesting Experiment.'" *MusicRadar,* March 9, 2022. https://www.musicradar.com/news/trevor-horn-dolby-atmos-seal

Sinclair, Paul. 2022. "Trevor Horn Remixes Seal's Debut." *Super Deluxe Edition,* March 19, 2022. https://superdeluxeedition.com/feature/saturday-deluxe-19-march-2022/

Stanfield, Geoff. 2020. "Lee 'Scratch' Perry: Super Ape Upsets The Studio." *Tape Op* 136: 24–30.

Tagg, Philip. 2012. *Music's Meanings: A Modern Musicology For Non-Musos.* New York: Mass Media Music Scholar's Press.

Warner, Timothy. 2003. *Pop Music—Technology and Creativity: Trevor Horn and the Digital Revolution.* Aldershot: Ashgate.

Williamson, Nigel. "Albums: SEAL." *Uncut (Archive: 1997–2000),* December 01, 1998, 110, https://ezproxy.mtsu.edu/login?url=https://www.proquest.com/magazines/albums-seal/docview/1771118348/se-2?accountid=4886

Zagorski-Thomas, Simon. 2014. *The Musicology of Record Production.* Cambridge: Cambridge University Press.

Zahlaway, Jon. "LiveDaily Interview: Seal." LiveDaily. November 01, 2001. http://www.livedaily.citysearch.com/news/3992.html

Zak, Albin. 2001. *The Poetics of Rock: Cutting Tracks, Making Records.* Berkeley, CA: University of California Press.

Discography

Seal. 1990. *Crazy* (Maxi-Single). Sire/Warner Bros./ZTT. 9 41003-2.

———. 1990. *Crazy* (Promotional Single). Sire/Warner Bros./ZTT. PRO-CD-4879.

———. 1991. "Crazy." *Seal.* Sire/Warner Bros./ZTT. 9 26627-2.

———. 1991. *Future Love EP.* WEA/ZTT. WPCR-2202.

———. 1994. *Seal.* Sire/Warner Bros./ZTT. 9 45415-2.

———. 1996. "Fly Like an Eagle." *Seal.* Atlantic/Warner Sunset/ZTT. 2 87046.

———. 1998. *Human Being.* Warner Bros. 9 46828-2.

———. 2001. *Togetherland*. Warner Bros. Unreleased.
———. 2003. *IV*. Warner Bros. 47947-2.
———. 2004. *Best 1991–2004*. Warner Bros. 9362-48943-2.
———. 2005. "Killer 2005" (Maxi-Single). Warner Bros. 42777-2.
———. 2007. *System*. Warner Bros. 279868-2.
———. 2010. *Seal 6: Commitment*. Reprise. 525056-2.
———. 2015. *Seal 7*. Warner Bros. 552673-2.
Yes. 1983. "Owner of a Lonely Heart." Atco. 7-99817.

Videography

"Crazy." Directed by Big TV. 1990. Warner Music Vision, 1991. VHS.
Seal: Acoustic. Directed by Gareth Roberts. 1990. Warner Music Vision, 1991. DVD.
Seal: Live at the Point. Directed by Gareth Roberts. 1992. Warner Music Vision, 1992. DVD.
Seal: Live in Paris. Directed by Martyn Atkins. 2005. Warner Music, Burbank, CA, 2005. DVD.
Seal: One Night to Remember. Directed by Volker Weicker. 2005. Toronto, ON: Warner Music Canada, 2005. DVD.

7

COME TOGETHER

Feeling the Distemper of Murk and Elation with the Beatles (1969) and with Sheila E. and Ringo Starr (2017; 2020)

Walter Everett and Katie Kapurch

Introduction

"Come Together" is the first track on *Abbey Road* (The Beatles 1969), the last album the Beatles recorded. Our chapter reveals the song's competing dynamic of invitational playfulness and hope versus violence, dark desire, and disappointment. We shall argue that an obscured distemper—a ghostly disease—suggests an off-color, failed pursuit of bliss born of intangible yet sultry passions. An examination of John Lennon's inspirations for the song leads to a close reading of its tracking process, lyrics, and music. Our discussion concludes with an ear toward the recording's afterlives in Black American artists' covers, including Sheila E.'s collaborations with Ringo Starr (Sheila E. 2017a and 2017b; Starr, 2020).

An Invitation

On July 21, 1969, John Lennon made his first studio appearance following a car crash three weeks previous, a "spinal cracker" that left him, the driver, with just a few facial lacerations, but pregnant passenger Yoko Ono with minor back injuries. The latter required a bed rest observed notoriously even at Abbey Road Studios, making for a rather disordered atmosphere surrounding the creation of "Come Together." The title phrase was provided by Timothy Leary, whom John and Yoko met on June 1–2 during the Montreal bed-in for peace. Leary was then projecting a 1970 gubernatorial bid in California and appealed to John for a campaign jingle composed around the rallying cry, "Come Together—join the party." The slogan was a punning invitation to both a political body and a social celebration (Leary familiar mostly as an advocate of the spiritual potential of LSD, then more widely a recreational turn-on than a devotional catalyst). Although the crusade was soon dissolved and the campaign's phrase commandeered by the Beatles in a wholly new direction, Lennon would a few years

DOI: 10.4324/9781003089926-11

later reassociate the resulting song with its political roots when he performed it in New York on August 30, 1972, ending the last verse with the line, "Come together, right now; stop the war!," suggesting a communal urgency in the mood behind the hopeful message.

The Recording

Figure 7.1 summarizes the recording's contents in a timeline of waveforms, presented in their presumed arrangement among the eight tracks of the working tape.[1] Surviving parts in the basic track, Ringo Starr's drums (with tea towels dampening the batter heads of the snare and all three toms, and a blanket muting the bass drum's one head), Paul McCartney's bass (in a blend of direct injection and some input from the miked amp), and George Harrison's rhythm guitar (initially and elsewhere with a clean Fender tone but in most places dirtied by a distortion pedal), were all led by John Lennon's guide vocal and handclaps in a live performance on July 21, 1969. (Although the guide vocal was wiped by a dub the next day, it can be heard along with bass and guitar bleeding onto the drum track.[2]) All remaining tracks received superimpositions on the 22nd, resulting in three captured passes:

- Firstly, an instrumental ensemble added to the break and coda, consisting of McCartney's electric piano on Track 5, and on Track 6, both Harrison's first lead guitar part and an inexpert maraca attributed to Starr (but perhaps shaken by Lennon, in the first halves of both break and coda—entering there at 3:34).
- Secondly, Harrison performed a second lead guitar part with a second maraca (certainly Starr's) for the break on Track 7. Underneath the piano break (at 2:02–2:13), the two new guitars have a repeated staccato passage shown in Example 7.1 (all-but-buried in all released mixes) before the high-register legato duet begins at 2:14. (Notably, Lead Guitars 1 and 2 are registrally intertwined: Guitar 1, the first of the two recorded, has a lower line than Guitar 2 in staccato, sings above it for the legato passage, and cadences below; see the last two bars of the duet in Example 7.2.)
- Thirdly, vocals were taped simultaneously by Lennon (leading, with live tape echo) on Track 4 and McCartney (lower harmony—an unheard assignment before "Hey Jude," but fairly common afterward) on Track 8.

On July 25, John double-tracked his lead vocal at the unison to reinforce parts of choruses and coda, punching in to Track 8; these and other such dubs are shown with lighter shading in Figure 7.1. Finally, additional guitar parts were similarly punched in on July 23, 29, and 30 (as per Lewisohn 1988, 181–83), resulting in John's grungy addition to Track 6 (a brief distorted double-stopped part in the chorus; muted in mixes, likely because it "steps on" the stop-time silence) and George's volume/tone-pedaled lead-in to the coda on Track 7.[3] Ten attempts at a stereo remix were made on August 7, with RS 1 mastered for releases on LP and reel-to-reel on September 26, on 8-track and cassette in 1970, and CD in 1987.[4]

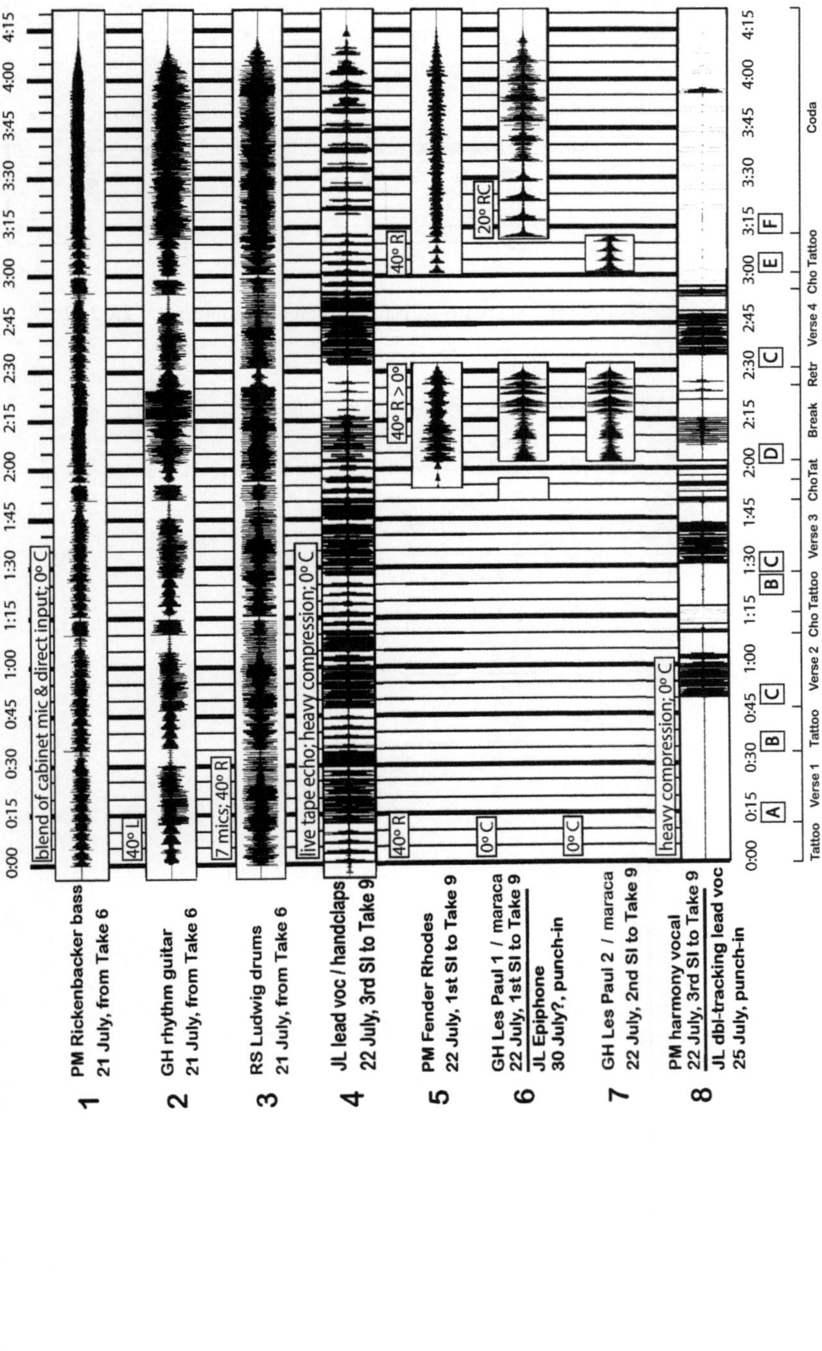

FIGURE 7.1 "Come Together": form-defining timeline of multitrack contents.

EXAMPLE 7.1 "Come Together": staccato guitars and electric piano, @ 2:02–2:13.

EXAMPLE 7.2 "Come Together": ending of instrumental break, @ 2:19–2:30.

This familiar mix has reverb produced by Harrison's Fender Twin and the tape echo printed to Lennon's main vocal but is otherwise dry except for the piano's sustain pedal and some reverb applied to each chorus-ending vocal statement of "over me" except the third. ("Over me" can be taken as Lennon's offer to personally unite whomever wishes to join—with "my direction," as previously offered in "And Your Bird Can Sing.") Vocal parts are compressed, but all instruments maintain a wide dynamic range.

Getting Back to Berry

While the song's title had been provoked by Leary, who was a recent acquaintance, much of its emerging substance confessed a boyhood hero: Chuck Berry, whose 1956 single, "You Can't Catch Me," is so obviously the source of the opening line of Lennon's first verse, the copyright owner of Berry's song, Morris Levy, sued the Beatle, who settled by promising to record and release three of Levy's holdings, thus terrifically boosting their future earnings. To comply, Lennon recorded Berry's alleged model and "Sweet Little Sixteen" plus two of Levy's other Big Seven publications, "Ya Ya" and "Angel Baby," in 1973–74. The last-named was released only posthumously.

The Berry connection in the opener of the Beatles' final LP bookends the launch of their first, in which "I Saw Her Standing There" had opened with a bass line lifted from Berry's "I'm Talking About You." But the pioneer of rock 'n' roll inspired much more than the "Come Together" kickoff. Other more subtle allusions abound: Lennon's vocal throughout all verses is dominated by a reciting tone on F♮, the lowered third scale degree (see the repeated pitch in the final bar of Example 7.3), also characteristic of the "You Can't Catch Me" melody. Harrison's rhythm guitar adorns power chords with their fifth's upper neighbor in a muted boogie figure that evokes Berry's signature rhythm-guitar playing (see the embellished I5 chord through the last bar of Example 7.3). Lennon's highly ambiguous lyrics (which he characterized in 1980 as "gobbledygook"; see Golson 1981, 210), often inspired by Berry's wordplay, specifically recall the model song's scansion: There's a ghostly retention of disrupted trochaic lines such as "I was—rollin' slowly 'cause of—drizzlin' showers" behind oddball attributes such as "He got—toe-jam football, he got—monkey finger." In fact, Lennon extends this borrowing in likely referring to his youthful admiration for Berry in an ad-lib sung just after Take 5 of "Come Together" breaks down by intoning, "He's got teenage lyrics."[5] An old-hat reversion to tape echo, applied to many Berry vocals, drenches Lennon's lines. The tape echo on the tattoo's downbeat handclaps timed to reflect six attacks per beat, perfectly answered by Ringo's subdivision on the hi-hat and a rack tom, may well have contributed to Grammys for Best Engineered, Non-Classical, Recording awarded to *Abbey Road*'s Geoff Emerick and Phil McDonald.[6] Note that the bottom line of Figure 7.1 shows the song to contain a recurring tattoo, four verses, three choruses, an instrumental break and retransition, and a coda. Lennon's rare lack of a contrasting bridge

EXAMPLE 7.3 "Come Together": count-in (later trimmed away), opening tattoo, and the first bar of Verse 1, @ 0:00–0:14.

("middle eight") section mirrors the typical Berry arrangement. Finally, Berry's Chess Records stablemate, Muddy Waters, even gets a name check in "Come Together," and he, in turn, likely motivates the reference to "mojo" (directly from Waters' "Got My Mojo Working" and more generally in the voodoo folklore of his "Manish Boy" and "I'm Your Hoochie Cooche Man"). The word "mojo" also seems to engender consonance with "ju-ju," "joker," and "toe-jam."

Coming Together?

Under Lennon's direction, the idea behind "Come Together" can be interpreted as having morphed from an invitation to join a social/political movement into a celebration of coordinated orgasms. This adoption was in line with John's puns such as in the "BED PEACE" ("PIECE [OF ASS]") posters in Amsterdam and Montreal (also appearing as "only trying to get us some peace" in "The Ballad of John and Yoko," the group's last release before *Abbey Road*). Just as Lennon used a gun's trigger and the line, "bang! bang! shoot! shoot!" as sexual metaphors in "Happiness Is a Warm Gun," the strong desire underlying the sustained but turbulent unvoiced fricative attack of "Shoot me!" in the sultry yet arresting, passion-arousing tattoo that introduces "Come Together" can easily be heard as a request for ejaculatory relief. (See the second bar of Example 7.3, especially for the shouts and the staccato echo's contrast against the slow, legato, lasciviously sliding up-and-down bass ostinato and understated though playfully syncopated guitar figure, followed by rapid cymbal then drum subdivisions.[7]) The glissandos' soggy gradation is matched by the sodden fade-ins provided by Harrison's first use in years of the volume/tone control in his guitar, giving nuance to the final tattoo that signals the coda at 3:00–3:11. Similarly liquid is Harrison's frequently wavering portamento in his Les Paul melodies, embellished by languid bends and pre-bends (some noted in Example 7.2), and a ubiquitous wrist vibrato (the instrument is not fitted with a whammy bar). Through the electric piano solo (at 2:02–2:12), Lennon breathes heavily on every backbeat while McCartney similarly repeats a wordless "oo" in his highest falsetto register (best audible in the isolated main vocal track).[8] They then both accent a half-time backbeat under the guitar duet (2:12–2:24), speak-singing a repeated "Come!," John with a sexy portamento (see the first two bars of Example 7.2). Of course, such meanings are veiled in obscurity—they, like the song's good-looking soul, are "so hard to see."

Part of the reason for the song's obscurity, its lack of definition, is that orgasm may be described therein, but it is apparently not achieved (at least not in a mutual sense). The early Beatles would culminate in successful climaxes through ecstatic retransitional dominants (lusty in "Twist and Shout," 1:24–1:36, "I Want to Hold Your Hand," 1:45–1:53, and "Day Tripper," 1:31–1:41; a bit less euphoric in "The Night Before," 1:05–1:07). Instead, the retransitional dominant of "Come Together" (beginning at **D**+5 in the Wise score; see The Beatles (1993, 154); it ends in the first two bars of Example 7.2) atypically fails, falling prematurely into tonic harmony in a closely collaborative but ultimately limp post-duet hocket (at 2:24–2:30, last two bars of Example 7.2). Here, suddenly, the pounding (thrusting) drums stop while an impotent post-coital stop-time texture meanders through the electric piano's falling thirds that syncopate over a half-length bass-and-guitar tattoo (now abandoning the hopeful "Shoot me!" command). The offbeat- and weak-beat-emphasizing piano is answered on beat two by guitar and four by bass, the collapsing lines clearly asynchronous. Sex without the rapture of simultaneous climaxing. This passage becomes the basis of the song's coda, which long repeats this failure under John's rising but colorless falsetto arpeggio, D5–A5–D6, expressing a more groundless hope than did the similar lead-guitar rise with which "Dear Prudence" had ended. Lennon's last-added vocal-fry moans sound more like disappointment than the bliss of a "little death."[9]

Murky Scales

The track's ambiguity, darkness, and sense of failure are all supported by the obscurity of its underlying musical scale. As in much rock music, the tonal texture is stratified in that soloistic melodic forces (principally main vocals and lead guitars) draw from a different collection of scale degrees than do more structural accompanimental ones ("fill" pitches largely chorded in the "rhythm" section: piano, rhythm guitar, bass).[10] But the constant lowered third in the bass groove, F♮, occurs so high in its register, it seems more a doubling of the vocal's minor-pentatonic F than a claim by a member of an elemental minor mode. Our notated examples use a key signature for D major, even though F♯s are shrouded. They are heard in the harmony-defining V chord in the rhythm guitar's boogie neighbor to E, but this appears only in two bars of the verse and under the four-bar lead-guitar duet in the failed retransition (as in the first two bars of Example 7.2). Of more structural value, F♯ is heard as the fifth of the vi chord with which the chorus begins, the B minor chord appearing just before the first ending (see The Beatles 1993, 153). The vi chord helps support the upper voice's structural scale degrees 3–2–1 for the cadence, thus clinching the foundational nature of D major.[11] Another indication of D major appears ever so slightly at 2:40 in the rhythm guitar, where the F♮ (E♯ in a functional ♯9 spelling) slides up to a barely articulated F♯ (like a pale version of the Fx hammer-on to G♯ in the lusty "Day Tripper" ostinato) as if inadvertently betraying a major-mode scaffolding.

C♮s also routinely defy the major key's signature in elemental ways. With its C, the "clean" rhythm-guitar part in the tattoo (see the second bar of Example 7.3) completes the minor-pentatonic set outlined in the bass. The sung C♮ (introduced in the last bar of Example 7.3) completes Lennon's (025) blues trichord, C–D–F, which accounts for all of his vocal pitches in verses but one (E); McCartney features C♮ in his lower harmony vocal through verses.[12] Both lead guitar parts in the duet (see its second half in the first two bars of Example 7.2) employ C♮s that clash against the piano's C♯s, as well as F♮ members of the V ♯9 chord, and both F and C are included in the four volume-controlled guitar chords that fade in to announce the coda.

So "Come Together" is in D major despite the predominance of embellishing tones that cancel its key signature. But B♭ (an important element of chords in D minor) doesn't make a single appearance, not even as an ornament. Instead, B♮ is prominent throughout, as in the grungy rhythm guitar's boogie on I5 (see the third bar of Example 7.3) and the chorus's structure-defining B minor and G major triads.[13] Perhaps of greatest interest in this regard is McCartney's Fender Rhodes part in the hocket retransition (last two bars of Example 7.2), in which he resolves tonic ninth and seventh (E4 and C4) to octave and sixth (D4 and B3), passing to seventh and fifth (C4 and A3) before passing through non-chord tones B3/G3 to A3/F3. (The connecting passage is notably similar to Billy Preston's stop-time syncopated, two-voiced pre-solo lick at 1:08–1:12 in the Beatles' "Get Back.") The piano's passing use of this B3, the major sixth scale degree in an otherwise minor context, exhibits a pungent dorian coloring idiomatic to pop keyboard playing in the 1960s: popularized on a sour Hammond organ in Booker T. & the M.G.s' "Green Onions" (1962), the lick then lent an inscrutable tinge to absence on Rod Argent's Hohner Pianet in the Zombies' "She's Not There" (1964); augured a dark D dorian thunder in the piano and brass intro to the Toys' "A Lover's Concerto" (1965) before being neutered as a dominant preparation for a pleasant C major rain; suggested a cuckolding in organ and rhythm guitar from 0:13 (not in the more prominent opening Wurlitzer part) in Marvin Gaye's "I Heard It Through the Grapevine" (1968); and was tied to libido by James Brown in his own piano riff at 1:03–1:20 in "Get Up (I Feel Like Being a) Sex Machine" (1970). The Beatles were no strangers to the dorian mode; a favorite exotic device of George Harrison, it added a bitter, heartbreaking intensity to McCartney's vocal melody in "Eleanor Rigby" (1966) ("wearing the face *that she keeps* in a jar by the door …").[14] These dorian portrayals seem well related to the murk and elation that pose united in "Come Together."

More Contradictions

Many rock recordings are close to single-minded in their execution (e.g., the Beatles' "All Together Now" or "I Want You (She's So Heavy)"). "Come Together," on the other hand, highlights many inner contradictions, some already suggested in the paragraphs above. Most singing is detached, often expressing the (025) trichord, but one line stands out in each verse with legato repetition of only one pitch, an anxious F♮ (e.g., "got to be good-looking 'cause he's so hard to see"); the mysterious intensity of these particular lines is amplified by the stop-time dropout of guitar and bass and a new menacing dotted rhythm in solo bass drum.[15] Lennon strains through the highest pitches of his belted vocal (his voice fails on A5 in the second and third choruses, 1:50 and 2:58, respectively, in his guide vocal to Take 1), whereas McCartney's vocal is always relaxed and well contained. Verses emphasize D minor pentatonic pitches, but choruses turn major until the cadence "over me" twists to the dark once again.

William Moylan (2020, 318) has described Lennon's vocal quality in this song as relatively distant, lacking many high frequencies and timbral detail. A cavernous tape echo and compression certainly contribute to this colorless impression, but the sense is at its strongest at a moment actually free of compression and echo, when John's shouted warning, "look out!," bleeds from the far-off guide vocal onto the drums track, clearly audible at the point of retransition while the reduced instrumentation begins to decay, at 2:30 (see the end of Example 7.2). Such distance is bridged by unison double-tracking in the urgent commands of "right now" and "over me" in the chorus and of the repeated title in the coda.

Continuing with such textural paradoxes, but in the instrumental forces, Harrison's rhythm guitar is clean and legato for the subdued tattoo (blending with the bass tone), but crunchy and staccato for aggressive verses and choruses (compare the second and last bars of Example 7.3). The tension of the unrelenting suggestion of a ♯9 chord in the substrate is relieved in the guitar duet's clearly major dominant, but this only supports minor-tinged lines while the creamy legato melodies are undercut by disjointed maracas. The stealthy, tip-toeing dual guitars of Example 7.1 would contribute further unease to the piano solo's "swampy and smoky" spirit (as understood by its creator, McCartney; see Goodman 1984, 87–88).

The Beatles' musical paradoxes bring to life the mysteriously obscure nature of Lennon's "gobbledygook" lyrics. In the past, Lennon's ambiguous lyrics have conveyed such disparate qualities as psychological portraiture in "Strawberry Fields Forever," psychedelic tripping in "I Am the Walrus," and both suspect and celebratory aspects of sex in "Happiness Is a Warm Gun." In "Come Together," sex—seemingly unrewarding—is contemplated among misfit personalities. But perhaps the sex is itself a metaphor for a more generalized set of relationships. After all, the four verses have been interpreted as portraits of the four individual Beatles themselves, not exactly the closest of friends in late July 1969 when invited to gather by a re-Beatled Lennon.[16] Coming together is not only a suggestion of the band reforming but also an album-opening invitation to its listening audience, an overture like that which begins *Sgt. Pepper's Lonely Hearts Club Band*. But it is a darker overture; McCartney's friendly "sit back and let the evening go" is now more like "look out! we plan to entertain you by gathering the diseased and other weirdos like Maxwell Edison, Mr Mustard, and Polythene Pam." On *Abbey Road*, "Come Together" is immediately followed by George Harrison's sublime "Something" (with which song it was paired for the album's single), transcendent sweetness supplanting the sour. But "Something" is similarly inharmonious because its hard-to-pin-down dynamics show a musician having a hard time being specific about what exactly he desires (Everett 2001, 250–51).

The Black (After)Lives of "Come Together"

"Come Together" responded to a discordant time—for the Beatles and the broader 1969 cultural-political milieu—through seemingly harmonious but still very inharmonious overtures. This thematic content and its attendant rhetorical potential carry on in the song's afterlives, especially in the recordings of Black Americans and during historical moments of political tension and public demonstration.

Given the obvious Berry lifts to which Lennon confessed, prompting his 1970s-era recording of earlier rock 'n' roll, it seems especially fitting that some of the most popular Black American artists have since covered the song at the heart of that controversy: Ike and Tina Turner, Diana Ross, Michael Jackson, and Prince, in addition to others, such as Prince's own drummer and collaborator from 1984 to 1989, Sheila E, the subject of the next section. When considered together and in light of the song's history, African Americans' interpretations aren't just about "Come Together." Collectively, they become emblematic reclamations of rock 'n' roll that raise the question of all the Beatles' many un-litigated borrowings, lifts, and appropriations from Berry and numerous other Black Americans *throughout* the band's tenure.[17] In a song that sees the Beatles returning to their Berry beginnings even though their fragmentation betrays a point of no return, Black artists' covers of "Come Together" participate in that ironic get-back loop, reconciling the irreconcilable. But there are more specific and recent reasons to consider the afterlives of "Come Together" in relation to rock 'n' roll's African American roots and US politics.

Drummers Unite: Ringo Starr and Sheila E.

On July 7, 2020, Ringo Starr performed the song—his favorite Beatles number—at the beginning of his virtual birthday bash, an annual party that went online due to the COVID-19 pandemic (Starr 2020). The show streamed live on his YouTube channel. "Come Together" followed the opener, "It Don't Come Easy," a pairing gesturing directly to the circumstances that motivated the format of his celebration.[18] Helping to kick off "Ringo Starr's Big Birthday Show," the unified fragmentation of "Come Together" was laid bare. We can't come together in person, Ringo seemed to be saying with the song selection, but we can still come together on this online platform; he allowed all the lonely people who could never make the trip to Los Angeles a chance to join the party. The virtual platform's very aesthetic, however, relies on visual framing that delineates users' screens and makes their isolation from each other palpable. At the same time, Starr also harnessed the refrain's coming-together motif to address the pressing issue of racial injustice in the United States. Starr's streaming celebration took place amidst heightened tension in the US and abroad, where #BlackLivesMatter demonstrations coalesced and intensified following the police murder of a Black Minneapolis man, George Floyd. This was also the summer that preceded a controversial presidential election, whose Republican incumbent and his followers were no supporters of BLM and are, to date, still contesting the outcome of a vote they lost, using violence (and its threat) and propaganda.

For the 2020 "Come Together" birthday performance, Starr was joined (virtually) by fellow drummer Sheila E. (Sheila Cecilia Escovedo), who sang and played congas and timbales; they were accompanied by several other Black musicians, including guitarist Ben Harper, well-known for his Beatles covers. Sheila E. was, in fact, no stranger to Starr and his drumming. In addition to performing with his All-Starr Band for much of the 2000s, she (2017b) featured Starr on the song three years prior for her *Iconic: Message 4 America*, an album responding to the 2016 US presidential election, whose outcome saw an uptick in violent crimes against people of color. When she promoted that record, Sheila E. made explicit the song's political potential, which she prefaced in a video in which she and Starr appeared together:

> I was telling people, I was like, the thing that Ringo continues to say constantly is 'peace and love' [performs peace gesture with her fingers] 'peace and love' and 'come together.' And that song says everything [Ringo says

> "yeah," nodding in agreement]. And if we can come together as a people and as this country and do something together— we can't keep dividing each other.
>
> *Sheila E., 2017a*

Listening to Shelia E. explain the song in 2017 is a reminder that people of color, artists especially, have long been calling attention to the problems of race in the United States. *Iconic* also includes her cover of Lennon and McCartney's "Blackbird," whose retroactive civil rights interpretation[19] comes through via Sheila E. in the context of her album's unambiguous racial themes in America (other songs sample speeches by Dolores Huerta and Angela Davis, for example).

In the hands of the Beatles, the lyrics of "Come Together" do not have the explicit civil rights message that Sheila E. articulates in the 21st century—but her theme acknowledges the song's inherent fracture and yearning for unity, which we have been addressing throughout this chapter. On the 2017 album track that features Starr, Sheila E. elevates "Come Together" to anthemic heights. She does this by repeating the title phrase so as to form a fuller chorus section than afforded by the song's composer. She also incorporates a selection from "Revolution," as well as the refrain from "All You Need is Love." "Revolution"[20] is credited in the song title, but the 1967 song is not, instead invoked only briefly (just as the Beatles' original quoted touches of "She Loves You" and more traditional numbers) as a transitional piece in her remix.

Sheila E. Calls Back to Hendrix and Prince

In Sheila E.'s "Come Together," the brief sample of "All You Need is Love" may gesture toward the flower-power psychedelia that may or may not have informed Prince's 1985 *Around the World in a Day*, on which Sheila E. drums. The album *Iconic* also includes her cover of Prince's "America" from that very 1985 record. Although Prince was always resistant to admitting a Beatles association, his own collaborators and critics have long drawn the comparison between *Around the World in a Day* and the Beatles' *Pepper.* Thus, even though she credits Ringo's messaging for her selection of "Come Together," Sheila E's song choice also nods to Prince, whose own lauded cover of "Come Together" might be his second most famous Beatles interpretation, next to the better-known, scene-stealing guitar solo on "While My Guitar Gently Weeps" at the Rock and Roll Hall of Fame's posthumous induction of George Harrison in 2004.[21]

As these examples help illustrate, when Black artists remix the Beatles, they often call attention to other Black artists, especially those who initially inspired the Fab Four or those who have subsequently remixed the Beatles.[22] Along these lines, in the 2017 video of "Come Together," Sheila E. displays an image of Jimi Hendrix, a Beatles contemporary who remixed the band ("Day Tripper," "Sgt.

Pepper's Lonely Hearts Club Band") and inspired the Fabs, too ("A Day in the Life," "It's All Too Much"). Similar to how Hendrix awed McCartney with an interpretation of *Pepper's* title track days after the album's release in 1967, decades later, Prince—an admirer of Hendrix—amazed those in attendance at the aforementioned live rendition of "While My Guitar Gently Weeps" in 2004. Along with Prince, the performance featured an all-star collective of Tom Petty, Jeff Lynne, Steve Winwood, Marc Mann, and Steve Ferrone, as well as Dhani Harrison, George's son. Prince stunned the musicians and the audience that night: after materializing onstage mid-performance, he issued an unrehearsed three-minute solo that culminated in him throwing his guitar up into the air (from where it never re-appeared) and coolly gliding offstage.

Of Prince's almost otherworldly contribution, the younger Harrison remembered his own "nervous" laughter, going on to explain, Prince's "solo was so incendiary that I started really laughing and enjoying it. But because they put me on stage in the middle, I'm in every shot right behind Prince, and it looks like I'm hysterically laughing while trying to play the end of 'While My Guitar Gently Weeps' and sing the backing vocals" (Dhani Harrison quoted in Light 2021). Prince's unsettling presence upstaged everyone else in the ensemble—throwing the performance into an ironic meta-territory, where the guitar and the guitarist being honored were made to weep. When Prince ultimately made his own guitar fly (a gesture that recalls his signature doves), he freed the instrument from the Beatle baggage that Prince has so profoundly interpreted and that plagued Harrison himself. Thinking again of Hendrix's influence on Prince, Dhani Harrison's use of "incendiary" couldn't be more apropos: the destruction or disappearance of the guitar—the very object at the heart of the song being performed—recalls Hendrix's own shocking onstage gestures whenever he lit his guitar on fire or smashed the instrument into pieces.

Seeing these Black artists—Hendrix, Prince, and Sheila E.—in a trajectory with each other demonstrates how Black artists call back to each other using the Beatles as a vehicle for musical conversations. According to Mike Alleyne, "Hendrix represents an axis point between The Beatles, Prince, and Sheila E., allowing them to virtually intersect and musically come together. Representing a genetic blend of Cherokee Nation and Black American, along with his global creative presence philosophically centered on transcending humanity's fatal flaws, Hendrix physically and aurally embodies integration on several levels" (Alleyne forthcoming).[23]

"Come Together" and #BlackLivesMatter in 2020

In addition to the gestures toward Prince and Hendrix, during Ringo's 2020 birthday performance of "Come Together," Sheila E. performs with Bob Marley displayed prominently in the background, offering a visual gesture toward

1970s Black uplift movements consistent with other choices in her and Starr's "Come Together." Thus, in 2020, Ringo only appears to be using "his" band's song at the beginning of the birthday bash. To signal his solidarity with #BlackLivesMatter, Starr invokes Sheila E.'s intentions and her arrangement. Although they do not include the 2017 song's latter two minutes (which include "Revolution"), they do maintain the preambulatory remarks included at the beginning of the remix. "Come Together," then, comes almost full circle, back to another 20th-century political moment with a sample from the Reverend Jesse Jackson's speech at Wattstax, the festival organized by Stax Records in the summer of 1972:

> This is a beautiful day …
> … We are unified and on one accord …
> … Today we are together …
> … When we are together we got POWER.

Jackson's words honored an occasion that commemorated the Watts Uprising in the previous decade when African Americans publicly demonstrated after the mistreatment of a Black Los Angeles man (and his mother) by police who had stopped Marquette Frye for intoxicated driving. In the context of "Come Together" during Starr's birthday performance, Jackson's words draw a direct parallel between that 1965 incident and the arrest and murder of Floyd in 2020—and the racism underlying the structural oppression of Black people that remains unresolved over that half-century in the United States.

Available in *Wattstax*, a 1973 documentary of the summer festival, the Jackson sample, in fact, recalls yet another "Come Together" track: Andy Weatherall sampled Jackson's speech for a Primal Scream "Come Together" remix (1990). But Sheila E.'s and thus Ringo's use of Jackson in their track retroactively binds Jackson's Black Power message to the Beatles' 1969 song—uniting those nearly contemporaneous historical moments. Together, Sheila E. and Starr demonstrate how the song's tensions still apply to moments of political and social unrest while binding "Come Together" to a civil rights event that postdates the song's original recording but very much informs the history leading up to #BlackLivesMatter in the 21st Century.

Notes

1 Tracks 1–3 represent the instrumental content of Take 6, the best of the band's eight attempts at a basic track, all originally recorded on a four-track Studer machine in EMI's Studio Three, presumably because all eight-track machines were in use elsewhere in the building. These recordings, along with Lennon's live guide vocal, were transferred the same evening to Tracks 1–4 of the eight-track copy illustrated in Figure 7.1 in Studio Two; the bounce-down was henceforth referred to as Take 9. The Figure's indexical numberings of all eight tracks are confirmed on p. 26 of the book accompanying The Beatles (2019). Timestamps and sound staging are as coded on The Beatles (1987). Two outtakes from this session have been released: Take 1 on The Beatles (1996) and Take 5 on The Beatles (2019).

2 Approximations of the eight isolated tracks can be heard in the mogg stems taken from three different latter-day 5.1 mixes made for The Beatles (2000, 2009, 2019).
3 Lennon's part on Track 6 and the staccato portions of Tracks 6 and 7 are displayed by Steve Levine in Allinson (2009). The order of recording history shown in Figure 7.1 is evident from a study of room bleed onto Tracks 5 and 6, and headphone leakage onto Tracks 4 and 8. It has been said elsewhere that McCartney added his vocal *after* Lennon perfected his, or did not sing at all (as asserted by the singer himself!), but these are not borne out by the recorded document.
4 In addition to the latter-day mixes listed in note 2, Giles Martin produced a stereo remix of "Come Together" for The Beatles (2019).
5 The performance halts after a bad bass note from Paul, who then refers to a just-sung phrase by asking, "early warning"? The expression is a British euphemism for bad breath, according to an unsigned essay at the "Beatles Music History!" page.
6 The tattoo is described as a pop/rock recording's short introductory riff that periodically reappears "as if to bring the song back into focus, perhaps to call extra attention to the following verse or, if the phrase had functioned as the song's introduction, to make it seem as if we are off to a fresh start" (Everett 2009, 151).
7 More on the Beatles' preoccupation with guns and sex in the *White Album* is presented in Kapurch and Everett (2020). The repeated "shoot me!" is a fragment taken from Lennon's draft, "Watching Rainbows," of January 14, 1969 (see transcription and discussion in Everett 2001, 232, 246–47). There, the request is for heroin, but of course, the command in "Come Together" has been taken as an eerie prediction of the Beatle's eventual murder (see Fink 2019). The lustful nature of such bass glissandi could be related to the priapic slides of the trombone in older recordings, as in Ada Jones' "I'm in Love with the Slide Trombone" (1907), its responding spurts in Eddie Cantor's "If You Knew Susie (Like I Know Susie)" (1925), or the double entendre in Dinah Washington's "Big Long Slidin' Thing" (1954).
8 The heavy breathing is reminiscent of the sex-depicting coda of "Lovely Rita," while John's and Paul's combination of nonsense vocalizations reminds us of the three-part chanting through the retransition of "A Day in the Life" (2:48-3:13), also best heard in its isolated vocal track. These isolated tracks also allow for more accurate transcriptions for this essay than have appeared elsewhere.
9 The French often refer to the orgasm as "le petit mort."
10 Everett (2004, 2009, 168–70) discusses the pitch-collection stratification of melody and accompaniment as a distinct system (fourth among six) of tonal structure; Temperley (2007) terms this the "melodic-harmonic divorce."
11 We label this passage—which carries the song's core message, "Come Together right now, over me"—as a chorus rather than a refrain (which term would befit its brevity) because it is not contained within the first verse. These concepts are covered in Everett (2009, 145–46).
12 The (025) blues trichord is covered in Everett (2001, 17–18, 2009, 169).
13 In the third chorus of Take 1 (2:59+), Harrison and McCartney accent the Bm–A–G expansion of the dominant preparation in harsh triplet eighths but discard this obtrusive rhythm in later run-throughs.
14 The dorian mode persists long beyond 1970 in pop music; its tangy appearance in the rhythm guitar of Janelle Monáe's "Tightrope" (2010) helps suggest this song's risk of imbalance while featuring as one of several references to Brown's "Sex Machine."
15 The sudden smoothness of this vocal line can be inferred from the density of Track 4 (at, e.g., 0:30–0:34, and Tracks 4 and 8 when double-tracked at c. 2:48–2:53) in the timeline of Figure 7.1.
16 The individual portraits are offered in Everett (2001, 246). Other Beatle songs with four fully developed verses are typically story narrations, though John's "Cry Baby Cry" is more episodic, less linearly sequential, than Paul's "Paperback Writer," "She's Leaving Home," or "Rocky Raccoon." The Beatles' one five-verse song is the folk ballad, "The Ballad of John and Yoko," a legend told in historical reportage.

17 The Beatles' deliberate career-long preoccupation with Motown soul and Stax R&B, perhaps most concentrated in *Revolver*, is the topic of Everett (2002).
18 Starr thought so much of "Come Together" and its relevance to his 2020 birthday that the setlist included yet another cover of the song by Gary Clark, Jr., whose 2017 cover amplifies the song's sexy groove.
19 For historically situated discussions of McCartney's attachment of a civil rights message to "Blackbird," as well as the implications of this interpretation in Black artists' covers of the song, see Kapurch and Smith (2016, 2020, forthcoming).
20 In 1969, Nina Simone famously responded to the Beatles' "Revolution" with her "Revolution," separated into two parts that call direct attention to the struggle of civil rights, including imagery of a foot on the back. Although Sheila E.'s cover does not take direct inspiration from Simone, we can still interpret the presence of "Revolution" as nodding toward Simone, who took the Beatles to task for their lukewarm engagement with the movement.
21 The Sheila E.-Prince "Come Together" associations present an opportunity for further analysis; indeed, there is a dearth of scholarship that addresses Prince's conversations (even resistant as they were) with the Beatles, a trend consistent with the lack of scholarly treatment on Prince relative to the Beatles (see Alleyne and Fairclough 2020). Kapurch and Smith's (forthcoming) *Blackbird Singing* addresses Prince's Beatle-esque moments in *Around the World in a Day* in the context of a discussion of bird and flight imagery.
22 See Kapurch and Smith (forthcoming) for a sustained treatment of this remixing.
23 For a comprehensive treatment of Hendrix's recording career, see Alleyne (2021).

References

Alleyne, Mike, 2021. *The Essential Hendrix*. San Diego: Cognella.

Alleyne, Mike. Forthcoming. "The Hendrix Axis: Crossroads of Connection" [essay in progress].

Alleyne, Mike, and Fairclough, Kirsty. 2020. "Introduction." In *Prince and Popular Music: Critical Perspectives on an Interdisciplinary Life*, edited by Mike Alleyne and Kirsty Fairclough, 1–8. New York: Bloomsbury.

Allinson, Richard. 2009. "The Record Producers." Two-hour audio documentary episode originally airing September 5. https://www.mixcloud.com/charliemouse/the-record-producers-george-martin/, accessed on December 24, 2020.

Beatles, The. 1969. *Abbey Road*. Parlophone PCS 7088.

Beatles, The. 1987. *Abbey Road*. Parlophone CDP7464462.

Beatles, The. 1993. *Complete Scores*. London, New York, Paris, Sydney, Copenhagen, and Madrid: Hal Leonard Publishing Corporation.

Beatles, The. 1996. *The Beatles Anthology, vol. 3*. Apple CDP724383445127.

Beatles, The. 2000. *Love*. Apple CD 09463-79810-2 3.

Beatles, The. 2009. *Rock Band*. Harmonix/MTV Games/Electronic Arts.

Beatles, The. 2019. *Abbey Road Anniversary Edition*. Apple CD 602577 92112 4.

"Beatles Music History!" http://www.beatlesebooks.com/come-together, accessed on December 24, 2020.

Everett, Walter. 2001. *The Beatles as Musicians: Revolver through the Anthology*. New York: Oxford University Press.

Everett, Walter. 2002. "Detroit and Memphis: The Soul of *Revolver*." In *"Every Sound There Is": The Beatles' Revolver and the Transformation of Rock and Roll*, edited by Derek Scott and Russ Reising, 25–57. London: Ashgate.

Everett, Walter. 2004. "Making Sense of Rock's Tonal Systems." *Music Theory Online* 10, no. 4 (December). http://www.societymusictheory.org/mto/issues/mto.04.10.4/toc.10.4.html, accessed 1 August 2021.

Everett, Walter. 2009. *The Foundations of Rock: From Blue Suede Shoes to Suite: Judy Blue Eyes*. New York: Oxford University Press.

Fink, Robert. 2019. "He Bad Production: Abbey Road as Timbral Journey." Presented at the Abbey Road Conference A Symposium Hosted by the University of Rochester Institute for Popular Music and the Eastman School of Music, September 27–29, Rochester, NY.

Golson, G. Barry, ed. 1981. *The Playboy Interviews with John Lennon and Yoko Ono: The Final Testament*. New York: Berkley.

Goodman, Joan. 1984. Interview: Paul and Linda McCartney. *Playboy* 31 (December).

Kapurch, Katie, and Smith, Jon Marc. 2016. "Blackbird Singing: Paul McCartney's Romance of Racial Harmony and Post-Racial America." In *New Critical Perspectives on the Beatles: Things We Said Today*, edited by Kenneth Womack and Katie Kapurch, 51–74. New York: Palgrave Macmillan.

Kapurch, Katie, and Smith, Jon Marc. 2020. "Blackbird Fly: Paul McCartney's Legend, Billy Preston's Gospel, and Lead Belly's Blues." *Interdisciplinary Literary Studies: A Journal of Criticism and Theory* 22, no.1–2: 5–30.

Kapurch, Katie, and Smith, Jon Marc. Forthcoming. *Blackbird Singing*. University Park, PA: Penn State University Press.

Kapurch, Katie, and Everett, Walter. 2020. "'If You Become Naked': Sexual Honesty on the Beatles' *White Album*." *Rock Music Studies* 7, no. 3: 209–25.

Lewisohn, Mark. 1988. *The Beatles: Recording Sessions*. New York: Harmony.

Light, Alan. 2021. "What Dhani Harrison Learned About His Dad, George Harrison, While Revisiting *All Things Must Pass*." *Esquire* (August 6) https://www.esquire.com/entertainment/music/a37231320/george-harrison-all-things-must-pass-rerelease-dhani-harrison-interview/, accessed January 16, 2022.

Moylan, William. 2020. *Recording Analysis: How the Record Shapes the Song*. New York: Routledge.

Sheila E. 2017a. "Sheila E. featuring Ringo Starr–Come Together/Revolution—Preview." *YouTube*. Posted September 14. https://youtu.be/rx4G7au123E, accessed 1 August 2021.

Sheila E. 2017b. *Iconic: Message 4 America*. Trier, Germany: Mooscious Records, Stilettoflats Music.

Starr, Ringo. 2020. "Ringo's Big Birthday Show." *YouTube*. July 7. Livestream [no link available]. Accessed July 7, 2020.

Temperley, David. 2007. "The Melodic-Harmonic 'Divorce' in Rock." *Popular Music* 26/2: 323–42.

The Late Show with Stephen Colbert. 2021. "Ringo Starr Takes The Colbert Questionert." *YouTube*. Posted May 1. https://youtu.be/JZKZY2pyt4c.

8

"A TSUNAMI OF VOICES"

10cc's "I'm Not in Love" (1975)

Mark Spicer and Stephen Spencer

This chapter is supported by a webpage.

Although the Beatles' recording career spanned less than a decade, the influence they exerted on subsequent generations of pop and rock musicians can hardly be overstated. This chapter will focus on one such British group that emerged in the early 1970s and whose recorded music cannot be fully understood without taking the Beatles' influence into account. Like the Beatles on their later records from *Revolver* onward, 10cc learned to use the multitrack recording console as an instrument in itself, the consummate example being their remarkable song "I'm Not in Love," which topped the UK singles chart for two weeks in July 1975.[1] With reference to firsthand accounts from the band members, we will trace the song's genesis and offer a detailed analysis of its unusual formal structure and harmonic design. Then, using 21st-century audio processing tools in tandem with William Moylan's (2020) system of "timbral balance," we analyze the sonic content of "I'm Not in Love" more precisely, revealing further just what makes this tour de force of analog recording so special.

From Hotlegs to 10cc

Hailing from Manchester, 10cc were something of a supergroup when they released their debut single, the doo-wop pastiche "Donna" (UK #2), in September 1972, with all four members having enjoyed their own share of success in various intertwining contexts as songwriters or performers (or both) since the mid-1960s. Eric Stewart was the guitarist for Wayne Fontana and the Mindbenders, who had scored a US #1 hit with "The Game of Love" in April 1965 at the height of the British Invasion. Stewart took over as lead vocalist after Fontana left the band and the Mindbenders continued as a three-piece, achieving one further top-five single, "A Groovy Kind of Love" (US and UK #2), in 1966.

DOI: 10.4324/9781003089926-12

Graham Gouldman briefly joined the Mindbenders as bassist before the group disbanded in 1968, yet, along with a stint with his own earlier Mancunian band, the Mockingbirds, he had already won recognition as an accomplished songwriter for other artists, having penned such hits as "Bus Stop" (1966, US and UK #5) for the Hollies, "Listen People" (1966, US #3) and "No Milk Today" (1966, UK #7) for Herman's Hermits, and, most famously, "For Your Love" (1965, UK #3, US #6) for the Yardbirds. Childhood friends and former art school students Lol Creme and Kevin Godley (previously the drummer for the Mockingbirds) began writing and recording songs as a duo in 1967, releasing a one-off psychedelic pop single, "Seeing Things Green," as the Yellow Bellow Boom Room in January 1968.[2]

Following the breakup of the Mindbenders, Stewart became a partner in a small recording studio at 3 Waterloo Road in downtown Stockport, which he named Strawberry Studios after his favorite Beatles song, "Strawberry Fields Forever." The equipment at Strawberry was initially very basic—just a pair of stereo TEAC tape machines slaved together—but it was enough for Stewart to begin developing his skills as a recording engineer. Gouldman, Creme, and Godley came on board shortly thereafter, and soon the four of them were using the makeshift facility to record their own experimental songs and demos. In 1970, after purchasing a new four-track Ampex tape machine for the studio, Stewart was eager to explore its capabilities and asked Godley to lay down a continuous drum pattern while he toyed with getting the best drum sound in the control room; he then "started experimenting to see how many drum tracks we could physically get on that tape by jumping stereo tracks backwards and forwards" (Newton 2020, 52), foreshadowing the kinds of multitrack experiments 10cc would later employ on records like "I'm Not in Love." The resulting multilayered drum groove became the rhythmic foundation for a novelty song, "Neanderthal Man," released under the name Hotlegs and which earned Godley, Creme, and Stewart a surprise UK #2 hit in August 1970.[3] The income from the song allowed for further upgrades at Strawberry: "With the success of 'Neanderthal Man' we bought an eight-track for the studio and got a real control desk in and we started letting the studio out to people in Manchester," Stewart recalls. "Before that it was basically a private studio for me and the other writers. So, now it became a successful, working studio" (Newton 2020, 56). In fact, Strawberry quickly gained a reputation as the only full-fledged professional recording studio in England outside of London, where over the next two years, the quartet would serve as the house band on a wide variety of recording projects, most notably with British psychedelic artist Ramases on his 1971 debut album *Space Hymns,* and as the backing musicians (Creme, Godley, and Gouldman) and recording engineer (Stewart) for Neil Sedaka's 1972 comeback album *Solitaire.*

By the time the foursome decided in 1972 to have a go of it as their own band, rechristened as 10cc, their combined skills as vocalists, multi-instrumentalists, and songwriters were unmatched in the British music industry, and in Eric Stewart—who was like Paul McCartney and Geoff Emerick rolled into one—they even had their own engineer extraordinaire as a member of the band.

It is useful to compare 10cc with another British group who also released their self-titled debut album in 1973, namely Queen. Both groups managed to successfully straddle the albums and singles markets during this remarkably eclectic period in British chart history, producing what Andy Bennett (2020) has labeled "progressive pop."[4] 10cc and Queen achieved their creative zeniths in 1975 with "I'm Not in Love" and "Bohemian Rhapsody," respectively, two songs that became huge UK #1 hits despite exceeding the usual length of a radio-friendly single (each clocking in at around six minutes). Both songs, now iconic, have since secured their place in rock history as extraordinary feats of analog recording wizardry, especially in the way multitracked background vocals created the effect of virtual choirs.[5]

Lyrics and Formal Structure

The lyrics of "I'm Not in Love," according to Stewart, were inspired by a conversation he had with his wife Gloria, who after eight years of marriage felt disappointed that he did not tell her often enough that he loved her.[6] In an attempt to write her a love song while avoiding clichés, Stewart explains, "I chose to say 'I'm not in love with you' while subtly giving all the reasons throughout the song why I could never let go of this relationship" (Buskin 2005).[7] He presented the lyrics to Gouldman, his usual songwriting partner in the band, and together they began work on fleshing out the melody and chords. A formal synopsis of the finished song is provided in Example 8.1A.

The formal structure of "I'm Not in Love" is best described as what John Covach (2006) calls an AABA form with abbreviated reprise—i.e., AABABA, in which the A section consists of a verse with its tail refrain ("I'm not in love, no no, it's because …") and the B section represents the contrasting bridge or "middle eight" (see the structural timeline in Example 8.1B). Inherited from the 32-bar AABA form common to Tin Pan Alley pop from earlier in the 20th century, this was the formal template used by the Beatles on the majority of their early hits through 1965 (e.g., "From Me to You," "I Want to Hold Your Hand"), after which their emphasis shifted toward contrasting verse–chorus and other more experimental song structures. At first glance, then, the formal design of "I'm Not in Love" might be seen as harking back to an earlier era, but 10cc was a band who constantly strove to raise their own bar and try out fresh ideas. Unusually, the song features two *different* bridges. The second, more conventional, bridge is the standard eight bars long (i.e., a true middle eight, consisting of a four-bar phrase and its repetition) with a change of key and ending with a retransitional dominant leading back to the fourth and final verse.[8] The first bridge, however, is better described as an extended musical interlude: it begins with the piano melody as notated in the example, then moves into a bass solo, accompanied by the repeated whispered voice, "Be quiet, big boys don't cry … big boys don't cry …" (we will examine this musical interlude in greater detail with the pitch density and timbral balance graph in Figure 8.2). Furthermore,

EXAMPLE 8.1 10cc, "I'm Not in Love" (1975): (A) Formal synopsis and harmonic design; (B) Structural timeline.

the piano melodies of the first and second bridges have been rhythmically displaced by half a bar, and the key signature has changed from four sharps to one sharp (E major to G major) with the head note E of the melody preserved. In this clever type of thematic transformation, known as *inflected repetition* (see Satyendra 1997), a melodic segment is repeated with one or more of its pitches inflected by semitone (in this case, E–D♯–C♯ becomes E–D–C).[9] While inflected repetition had been employed as an expressive effect in chromatic music by Franz Liszt and other 19th-century classical composers, one would be hard pressed to find another such example in a 1970s pop song.

Harmonic Design and Two Precursor Songs

Let us look more closely now at the song's harmonic design. The eight-bar intro is based on a two-chord shuttle—in this case, close position A major and B major triads alternating over a B pedal. (This two-chord shuttle returns as the basis for the extended outro.) The shuttle repeats four times before moving to a G♯7 chord that we would expect to resolve to C♯ minor but instead resolves deceptively to another A major chord at the onset of the verse. Example 8.1A includes a Roman numeral analysis of the verse–refrain's chord progression. The verse–refrain or A section is organized into three four-bar phrases, with the first two (the verse) veering toward the relative C♯ minor or vi, and the third phrase (the refrain) serving cadential function by arriving on the tonic chord E major, embellished by IV–♭III–IV over a tonic pedal.[10] The harmonic trajectory of the introduction and verse–refrain thus exemplifies what I have called an *emergent* tonic (Spicer 2017): beginning with the distinctive sonority of a "soul dominant" (a close position IV chord over scale degree 5 in the bass), the two-chord shuttle of the intro prolongs the dominant, swerving to the IV chord at the onset of the verse, and then only in the refrain does it find its tonic. The song's final refrain is modified, as notated in the example, with Stewart singing the title lyric twice, the second time with heightened passion over an unresolved F♯ half-diminished seventh chord (Am6) in a melismatic flourish that emphasizes the blue note G-natural (the flatted third scale degree and apex of his lead vocal melody, drenched in tape delay) as his final vocal utterance before the song moves into the extended outro.[11]

Two precursor songs may have influenced the harmonic design of "I'm Not in Love." The first is Daryl Hall and John Oates' "She's Gone" from their 1973 album *Abandoned Luncheonette,* which opens in a strikingly similar way, with an extended intro featuring oscillating A and B chords over a dominant pedal, in the very same key.[12] It is possible that 10cc had heard "She's Gone" before, but the song had failed to chart and Hall and Oates were not widely known in the UK until receiving national exposure on the BBC2 program *The Old Grey Whistle Test* in May 1976. (They offered a stellar performance of the song, but by that point British audiences might have thought they were ripping off 10cc!) The other precursor song is perhaps more surprising. Adept at writing stylistic pastiches, Stewart and Gouldman had originally conceived of "I'm Not in Love" as a bossa nova, a style originating in the nightclubs of Rio de Janeiro in the 1950s and which became a global sensation in 1964 with the Getz/Gilberto recording of Antônio Carlos Jobim's "The Girl From Ipanema" (US #5, UK #29). Rather than learning the chords of "Ipanema" exactly from the original recording in the key of D-flat major, Stewart and Gouldman most likely knew how to play the song from its "fake book" lead sheet version in F major that was widely disseminated in the 1970s, in which the chords of the verse are as follows: Fmaj7 | G7 | Gm7 G♭7 | Fmaj7.[13] The central three chords of this progression map onto the opening three chords of the verse of "I'm Not in Love" (A | Am | G♯m), where we also have a major chord immediately followed by its minor counterpart and then slipping down to a third chord whose root lies a semitone below. Such chromatic moves are, in fact, a hallmark of Jobim's bossa nova songs (e.g., in "Corcovado," we find the progression Fmaj7 F6 | Fm6 | Em7). "I'm Not in Love" might therefore be better characterized not as an instance of strategic intertextuality (see Spicer 2009) but rather *stylistic* intertextuality in the way it references the harmonic language of bossa nova more broadly.

The Emerging Track and Arrangement

With their bossa-nova-style arrangement of "I'm Not in Love" now in place, Stewart and Gouldman were ready to present the song to their fellow band members and begin the recording process. This fledgling version, however, did not impress Creme and Godley. As Creme recounts in the 2015 BBC4 documentary celebrating the song's fortieth anniversary, "When we heard it back everyone was underwhelmed, I think is the best word … and nobody had the real enthusiasm to carry on with it, it sounded so underwhelming." Godley was more emphatic:

> [KG] I hated it. … There was something in there … but that treatment didn't really bring it to life and we weren't really sure what we could do with it. I think I said at the time, "Well, why don't we do it with voices—no instruments, just all a tsunami of voices, a wash of voices." Probably out of desperation really, trying to come up with something to bring this thing to life.

The other band members continue the story:

> [ES] I said, "What do you mean, a cappella?" … [Kevin] said, "No, the backing track. Let's do the whole backing track with voices, like a massive choir, the biggest choir you can imagine." I said, "Great, how do we do it? … There's only four of us." And Lol said, "Tape. Tape loops." I said, "… alright, let's do it." And we spent three weeks recording three of them in the studio, Kevin, Lol, and Graham, singing "ah" as long as they could hold their breath. Then they did that sixteen times on the 16-track machine, copied all that across to a stereo machine, and we had thirteen notes to choose from, a chromatic scale with a top C and a bottom C, and 624 voices were there at our disposal. I then played them back through the 16-track machine, through the control desk, and I gave each of us a set of three faders to go up and down when the chords were changing.
>
> [LC] All of us worked the desk and played it like a keyboard, and when we had all the notes playing together you had that lovely abstract [whispered "ah"] 'cause it's a mass of harmonics bleeding together. So we kept them, all of the notes, in to a certain degree with a piece of masking tape

> so the faders were all at a certain level. And that's what made that in-tune, harmonic sound but with an extra something, and it made something really special.
>
> [GG] Originally, I think we were going to do it with just voices, nothing else. So we recorded a rhythm track which we thought was going to be a temporary and ... the idea was we'd sing to that and do all the vocals then we'd take that off and we'd be left with this song recorded with just the choir and the vocals. But there was something ... magic about it, and of course once we'd put the other voices on we didn't want to touch anything, we didn't want to take it out.

By the time they were finished recording, the only thing left of their original bossa nova arrangement was a syncopated rhythm guitar part played by Gouldman, buried deep in the mix. The rest of the rhythm track consisted of Stewart on a Fender Rhodes electric piano, with Godley, rather than using a conventional drum set, tapping out a repeated bossa-nova-style rhythm (dotted-8th note–16th note–quarter note) on a Moog synthesizer, programmed by Creme to sound like a bass drum and which thuds like a heartbeat throughout most of the song. These effects are innovative and striking, but the real magic of "I'm Not in Love" lies in the multitracked virtual choir. Listening to the band members fondly recall how they fed the tape-looped "ahs" into the 16-track control desk and "played it like a keyboard," using the volume faders to bring the pitches of the virtual choir in and out of the mix as the song unfolded, one cannot help but be reminded of the Beatles' similar experiment with tape loops some eight years earlier on "Tomorrow Never Knows."[14] The Beatles were still using four-track equipment at Abbey Road when they recorded the groundbreaking *Revolver* album, yet by the time 10cc were working on their third album, *The Original Soundtrack* (which includes "I'm Not in Love") in late 1974, Strawberry had recently upgraded again to a 16-track machine, quadruple the number of tracks that the Beatles and their maverick engineer Geoff Emerick had available to them in 1966. Yet even with these advances in equipment, what 10cc was able to achieve with "I'm Not in Love" pushed the boundaries of analog recording beyond anything known at the time.

The Sizzling Hum of the Virtual Choir

In what follows, we will use 21st-century audio-imaging software to analyze the pitch content of the tape choir part more precisely, revealing along the way some possible discrepancies in Stewart's recollection of how the virtual choir was recorded. Figure 8.1 shows an annotated spectrogram and pitch representation of the tape choir cluster that we hear in isolation for one measure at the song's opening, accompanied by just the Moog bass drum in steady quarter notes before the rest of the rhythm track enters.[15] In Figure 8.1A, separate regions of the sound spectrum are annotated, corresponding to different components

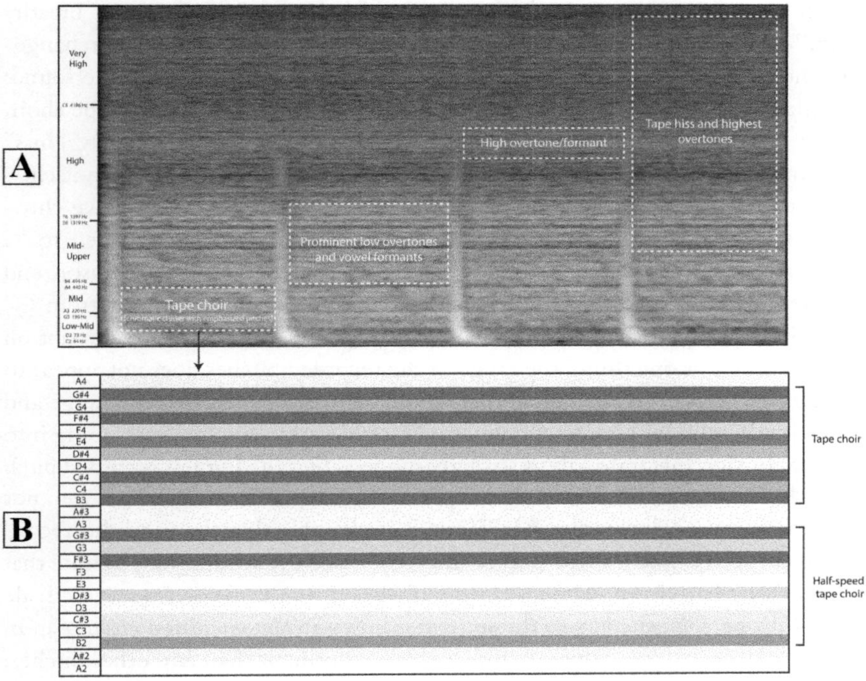

FIGURE 8.1 10cc, "I'm Not in Love" (1975): (A) Spectrogram representation of the opening tape choir cluster; (B) Weighted pitch map of the opening tape choir cluster.

of the overall timbre: the actual sung pitches in the mid-range, the prominent harmonic partials in the mid-upper range, and—particularly noticeable in this representation—the high-frequency content of the recording, resulting from the hiss of the tape and the ringing overtones of the voices reinforcing one another (or "bleeding together," as Creme puts it). In Stewart's words,

> Sound degradation caused by all the bouncing didn't matter at all because, when each of us were using control desk faders to mix the voices, there was a piece of gaffer tape across the bottom of the fader paths to stop them ever going to the bottom. That meant we had a chromatic scale sizzling underneath the track all the time, a hiss just like the hum you sometimes hear at a football match when nobody's shouting. If you listen to the opening of the song, where the bass drum beats us in, you will hear a sizzling hum there that continues all the way through the track. We actually created "hiss" on the track, when we would normally have been fighting to get rid of hiss!
>
> *Buskin 2005*

As Stewart explains, the continual presence of the entire "chromatic scale" serves both to mask unwanted sonic artifacts and lend the track a distinctive timbre, which he describes variously as "hum," "hiss," "sizzling," or simply a "breathy sound" (Allinson and Levine 2009). While Stewart uses these words interchangeably, he seems to be picking out at least two distinct components of the sound: the mid-range "hum" resulting from the cluster of pitches sung by the tape choir, which he likens to a crowd of people talking, and the high-frequency "hiss" resulting from both the overtones of the tape choir and the tape medium itself.

Several accounts by the band members refer to a complete one-octave chromatic scale—Stewart, for example, in different interviews has referred to "a chromatic scale of twelve voices" without specifying its range (Allinson and Levine 2009), and a 13-note "chromatic scale with a top C and a bottom C" (BBC4 2015). Spectral analysis of the recording, however, casts some doubt on these details. For one thing, the range of the recorded pitches does not appear to extend to a "top C" (C5), a pitch that would be firmly in the soprano range and exceedingly high for Creme, Godley, and Gouldman to sing without going into falsetto. In fact, spectral analysis strongly suggests that the highest recorded pitch was G♯4. The lower boundary, too, appears not to be G♯3, one octave lower, nor C4 ("bottom C"), but rather B3. While it is plausible that the band did record all ten chromatic pitches within this range (B3–G♯4), it is important to note that this opening bar clearly emphasizes the pitches of the E major pentatonic scale (B, C♯, E, F♯, G♯), the key of the song. As shown in the weighted pitch map in Figure 8.1B, these five notes are much more prominent than any other pitches. Some pitches (e.g., D4, A3, A♯3), if present at all, are barely detectable.[16] The computer-assisted pitch analysis shown in Figure 8.1B also reveals fundamental frequencies that extend down to B2, a full octave below the lowest note recorded by the tape choir. These low pitches were not recorded to tape in the same painstaking manner as the higher ones but rather were achieved through another experimental studio trick: half-speed playback. Playing the tape loops at half speed produces a sound one octave lower than the pitch at which it was recorded, bringing the tape choir into the low register and, for Stewart, producing an alluring timbre: "If you slow the loop down to 7.5ips, a human voice sounds amazingly just like a 'cello. It's got the rasp from the throat that sounds like the rosin on a bow swiping across the strings" (Buskin 2005). Again, one cannot help but be reminded of a similar trick the Beatles employed—with George Martin's help—on several of their songs, where a part was deliberately recorded at half speed an octave lower, then doubled in speed for the final mix so that it sounds an octave higher, noticeably distorting the timbre in the process; the most famous example of this is the "harpsichord solo" on "In My Life" (1965), achieved by Martin playing the part at half speed an octave lower on piano.[17]

The Virtual Choir as Accompanist

The virtual choir, in both its registers, is used in two distinct ways throughout "I'm Not in Love." The first, discussed above, provides the unique "sizzling hum" that permeates the track. In the second and much more conspicuous role, it serves as *accompanist*: to outline the harmonic progression of the song and provide countermelodic interjections. During the first verse, for example, the tape choir doubles a salient inner voice of the harmonic progression (A | Am | G♯m G♯7 | C♯m), highlighting the descending and ascending chromatic line that characterizes the progression (C♯–C–B–B♯–C♯). Then, upon reaching the C♯m chord in the fourth measure, a syncopated two-note motive (G♯–F♯) is passed from the left to right channel three times, hocketing back and forth in stereo (see Example 8.2B, first measure). Although the subsequent verses of the song follow the same general plan, the band was never one to settle for mere repetition. Indeed, there is near constant experimentation with subtle octave doublings and stereo effects throughout the remainder of the song. Example 8.2A provides a transcription of the tape choir part during Verse 2, with separate staves for the left and right channels. This time through the verse progression, the half-speed tape choir is added to the mix, doubling the chromatic inner voice an octave lower in the opposite channel.

Each of the four verses of the song features a subtly different approach to "orchestration" (i.e., octave doubling and stereo placement of the tape choir). To illustrate this, Example 8.2B compares the same two-bar segment across all four verses, with annotations clarifying the differences between them. Verse 3 includes additional half-speed notes, this time doubling each note of the hocketed G♯–F♯ motive in the opposite channel and then switching places. Subtler yet is the orchestration in Verse 4, wherein the two-note motive alternates between doubled and solo treatment upon each repetition. The two main techniques featured here—octave doubling and stereo "hocketing"—are nowhere better exemplified than in the third measure of Bridge 1, shown in Example 8.2A. Here, the tape choir joins the acoustic piano for a syncopated four-note descending melody (G♯4–F♯4–E4–D♯4), which is doubled down the octave by the half-speed choir. Like the hocketed motive in Verse 3, the first two notes (G♯–F♯) switch places in stereo space, with the high voice moving left to right and the low one moving right to left. Over the next two notes, the process is reversed, the voices swapping places again and returning to their original stereo positions for the held D♯. When this leading tone resolves upward, doubled by the acoustic piano two octaves higher, the voices exchange positions one final time before dropping out of the mix for the remainder of the first bridge.

"Be Quiet, Big Boys Don't Cry …"

As noted above, this "first bridge," although it began life as a middle eight, is in its final version more aptly considered an extended interlude. Having completely stripped away the full-band arrangement and replacing Stewart's vocal melody with an acoustic piano, accompanied by just electric bass and a few other sound sources, we are left with a strikingly transparent texture and a haunting sense of timelessness. Woven seamlessly into the texture is a "child's music box," which, according to Stewart, was attached to a string so that Creme "could whirl it slowly

EXAMPLE 8.2 10cc, "I'm Not in Love" (1975): transcription of the virtual choir, showing stereo separation in the left–right channels: (A) Verse 2 leading into Bridge 1; (B) Comparison of a corresponding two-bar segment of the virtual choir part across all four verses, showing differences in the orchestration. To hear audio examples of these excerpts with isolated left and right channels, see Web Example 8.1 in the online companion to this chapter.

over his head" in front of the drum overhead mics, producing a high-frequency metallic sparkling (Buskin 2005). The final analysis of this chapter, which illustrates further the textural intricacy of this interlude, is indebted to William Moylan's *Recording Analysis*. We have adopted Moylan's notion of "timbral balance," which "represents the track's overall frequency spectrum ... the combination of all the pitch densities of all the record's sound sources" (Moylan 2020, 254). Figure 8.2 shows a pitch density and timbral balance graph of Bridge 1, plotting the sound sources onto an XY graph to indicate their position within the frequency spectrum through time (shown in measures and beats). The bottom boundary of each sound source represents its lowest pitch level, while the top boundary estimates its timbral spread, indicating the approximate height of its lowest significant overtone. From this, we can see that the background tape choir cluster (Item 1), with its upper partials and high-frequency hiss, continues to "sizzle" in the background of the interlude, as it does throughout the entire track, providing a broadband bed of sound onto which other elements are placed. The piano melody (Item 5) occupies the mid-upper register before being doubled by the foreground tape choirs in the mid and low-mid registers (Items 6 and 7), bouncing back and forth in stereo as described above. But perhaps the most important contributor to the ethereal nature of this section is the whirling music box, which appears simultaneously in two different guises—one in the very high register, with inharmonic partials that extend into the highest regions of the spectrum (Item 8), and another an octave lower, likely achieved through half-speed tape playback (Item 9). The actual pitches supplied by the high music box are themselves noteworthy: G♯6, A♯6, B6, and C♯7[18]—four pitches which, remarkably, are approximations of the tenth through thirteenth harmonics of E3, the tonic note of the song. Before the end of the interlude, the music box disappears, but it returns during the final fadeout, again providing a kind of spectral decoration of the tonic note.

At the same time, in the low register, syncopated stabs in the bass guitar (Item 10) provide a hint of earlier material, tracing the inner chromatic line (B–B♯–C♯) of the verse's chord progression. These impulses develop into a full-fledged bass solo that outlines the two-chord shuttle of the intro, as the tape choir gradually reemerges and leads us to the onset of the third verse (see the transcription of the bass solo in Figure 8.2). As if the overall effect of this extended interlude was not magical enough, during the bass solo we are treated to an additional sprinkling of fairy dust: at Creme's suggestion, Kathy Redfern, the secretary at Strawberry, was brought in to the studio to whisper the line "Be quiet, big boys don't cry" (Item 11), which, along with the tape choir, became one of the song's most memorable hooks.

Closing Remarks

Like "Bohemian Rhapsody," "I'm Not in Love" is one of those songs from the mid-1970s that will forever define a very special period in the history of analog recording, prompting Paul Gambaccini's claim that "both songs were dare I say

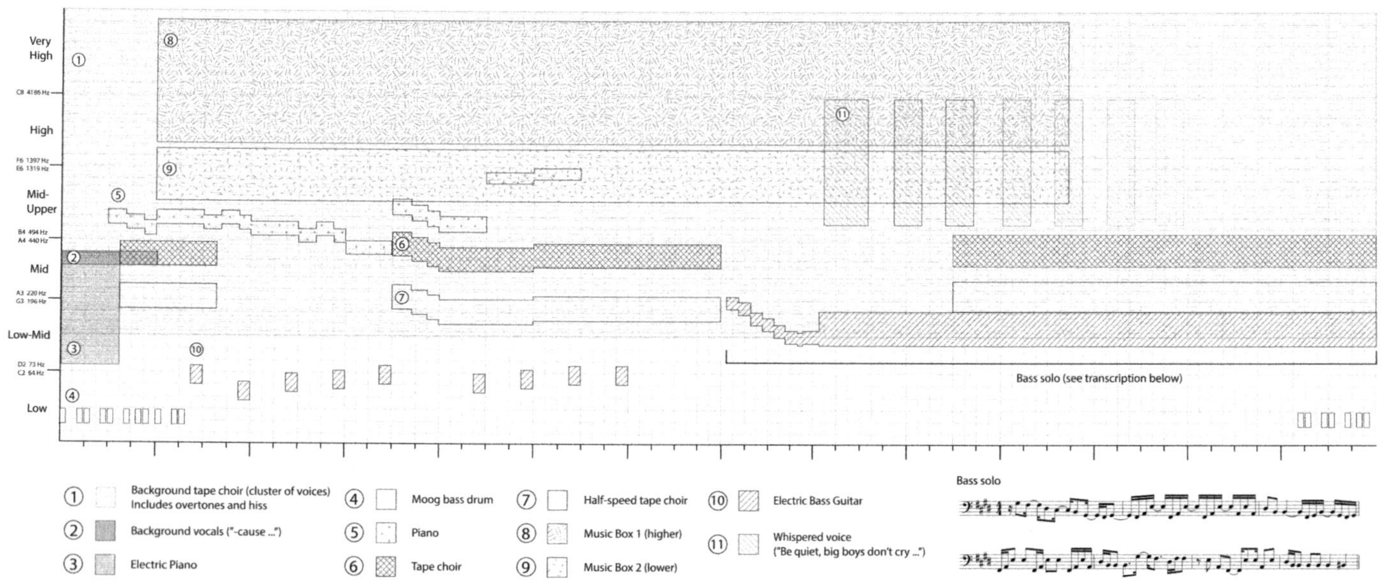

FIGURE 8.2 10cc, "I'm Not in Love" (1975): Pitch density and timbral balance graph of Bridge 1 (1:52–2:42). To listen to an audio excerpt of this section accompanied by a color version of the figure, see Web Example 8.2 in the online companion to this chapter.

it the greatest multitracked vocals of all time" (BBC4 2015). 10cc would record one more album with the original four members—*How Dare You!,* released in January 1976—before Godley and Creme, much to the surprise of Stewart and Gouldman, decided to leave the band and continue as a duo, their debut release being the ambitious triple-LP concept album *Consequences* (1977, UK #52).[19] Meanwhile, backed by an assembled roster of top UK session musicians, Stewart and Gouldman carried on as 10cc, releasing five studio albums and achieving three further top-ten singles—"The Things We Do For Love" (1976, UK #6, US #5), "Good Morning Judge" (1977, UK #5), and the reggae-tinged "Dreadlock Holiday," which earned 10cc their third UK #1 in September 1978—before finally calling it quits in late 1983, long after the hits had begun to dry up.[20] Like the Beatles before them, the original quartet of 10cc was always much greater than the sum of its individual parts.[21] Godley and Creme may not have felt so at the time, but looking back on this period, both agree now that "I'm Not in Love" was the group's masterpiece. We will let Creme have the last words: "[It was] the best track I've been involved with. You just look up to the gods and say thank you" (BBC4 2015).

Notes

1 For a close study of the Electric Light Orchestra (ELO), another British group that emerged in the Beatles' wake and who also demonstrated a profound anxiety of influence toward the Beatles, see Spicer (2018).

2 Echoing the sentiment of many aspiring pop musicians at the time, Gouldman says that "when the Beatles came along they were an inspiration. It was like, 'If they can write songs, I'll have a go as well.' I don't think I'd have started as a songwriter in the 60s if it hadn't been for them" (Newton 2020, 24). Godley adds, "I remember the day *Sgt. Pepper* came out. ... Everything changed. I didn't understand what I was listening to. Hang on, these guys have just stepped into the future and left us behind! We need to catch up real fast!" (ibid., 44).

3 Gouldman had temporarily relocated to New York to fulfill the remaining terms of his contract as a staff writer for Jerry Kasenetz and Jeff Katz's Super K Productions and therefore was not involved in the recording of the Hotlegs single. Dick Leahy of Philips Records (the company that had released all the Mindbenders singles) was in the Manchester area and happened to stop by Strawberry as the trio were working on "Neanderthal Man." Convinced the song would be a hit, Leahy offered them a £500 advance to release the single.

4 Bennett uses this genre label in contrast to "progressive rock," which was also enjoying its heyday in Britain during the first half of the 1970s. An important distinction for Bennett is that progressive rock bands like Genesis and Yes were focusing their energies on making albums and not singles (this would change, of course, when these bands themselves "went pop" around the turn of the 1980s). In addition to 10cc and Queen, other artists featured in his book include David Bowie, ELO, Roxy Music, and Steve Harley & Cockney Rebel. As Bennett puts it, "[i]n the context of the early 1970s, none of these artists could have been truly defined as either rock or pop as their musical output extended beyond the sonic boundaries commonly applied to each of these categories" (2020, 28).

5 For an in-depth analysis of "Bohemian Rhapsody," see Braae (2015). The single version of "I'm Not in Love" was edited down to under four minutes for release in the

United States and Canada only, skipping halfway through the second bridge directly to the final refrain and omitting most of the extended outro.

6 Married on March 10, 1966, Eric and Gloria recently celebrated their 56th anniversary.

7 For an analysis of the lyrics to "I'm Not in Love" as an exemplar of what has been termed linguistic negation, see Nahajec (2019).

8 Technically, the second bridge ends with a retransitional *secondary* dominant, where the E major chord serves as V of the IV chord that begins the final verse.

9 In the original take of "I'm Not in Love," at the onset of the first bridge, Stewart sings this melody set to the lyrics "Don't be let down, don't get hung up, we do what we can …," but they did not like the result and decided instead that Creme should play the melody on piano, discarding those lyrics entirely (Allinson and Levine 2009). For the second bridge, this piano melody works in counterpoint with the vocal melody ("Ooh, you wait a long time for me," not shown in the example).

10 The phrase design of the verse–refrain follows a common pattern in pop and rock music that Walter Everett (2009, 140–41) has identified as Statement–Restatement–Departure–Conclusion (SRDC), in this case with the D and C functions collapsed into one phrase (as in, for example, the Beatles' "A Hard Day's Night"). Additionally, the verse of "I'm Not in Love" begins off-tonic, highly unusual for an SRDC; see Nobile (2020, 47–50).

11 Interestingly, in the live recording of "I'm Not in Love" for 10cc's 1977 double-LP *Live and Let Live,* Stewart clearly plays a *fully* diminished seventh chord (vii^{o7}) at this juncture. Nahajec interprets this final utterance of the title lyric as "introduc[ing] a sense of incompleteness, even doubt, to the musical persona's assertions that he is not in love" (2019, 36).

12 "She's Gone" also features an emergent tonic; for an analysis of its harmonic design, see Spicer (2017, 2–4).

13 Not surprisingly, "The Girl from Ipanema" is also in AABA form with an abbreviated reprise (AABABA). For an illuminating documentary on "The Girl from Ipanema," including an analysis of how the song's actual chord progression is "far weirder" when compared to its fake book version, see Neely (2020).

14 For a detailed account of the recording process of "Tomorrow Never Knows," see Ryan and Kehew (2006, 412–13).

15 The spectrogram (Figure 8.1A) was created using Izotope RX8 software. The frequency content was analyzed using a short-time Fourier transform (STFT), with an FFT size of 4096 and the frequency scale set to the Mel scale. The amplitude ranges from −80 dBFS (low) to 0 dBFS (high). The pitch transcription (Figure 8.1B) was created in Sonic Visualiser using the plug-in "TIPIC Smoothed Pitch Representation," created by Chris Cannam from an algorithm developed by Sebastian Ewert and Meinard Müller. The plug-in divides the frequency spectrum into 88 semitone-tuned bands, thereby revealing how much energy is present at each equal-tempered pitch (Figure 8.1B includes only pitches below A4).

16 Perhaps the most remarkable aspect of the weighted pitch map is the blatant non-appearance of the note A in any of its octaves. Had the band recorded a complete one-octave chromatic scale and "kept all the notes in," one imagines that the note A (either A3 or A4) would be detectable in the audio signal. And, while we might be venturing slightly into speculation, if the note A was available to them it stands to reason that it would appear in some of the tape choir's more "foreground" material later in the song, perhaps as the root of the A major chord that begins every verse.

17 10cc had experimented with playback speed before, most notably on their first UK #1 single "Rubber Bullets" (1973), which features a double-speed guitar solo (sounding one octave higher and twice as fast as originally recorded, like the keyboard solo on "In My Life"). To the best of our knowledge, the only time the Beatles used the opposite trick that 10cc employed on "I'm Not in Love"—where a taped part is deliberately *halved* in speed to sound an octave lower—was the crashing final E major

piano chord that ends "A Day in the Life," for which the piano ensemble (John, Paul, Ringo, and Mal Evans) apparently overdubbed two additional performances of the chord with the tape machine running at 7.5 ips, mixed together with a slowed-down (half-speed) version of the original piano chord, to help create "a sense of weight and infinite sustain" (Ryan and Kehew 2006, 445).

18 The lower music box produces the pitches A♯5, B5, and C♯6.

19 Drawing upon their art school training, Godley and Creme also went on to enjoy a highly successful parallel career in the 1980s as pioneering music video directors, being responsible for some of the most innovative and memorable videos during the early years of MTV—e.g., Duran Duran's "Girls on Film" (1981), Herbie Hancock's "Rockit" (1983), the Police's "Every Breath You Take" (1983), Frankie Goes to Hollywood's "Two Tribes" (1984), and their own transatlantic top-20 hit "Cry" (1985), to name but a few.

20 After almost a decade apart pursuing other musical projects, Gouldman and Stewart recorded and released two more 10cc albums in the 1990s, … *Meanwhile* (1992) and *Mirror Mirror* (1995), the former of which was the closest we ever got to a reunion of the original quartet, featuring guest vocal appearances by Creme and Godley on several tracks. Unfortunately, both albums failed to chart. Since 1999, Gouldman has been performing regularly as 10cc—playing mostly the 1970s hits—with a five-piece band that includes Paul Burgess (drums) and Rick Fenn (guitars) from the late-70s lineup.

21 As Stewart puts it, "we carried on, and we sold some more records, but it wasn't the same anymore. I missed the wacky brains of my two friends and I think they missed me too" (Newton 2020, 440).

References

Allinson, Richard, and Steve Levine. 2009. "10cc." Episode of *The Record Producers*, BBC Radio 2, first aired on May 4.

BBC4. 2015. *I'm Not in Love—The Story of 10cc*. Documentary first aired on December 4.

Bennett, Andy. 2020. *British Progressive Pop 1970–1980*. New York: Bloomsbury Academic.

Braae, Nick. 2015. "Sonic Patterns and Compositional Strategies in Queen's 'Bohemian Rhapsody.'" *Twentieth-Century Music* 12/2: 173–96.

Buskin, Richard. 2005. "Classic Tracks: 10cc 'I'm Not in Love.'" *Sound on Sound* (June). www.soundonsound.com/techniques/classic-tracks-10cc-not-love

Covach, John. 2006. "From 'Craft' to 'Art': Formal Structure in the Music of the Beatles." In *Reading the Beatles: Cultural Studies, Literary Criticism, and the Fab Four*, ed. Kenneth Womack and Todd F. Davis, 37–53. Albany, NY: State University of New York Press.

Everett, Walter. 2009. *The Foundations of Rock: From "Blue Suede Shoes" to "Suite: Judy Blue Eyes."* New York: Oxford University Press.

Moylan, William. 2020. *Recording Analysis: How the Record Shapes the Song*. New York: Routledge.

Nahajec, Lisa. 2019. "Song Lyrics and the Disruption of Pragmatic Processing: An Analysis of Linguistic Negation in 10cc's 'I'm Not in Love.'" *Language and Literature* 28/1: 23–40.

Neely, Adam. 2020. "'The Girl From Ipanema' Is a Far Weirder Song Than You Thought." YouTube video. https://www.youtube.com/watch?v=OFWCbGzxofU

Newton, Liam. 2020. *10cc: The Worst Band in the World*. London: Rocket 88 Books.

Nobile, Drew. 2020. *Form as Harmony in Rock Music*. New York: Oxford University Press.

Ryan, Kevin, and Brian Kehew. 2006. *Recording the Beatles: The Studio Equipment and Techniques Used to Make Their Classic Albums*. Houston: Curvebender Publishing.

Satyendra, Ramon. 1997. "Conceptualizing Expressive Chromaticism in Liszt's Music." *Music Analysis* 16/2: 219–52.

Spicer, Mark. 2009. "Strategic Intertextuality in Three of John Lennon's Late Beatles' Songs." *Gamut* 2/1: 347–75.

———. 2017. "Fragile, Emergent, and Absent Tonics in Pop and Rock Songs." *Music Theory Online* 23/2.

———. 2018. "The Electric Light Orchestra and the Anxiety of the Beatles' Influence." In *The Pop Palimpsest: Intertextuality in Recorded Popular Music*, ed. Lori Burns and Serge Lacasse, 106–36. Ann Arbor, MI: University of Michigan Press.

9

COUNTERPOINT AND EXPRESSION IN THE MUSIC OF U2

"Gloria" (1981)

Timothy Koozin

This chapter explores counterpoint and relationships that arise through the combination of musical parts in "Gloria," a song by the Irish band U2, showing how gestural qualities in the voice and guitar contribute to U2's signature sound while projecting temporal processes that shape overall form and implications of meaning in the lyrics.[1] Previous studies of counterpoint in rock music provide a starting point in considering how the interaction of musical gestures in combination can leverage the expressive potential of individual parts. To illustrate enactments of embodied musical gesture as hierarchically organized actions, I identify basic gestures and trace their role in projecting large-scale gestural motions that span the entire song. Gestural analysis also provides an interpretive tool in showing how guitar and vocal gestures, heard in combination within the song, can be interrogated as agents of meaning that project a web of cultural references and associations.

Beginning with close listening to the musical track as a starting point, I engage with the song from the multiple perspectives of scholar, fan, and performer, interrogating my own gestures in singing and playing the song. In this way, the essay documents my own process of discovery, as performance and analysis inform one another mutually. My approach to musical gesture draws from Middleton (1993):

> an advantage of the method being sketched here is that in principle the analyst double as 'informant' from within the culture—laying out the gestures through participation—and as 'critical outsider', cross-checking the information against schemas drawing on a wider body of musical data. The role of the 'scholar-fan' becomes vital. (180)

This approach also resonates with core principles articulated in William Moylan's *Recording Analysis* (2020, 28–30): that *every record is unique*; analysis and

DOI: 10.4324/9781003089926-13

interpretation are an outcome of *listening with attention and intention*. The studio recording of "Gloria," the opening track from U2's 1981 *October* album, is the primary object of study. If, while listening to the song, I might sing along or hold a guitar in my hands to try fingerings, it is because I seek to actively engage the body and imagination, developing a perspective of how the music may have been played and sung. This *embodied knowledge* provides grounding for observations leading to analysis and interpretation of meaning in the song.

As a fairly early U2 song from their second studio album, "Gloria" exemplifies musical traits and creative strategies that would continue to be important throughout the decades of the band's musical output. I argue that U2 employs a unique approach in songwriting and performance that correlates interplay between loose and tight counterpoint with blendings of sensual and spiritual coding in their lyrics. In "Gloria," as in many other U2 songs, the band leverages gradations in musical texture and cultural coding to artistically engage with complex issues of musical style, identity, politics, and religion.[2] Aligning with folk and punk traditions that influenced the band, U2's creative strategies can be interpreted as a romantic impulse to express a higher truth in active resistance to an oppressive societal world.[3]

The analysis also explores how spatial settings of individual sound sources on the "Gloria" track form a combined sonic signature that idealizes the multivalent sonic intensity of a live arena concert. This has been aesthetically important to the band, as U2 guitarist The Edge explains:

> Well, you have to remember that U2 is fundamentally a live group. When you perform live certain things work and others don't—certain things get lost. But there's something really powerful about the live combination of guitar, bass, and drums, and in the early days we disciplined ourselves to use only those primary colors of rock 'n' roll.
>
> *Hutchinson 1986, 33*

Moylan observes that "All sounds come from places, whether natural spaces or created environments" (2020, 333). In U2's music, a realistic sonic environment perceived as a virtual stage is superimposed against an enveloping, seemingly limitless sonic landscape. This quality of space within space, literally enacted in a live arena concert, metaphorically supports an artistic impulse to locate human action and experience within a higher meaningful framework of social and spiritual ideals.

In examining guitar gesture in the song, the analysis shows how Edge maximizes the sonic effect of relatively simple guitar moves derived from rhythm & blues traditions, using delay guitar effects that he helped pioneer. As Edge incorporates delayed echoes in real time to craft his signature guitar sound, they become part of the rhythmic fabric of the song, blurring perceived boundaries of what is "played" and what is "recorded."[4] Since reverb and echo are perceived as attributes of environment, actions in performance that dynamically shape reverb and echo actively animate the perceived sonic space. Because this mode of spatial shaping is rendered "live" as part of the rhythmic impulse in playing the guitar part, it projects sonic qualities associated with a live concert that are preserved on the recorded track.

Observations on musical form in this study are correlated with shifting spatial qualities of sound on the recorded track: sounds located in highly reverberant spaces that a listener might associate with a live arena concert are organized and energetically shaped in a gestural arc that spans the song. But it is not only the sonic grandeur of U2's music that gives it vast spatial dimensions; it also results from the epic scope of the subject matter. Analysis of vocal persona and musical form in the song shows how U2's "Gloria" is framed as an expansive romantic journey. The song takes as its theme nothing less than the human need for self-expression and love. What emerges at the end of the song is a hopeful utopian realization of this creative impulse, fueled with the powerful sound associated with U2's arena concerts and exaltations in Latin to the glory of God.

U2's "Gloria" is also notable as one of the first MTV videos to receive heavy airplay. Filmed in 1981, the video features the band playing "Gloria" on a floating barge in an industrialized area of the Grand Canal Basin in Dublin while a crowd of onlookers watch and dance. The video is noteworthy as an important part of the song's reception history. As Bono remembers:

> When we went on tour in America, something strange happened. In some cities we would still be playing clubs. "Gloria" was not getting on the radio, and things were looking a bit wobbly. But in other cities, there would be a thousand people there, or two thousand. We played the Hollywood Palladium in Los Angeles, to over four thousand. It turned out these cities were test markets for a new idea, which was music television, MTV. There weren't many music videos around but one of them was "Gloria," and they played it all the time.
>
> *Bono et al. 2006, 125*

The music video features dramatically contrasting perspectives of visual space. The beginning gives the impression of viewing the band as if looking up from underwater as the music fades in. Throughout the video, wide-angle aerial views of the band positioned within a vast industrialized landscape are contrasted against close-ups of the performing musicians. While a detailed reading of the video is beyond the scope of this essay, it is worth noting that shifting perspectives of visual space created in the video—from the panoramic to the personal—do parallel the changing aural representations of sonic space that are explored in the study.

Gesture, Counterpoint, and Agency in Two "Gloria" Songs

Gesture can be understood as the intersection of human movement and expression. I examine gestural qualities of guitar and vocal performance in U2's "Gloria" as a means to contextualize syntactical elements such as harmony and form as embodied movement and expression. The larger aim is to contribute to our understanding of how popular music communicates meaning, exploring how gestural actions in "Gloria" are empathically grounded in the physicality of human activity, implying virtual actors that navigate trajectories through the virtual world of the music.

Counterpoint refers to musical relationships that arise through a combination of one note or musical line against another. Conceiving of counterpoint broadly as the combination of musical parts, I argue that the strategic juxtaposition of material projecting different and even conflicting gestural, topical, and structural implications is a recurring creative strategy employed by U2 that leverages relatively simple materials rooted in rock and blues traditions to create the rich textures of their signature epic sound. The analytical lens of counterpoint also provides a framework for exploring complexities of *virtual* space that are crafted artistically in U2's music. This involves interrogating spatial metaphors in multiple domains, including

1. The virtual sonic environment that positions sounds spatially in the recording,[5]
2. The embodied actions of the performers that project instrumental and vocal virtual space, and
3. An imagined virtual space in which the narrative arc of the song unfolds. Imagined actors inhabiting that fictive narrative space project human-like attributes of *virtual agency*.[6]

It has been observed that blues and rock songs often have contrapuntally free or "loose" verses and more tightly aligned choruses (Temperley 2007, 323). We will see that something like this happens in U2's "Gloria." But first, we briefly turn to an earlier "Gloria," written by another Irish songwriter (from Northern Ireland specifically), for an example of what Allan Moore first termed the "Melodic-Harmonic Divorce."[7] Example 9.1 shows a transcription from Van Morrison's song, "Gloria," which he recorded with his band Them in 1964.[8] The solo and backing vocal lines do not participate in the harmonic moves to D major and A major shown in the repeated guitar riff. Van Morrison's lead vocal is persistently centered on the tonic E, decorated with bluesy coloration on G natural. This modal clash, with E major chord changes heard against the minor inflection of G natural in the melody, is a common feature in songs that draw from blues traditions.

The neighboring motions to D natural in the vocals do not align with the harmonic changes. Embellished with E minor pentatonic coloration, the vocals are aligned with a deeper hierarchical level of tonic harmony on E but dissonant at times against the local harmonic changes. This results in structurally loose counterpoint or a so-called "divorce" between melody and harmony. Van Morrison's "Gloria" uses a repeated progression of four chords throughout, a "chord loop" that leads back to its own beginning. Because the continuous groove of the chord

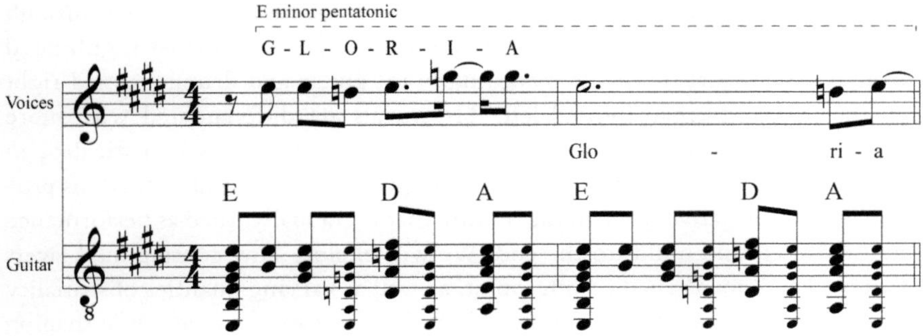


```
    E     * D * A *   E     * D * A *
E|-0-0-0-0-2-0-0-0-|-0-0-0-0-2-0-0-0--|
B|-0-0-0-0-3-0-2-0-|-0-0-0-0-3-0-2-0--|
G|-1---1-0-2-0-2-0-|-1---1-0-2-0-2-0--|
D|-2-----0-0-0-2-0-|-2-----0-0-0-2-0--|
A|-2-----0-----0-0-|-2-----0-----0-0--|
E|-0-----0---------|-0-----0----------|
```

* Open-string linear chord

Double-plagal progression:	D	A	E
	IV/IV	IV	I

EXAMPLE 9.1 "Gloria," Them (1964), 2:01.

loop lacks goal-oriented harmonic motion, the melody has an added burden in delineating phrases independently of the underlying chords. Drew Nobile labels this kind of loose counterpoint a "loop divorce" (2015, 189). We will discuss later how U2's "Gloria" ends with a final section that uses the same "double-plagal" chord progression found in Van Morrison's "Gloria," albeit in a different key.[9] In the earlier song, the progression in descending fourths provides for a clear gestural trajectory in the guitar part with chord roots played on successively lower open strings, D–A–E.

I would like to underscore the element of embodied expression that loose or "divorced" counterpoint brings to the music. In Van Morrison's "Gloria," it falls particularly to the singer to vocally shape phrases so that we hear phrase beginnings, highpoints, and endings, all while the repeated chord loop continues with unbroken circularity. It is the lead and backing singers that take this role until that task is passed to the guitarist during the guitar solo. In moments of loose, nonaligned counterpoint, we hear the vocal line or guitar solo resisting implications of control manifested in the chord changes. This enriches elements of virtual agency in the music, in which we empathetically identify with human-like attributes that are projected through purely musical means. The implication of willed intention in overcoming musically projected constraints can be important in representing a virtual agent that might be identified with a song's protagonist moving through a virtual environmental space.

The tradeoff between vocalist and guitar soloist in Van Morrison's "Gloria" could further be described as a staging of *multiple agency*, with attention drawn to the perceived exchange of roles and musical ideas that is expressed through interaction in the musical parts.[10] The musical exchange is further highlighted through stereo location in the recording, with guitar and drums panned right and the bass and organ located left. The vocals are differentiated with more reverberative and centered sound. After the guitar solo there is a dramatic drop in textural density (1:18), with instruments sharply panned left and right. This provides a gap or negative space in the texture that is gradually filled as performance intensity resets and builds to the end. We will explore how contrapuntal layers of aligned and nonaligned counterpoint, as well as varying qualities of spatiality in the recording, contribute importantly to U2's signature sound while shaping large-scale textural changes that delineate musical form.

Van Morrison's "Gloria" leverages the open strings of the guitar in rendering harmonies but also in creating an element of artistically shaped, rhythmicized, guitar-idiomatic sound that is not necessarily locked in alignment with harmonic changes. Dissonant "ghost" notes on open strings sound while the player is shifting hand position on the fretboard to grasp each chord.[11] The kinesthetic organization of the unbroken down-up strumming motion keeps the unflagging rhythmic energy of the chord loop on track. The result is a "rhythm guitar" part that powerfully asserts embodied musical expression, drawing attention to the sound of human movement as the guitarist moves through the chord changes. We will see how Edge does something like this in his highly inventive guitar work.

U2's "Gloria": Sonic Space and Narrative Trajectory

U2's "Gloria" overtly navigates the liminality between sensual and religious expression, providing an early prototype for later U2 songs that will do this including "Mysterious Ways" (1991) and "American Soul" (2017). Intersections of the spiritual and the sensual in the lyrics correlate with contrasts in sonic spatiality, vocal quality, and guitar technique, creating a dynamic virtual space that can imply both the grandeur of a vast sonic landscape and the intimacy of personal space. Example 9.2 shows an overview of musical form and phrase rhythm, along with transcribed passages. The protagonist in the song seems to

Instability *Stability*

Fade-in Pre-Intro 4 + 4 (D minor)	Guitar Intro 4 + 4 E minor	"I try to sing this song..." Verse 2 + 4 + 4	"...I'm complete" Extension 2	"Gloria in te Domine" Chorus 1 4 + 4 + 4 + 4 B minor

(2x)

Musical offering *Exaltation*

"I'd give it to you" Instrumental Break 4 B minor	Echo (guitar 1) 4 + 4 + 4	Slide (guitar 2) 4 + 4	Slap bass Transition 4 + 4	"Gloria in te Domine" Chorus 2 4 + 4 + 4 + 4 + fade B major

EXAMPLE 9.2a "Gloria," U2, October (1981). Formal outline.

EXAMPLE 9.2b "Gloria," U2. Verse 1 (0.27).

be singing to a woman in a spiritual way while also perhaps singing to God in a sensual way. Bono's comments on the lyrics indicate that he himself valued this double meaning and also enjoyed that the song shares its name with the famous Van Morrison song. In John Waters' book, Bono is quoted:

> … of course Gloria is about a woman in the Van Morrison sense. Being an Irish band, you're conscious of that. And I think that what happened at that moment was very interesting: people saw that you could actually write about a woman in the spiritual sense and that you could write about God in the sexual sense.

I contend that reversals and juxtapositions in contrapuntal alignment provide a means to animate varied implications of meaning in the lyrics with changes in the spatial fabric of the music. In the Pre-intro fade-in and Intro that follows, the guitar and vocal lines dovetail one another, never aligning metrically. The fade-in gives the impression of music, having already begun, now moving toward the listener from a vast distance. There is increasing timbral detail as the full instrumental sound comes into focus and perceived distance decreases (Moylan 2020, 306). This is the sonic environment in which we hear the protagonist literally call out to his "Gloria," as if projecting the solo voice across a vast sonic landscape.

EXAMPLE 9.2c "Gloria," U2. Chorus 2 (3.05).

Here in this brief opening passage, Bono sings the highest vocal notes in the song. Bono has stated that his father once said, "You, Bono, are a baritone who thinks he's a tenor" (Boyd 2015). In listening to Bono's opening vocal delivery, this quality comes through as passion and commitment as the singer projects audible effort in starting off the vocal in his high range with full voice.[12]

There is a personic shift as the song's enacted vocal protagonist, now Bono the front man, counts off "two-three-four" (0:12). This initiates the main instrumental fabric of the Verse, which juxtaposes two sonic host environments: the highly reverberant guitar projects sizable environmental width and depth (Distant Zone), while the drums and bass with more crisply articulated sounds are perceived as located in closer proximity to the listener (Public Zone).[13] As the motoric energy of the Intro kicks in, Bono calls out as if energizing the crowd at a live concert, "Get up, up, up" (0.19).

Figure 9.1 diagrams a large-scale narrative arc that traverses changes in sonic space and the enactment of vocal persona. Focus shifts from the singer as performer to the imagined protagonist depicted in the song as the Verse begins (0:27). Expression is turned inward, depicting a protagonist who is alone and aware of his limitations ("I try to sing …, I try to stand up …"). Conveying the isolation of artistic self-awareness, the Verse objectifies the song as a search for expression, a search that is metaphorically realized by the end of the song. The protagonist's search assumes epic proportions as it is represented through unstable rhythms, overlapping phrases, loosely aligned counterpoint, and a guitar-driven sonic landscape.

The musical parts transcribed in Example 9.2b do not quite align metrically or harmonically, exemplifying what could be termed as "loose" counterpoint. The first measure shown seems to superimpose G major over an A bass, with dissonant guitar harmonics added in the second measure. The guitar and bass

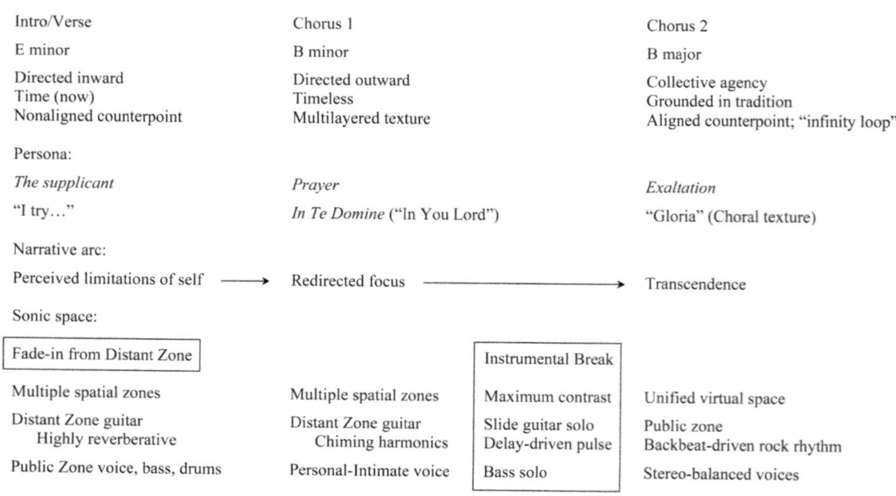

FIGURE 9.1 "Gloria," U2. Persona, narrative, and sonic space.

only shift into the Verse rhythmic patterning and E minor harmony at the third measure. The resulting effect is that Bono seems to begin his vocal phrase a full two measures before the instruments fully engage with motoric accompaniment. At that point, The Edge begins polyrhythmic patterning in a continuous stream of eight notes derived from the song's Introduction. At the fifth bar in the Verse, on the lyric "stand up," the voice and guitar are dissonant against the harmonic shift to subdominant A in the bass. This nonalignment or musically expressed confusion nicely coincides with a protagonist depicted in the lyrics that is confused and restrained, unable to "find his feet."

While the parts are at odds with one another, they are each very grounded as embodied musical gestures, maintaining stable *gestural* rhythms. Much of the guitar work can be played in a single position on the fretboard, with barre chords, followed by two-note figures that can be executed on the fretboard while moving a single finger, along with shifts from the B string to neighboring high E string, combined with pedal-like repetitions on the open E string. Highly characteristic of Edge's guitar work, the passage is very idiomatic to the guitar, efficient, and motivically organized, providing for abundant sonic activity with minimal motion.

Analyzing the Verse as an example of the sort of "melodic-harmonic divorce" that gives rise to "loose" counterpoint, I would argue that the vocal melody is aligned to a deeper level of structure elaborating the E minor triad. The Edge's pedal tone repetitions of the open E string connect with the E minor tonic prolongation. The moves to A and D in the bass support surface-level neighboring harmonies. The embodied musical actions—the musical gestures enacted in the vocal, guitar, and bass—are really quite independent rhythmically and harmonically, while Larry Mullen's steady drum beat keeps the driving groove on track.

While the voice, bass, and drums seem to occupy the Public Zone of a virtual stage, the guitar continues to project a more vast and distanced sonic dimension. Bono's spacious vocal gesture personifies the perceived musical agent inhabiting and moving through that virtual world. He sings broadly phrased lines, like an orator. This will connect to the hymn chorale topic found in the two different choruses that use the Latin text, "Gloria in te Domine" (Glory in you, Lord). Bono himself described the ending as "this Latin thing, this hymn thing. It's so outrageous at the end going to the full Latin whack. That still makes me smile. It's so wonderfully mad and epic and operatic" (Waters 1994). Bono's statement indicates his awareness of the religious hymn topic as a source of epic "operatic" dimension, while his description of it as "wonderfully mad" and "outrageous" suggests an awareness of the potential to hear the music as transgressive and ironic, with multiple interpretations of meaning possible.[14]

The Quiet Chorus

After the Verse dramatizes the enactment of a high-energy band playing *loud*, there is a shift to quiet intimacy in Chorus 1 (0.48). If we trace a spatial trajectory of the vocal, from the fade-in through the public sphere of the Verse, then here in Chorus 1, the vocal becomes increasingly intimate and introspective, as if addressed directly to the loved one or perhaps in prayer to God. This formal juncture is not so much a shift in the actual loudness of sounds but rather a shift in prominence.[15] After the driving rhythms and dense contrapuntal interactions of the Verse, the music here opens into a spacious environment that is resonant with sustained reverberant sound. The reflective and spacious Chorus 1 creates a formal instability, reversing the normal prototype of a full-sounding chorus that could serve as a song's climactic highpoint. This musically telegraphs that there is more musical material yet to come.

Chorus 1 previews the Latin verse used later in the ending of the song while expressing longing and the protagonist's perceived inadequacy in having nothing to give. With ironic friction in the invocation of biblical Latin, the vocal quality is overtly physical and sensual, especially toward the end of the verse as Bono vocalizes in his low range with audible glottal fry, "Oh Lord, loosen my lips" (1.11). The containment of voice, bass, and drums in a more localized sonic space within an enveloping guitar-driven environment continues, as chiming guitar harmonics are shaped with noticeable reverb, echo, and vibrato. High notes on piano blend with the guitar harmonics, contributing to the otherworldly atmosphere. The extended reverberant energy of vocal and instrumental sounds in Chorus 1 is enhanced by the positioning of sounds in a more texturally sparse context, which foregrounds and marks the reverberant qualities of sound as expressively significant. The listener's perspective of being drawn in closer at Chorus 1, as if invited into the more delicate sound world, is abruptly redirected in another sonic shift, with the return to public spatiality for the second verse.

Instrumental Break and Chorus 2

After two cycles through the Verse and Chorus 1, there is a shift in personic voice, from enactment of the song's protagonist to a focus on the multiple agency of the performers. The vocal lyric that prepares for this shift is "I'd give it to you" (2:14), which sets up the Instrumental Break to function in the song's narrative as a kind of "musical offering" from the artist-protagonist. The ending of the vocal is processed with a pronounced reverb tail that enhances the transition into the more atmospheric Instrumental Break.

The Instrumental Break clearly exemplifies a recurring technique that Allan Moore has noted in the music of U2: "the creation of the illusion of space by a divergence between fast, intricate surface movement and slow, underlying harmonic change" (1998, 21). A rapid and percussive delay-driven 16th-note pulse in one guitar part, matched by hi-hat drums, provides contrast for a slide guitar solo with continuously sustained sound and thick reverb, creating the effect of rhythmic activity within a vastly spacious environment. The slide guitar incorporates motivic material from the Verse vocal melody, forming a kind of instrumental continuation of the singing voice. Very quiet high notes on piano overdubbed by Edge provide a steadily recurring pattern that is motivically related to Bono's opening "Gloria" vocal, perhaps further suggesting the disembodied presence of the vocal protagonist, as if his longings for an exalted state are here channeled into wordless artistic expression. U2 bassist Adam Clayton introduces timbral contrast with a bright and percussive slap bass solo (2:59). Here we enter into the most atmospheric part of the sound recording, with metallic percussion and sonic layerings panned left and right in the stereo mix that show the influence of producer, Steve Lillywhite. As the Instrumental Break suspends the fuller vocal-harmonic texture of the song, it functions as a transitional pause to gather energy for the final Chorus 2 section.

As shown in Example 9.2c, Chorus 2 brings the voices and instruments together in close alignment or "tight" counterpoint along with a modulation to a major key. This provides a large-scale resolution of the rhythmically unstable contrapuntal juxtapositions heard earlier in the minor key. The repeated chord loop comprising the Chorus 2 section is introduced with the "Gloria" melodic motive in open fifths on guitar (2:59). I cannot help but notice that this sounds rather like a bit of medieval parallel organum.

Chorus 2 is a clearly separated new section not relying on previously heard music.[16] The multiple spatial zones heard throughout the song here coalesce into a single unified sonic environment. Guitar closely aligns with the harmonized voices, resulting in a rock-energized choral texture, complete with borrowed Latin text, "Gloria in te Domine." The music here achieves grounding through different topical markers with rich cultural associations. First, there is the vocal-choral style with Latin text, which resolves the song's multilayered textures into a unified texture associated with the collective agency and uplifting expression of a choir. This quality is enhanced with a balanced stereo sound that positions Bono's vocal line higher and center, with harmonized vocals of the other band members located closer left and right. In addition, the varied patterns and timbres of drums and percussion provided by Larry Mullen Jr here give way to a straightforward backbeat-driven rock rhythm. The strong and unflagging backbeat on snare is an iconic marker of rock style.

In summary, the shaping of sonic space is crucial in articulating the overall narrative trajectory of the song: the music comes from a far-off place (Fade-in), stages a quest for love and spiritual connection as public spectacle (Verse), provides a window into an intimate and personal space (Chorus 1), sublimates vocal expression in wordless sound (Instrumental Break), and finally achieves connection in a transcendent participatory space that powers the collective exaltation of choral voices with driving rock rhythm (Chorus 2).

Guitar Work

The Edge is known for his extensive use of analog and digital guitar delay effects that became available in the late 1970s and 1980s, replacing earlier tape-driven echo effects. Edge's elaborate sonic treatments sometimes involve sending the split guitar signal, processed through two different delay settings, into dual amplifiers that feed into microphones in live performance to produce added ambient spatial and temporal effects.[17] Without getting into the extensive lore that has been documented on Edge's playing technique and equipment, it is important to note that his use of delay effects is not only an alteration of tone quality. It produces virtual sonic images in real-time that impact the mode of performance of all the performers. As Edge explains, "… I became the timekeeper with the band for a while, and Larry would play to me, because everything had to sync with my echo" (Hutchinson 1986, 33). The delay machine is, in effect a virtual conductor, cuing the live performers through its synthetic timing of multiple virtual reiterations of the humanly rendered guitar sound. In "Gloria," we hear this especially in the Instrumental Break before the ending (2:19), but it contributes a spatial dimension to the music throughout the song.

Example 9.3 provides a summary of motivic materials in the guitar work, showing how they are patterned through embodied movement that invests the guitar sound with a clear gestural logic and material presence. The Edge's guitar sound is sometimes described as having a "chiming" quality. The ringing sound of the opening guitar figuration in the Pre-intro provides an example. In what could be described as parallel gestures, the two chords are formed with the same hand shape on the guitar fretboard, with a down-string shift in moving from the D minor to A major chords, as diagrammed in the guitar tab. The guitar voicing maximizes the effect of ringing open strings, with added ninths first on the high E string and then on the B string. Edge has commented that he considers his blending of melodic notes against droning open strings to be instinctively derived through Irish musical influence. (Hutchinson 1986, 33–34) Notes sounded on open strings in Example 9.3 are shown numerically as "0" in the guitar tab.

Each of the guitar motives shown in Example 9.3 involves parallel gestures in pairs. The main riff for the Verse (0:30) is a two-note figure that shifts from B to E string in the same fretboard position, fusing with pedal tone on the open E string to create a 3/8 cross rhythm. The Chorus 1 guitar work (0:47) features natural harmonics that shift in parallel movements between the seventh and fifth frets, providing an ambient atmosphere for the prayerful vocal delivery of the Latin "Gloria in te Domine." The Instrumental Break (2:19) with slide guitar solo dramatically expands the spatial dimensions of the music while building expectation for the Chorus 2 final section.

The Edge's "organum" style open fifths (2:59) are processed with a pulsing delay echo timed to the quarter beat, which serves to announce the driving rhythm of Chorus 2. At the repeated loop of Chorus 2 ending (3:05), really for the first time in the song, the guitar locks in rhythmically, melodically, and harmonically with the voices and other instruments. Slides along the fretboard in moving through open-fifth chords enhance the embodied materiality of the guitar sound. The parallel open-fifth barre chords high on the instrument fuse with the vocals, enhancing the power of a few voices to emulate a full choral texture. This illustrates another general characteristic of The Edge's guitar work;

a. Pre-intro:
Chords with added ninth on open string
Basic hand position on two frets
Parallel shift down-string to form second chord

b. Intro–Verse (0.14):
Two-note motive in parallel shift up and down-string

c. Chorus 1 (0.48):
Natural harmonics on seventh fret
Natural harmonics on fifth fret

d. Chorus 2 (3.03):
Parallel "organum" in fifths and octaves
Barre chord with fixed hand position
Shifts up and down the fretboard dramatized with echo delay
Engages with choral-style vocal

EXAMPLE 9.3 "Gloria," U2. Gestural parallelisms in guitar work

rather than playing barre chords on low strings for a heavy sound typical in rock, he more often saturates higher regions of the sonic texture.

The Ending: A Transcendent "Infinity Loop"

If we regard the Intro, Verse, and Chorus 1 as the conventional formal units of the song, then the Pre-intro fade-in and the repeated Chorus 2 chord loop with fade-out at the end constitute elements that are outside generic formal boundaries

TABLE 9.1 Three related chord loops

"Gloria" (Them, 1964)	E	D	A	E
"Hey Jude" (Beatles, 1968), ending	D	C	G	D
"Gloria" (U2, 1981), ending	B	A	E	B
	I	IV/IV	IV	I

and are therefore marked for special expressive significance. The fade-in pre-intro section, while only a few seconds in duration, conveys a vastness as if we are entering into an immense sonic landscape that is somehow already "out there." The ending Chorus 2 section further expresses the feeling that the song cannot be contained, as it energetically bursts out of its perceived frame, ending in an added "bonus" section of heightened energy.

The repeating eight-measure loop of Chorus 2, with its even hypermeter, resolves the song's rhythmic instabilities, freeing the music from tensions that could be interpreted to represent the protagonist's longing and effortful search for expression. This repeated phrase, ending in a fade-out, exemplifies a device that Frank Samarotto (2012) has described as the "trope of expectancy/infinity," in which a song's conventional form builds in expectation for "an endless loop of a single repeating musical phrase not heard in the previous section." The archetypal example of this is the final repeated phrase at the end of the Beatles' "Hey Jude."[18] While there are other songs that end with the infinity loop, the connection to "Hey Jude" is especially close, as it uses a very similar chord pattern based on the double-plagal chord progression. Interestingly, the chord progression also corresponds to the chords of Van Morrison's "Gloria." (See Table 9.1.)

The ending Chorus 2 section has connotations of an infinitely extending expressive space. It does not further advance the narrative but rather offers the continuous momentum of an exalted state. The striving depicted in the lyrics and through embodied performative energy in the song is answered in the transcendent expression realized in the infinity loop. In this way, the entire song can be understood as an environment for the staging of virtual agency that projects a single unifying gestural motion: the enactment of a prayer answered. As a love song, whether earthly or spiritual in its devotion, or as a rock show that transgressively borrows the Latin hymn topic, the "glory" manifested musically in the infinity loop is its own reward.

The songs of U2 do not so much communicate a social message directly, as might be found in a folk song, but rather, through representation and identification with a singer's persona that *is* passionate, provoked to action, and positioned within a highly energized virtual space that metaphorically maps onto the energizing social context of an activated community. This examination of U2's "Gloria" through the analytical lens of counterpoint has shown how vocal and instrumental gestures are combined freely—without strict contrapuntal alignment—to form a unique gestural approach that engages with lyrics and spatial sonic design to project U2's distinctive sound. Vocal and instrumental sound sources on the recorded track are further individualized spatially in the virtual sonic environment, dynamically positioning musical actions within a vast sonic landscape. The strategic juxtaposition of material projecting different and even conflicting gestural, topical, and structural implications in U2's music provides a framework for creativity that the band would continue to leverage through their various changes in style as a means to expressively mediate between the romantic inner world of the artist and an oppressive societal world.

Notes

1 Portions of this study were presented at the U2 Conference, 2018, Belfast, Northern Ireland.
2 The role of religion in the lives of U2 band members and the reception of their music has been discussed and researched extensively. For an overview, see Rothman (2014).
3 McLaughlin and McLoone (2012) provide a history of Irish folk, rock, and punk traditions that have shaped U2's musical development.
4 Moylan (2020) discusses properties of echoes and reverberation (333–37) and, specifically, the rhythmic properties of echoes in recordings of the 1980s (336).
5 Moylan's comments on "Hearing Invisible Sounds in Virtual Space" are useful here (2020, 287–88). Allan Moore (2012) discusses virtuality in musical motion, performance space, and the construction of persona. For a study examining spatially surreal sound in virtual space with sonic features that could not occur in actual physical environments, see Brøvig and Danielsen (2013).
6 I draw from Robert Hatten (2004, 2018) in formulating an approach to musical gesture and virtual agency. Another important source, Seth Monahan (2013), draws from the seminal work of Edward T. Cone (1974).
7 Moore (1995, 188–89). For a comprehensive discussion, see Nobile (2015), "Counterpoint in Rock Music: Unpacking the 'Melodic-Harmonic Divorce.'"
8 The U2 band members were very aware that Irish fans, in particular, would associate their "Gloria" with the earlier song written and sung by Van Morrison; this can be appreciated as one of the many undercurrents of meaning in the song. U2 has incorporated Van Morrison's "Gloria" in their live shows, sometimes inserting the iconic "G-L-O-R-I-A" lyric into live renditions of their song. The Edge has also acknowledged the influence of Patti Smith's 1975 version of Van Morrison's "Gloria" (Bono et al. 2006, 37).
9 Chord motions by descending fourths, known as the double-plagal progression, are discussed in Biamonte (2010) and Everett (2004).
10 For a study of multiple agencies in the chamber music of Mozart, see Klorman (2017).
11 For more discussion of embodied expression in Van Morrison's "Gloria," see Koozin (2011).
12 McLaughlin and McLoone (2012) discuss this quality in describing Bono's unique vocal timbre (182).
13 See Moylan (2020, Chapter 8), and particularly his discussion of "Personal Space and Proxemics" (307–09).
14 According to U2 producer Steve Lillywhite, Bono's idea for the Latin chorus was inspired by listening to albums of Gregorian chants owned by U2's manager Paul McGuinness. (U2 official website. Retrieved at https://www.u2.com/music/singles/4023.)
15 Moylan (2020) discusses prominence and loudness in recordings (359).
16 Brad Osborn (2013) makes this distinction in analyzing songs that end with new material forming a "terminal climax."

17 For an extensive study of Edge's guitar delay techniques, see "A Study of The Edge's (U2) Guitar Delay" by Tim Darling at http://www.amnesta.net/edge_delay/. The Edge is known to have used the Electro-Harmonix Memory Man on the first two U2 albums, which was the first analog echo/delay unit with no moving parts.

18 In addition to "Hey Jude" (1968), songs that end with an infinity loop include: "Atlantis" (1968), Donovan; "Beginnings" (1969), Chicago; "Free Bird" (1974), Lynyrd Skynyrd; "On Every Street" (1991), Dire Straits.

References

Biamonte, Nicole. 2010. "Triadic Modal and Pentatonic Patterns in Rock Music." *Music Theory Spectrum* 32 (2): 95–110.

Bono, The Edge, Adam Clayton, Larry Mullen, and Neil McCormick. 2006. *U2 by U2*. London: HarperCollins.

Boyd. Brian. 2015. "Bono: The Voice of Innocence and Experience." Interview with Bono in *The Irish Times*, September 18. Accessed at https://www.irishtimes.com/culture/music/bono-the-voice-of-innocence-and-experience-1.2355501.

Brøvig, Ragnhild, and Anne Danielsen. 2013. "The Naturalised and the Surreal: Changes in the Perception of Popular Music Sound." *Organised Sound* 18 (1): 71–80. Cambridge University Press.

Cone, Edward T. 1974. *The Composer's Voice*. Berkeley, CA: University of California Press.

Everett, Walter. 2004. "Making Sense of Rock's Tonal Systems." *Music Theory Online* 10 (4). https://mtosmt.org/issues/mto.04.10.4/mto.04.10.4.w_everett.html.

Hatten, Robert. 2004. *Interpreting Musical Gestures, Topics, and Tropes: Mozart, Beethoven, Schubert*. Bloomington, IN: Indiana University Press.

———. 2018. *A Theory of Virtual Agency for Western Art Music*. Bloomington, IN: Indiana University Press.

Hutchinson, John. 1986. "U2's Leading Edge." Interview with The Edge in *Musician* magazine. Accessed at http://www.amnesta.net/edge_delay/musician_magazine_1986.html.

Klorman, Edward. 2017. *Mozart's Music of Friends: Social Interplay in the Chamber Works*. Cambridge: Cambridge University Press.

Koozin, Timothy. 2011. "Guitar Voicing in Pop-Rock Music: A Performance-Based Analytical Approach." *Music Theory Online* 17 (3). https://mtosmt.org/issues/mto.11.17.3/mto.11.17.3.koozin.html.

McLaughlin, Noel, and Martin McLoone. (2012) *Rock and Popular Music in Ireland: Before and After U2*. Dublin: Irish Academic Press.

Middleton, Richard. 1993. "Popular Music Analysis and Musicology: Bridging the Gap," *Popular Music* 12 (2): 177–90.

Monahan, Seth. 2013. "Action and Agency Revisited." *Journal of Music Theory* 57 (2): 321–71.

Moore, Allan. 1995. "The So-Called 'Flattened Seventh' in Rock." *Popular Music* 14 (2): 185–201.

———. 1998. "U2 and the Myth of Authenticity." *Popular Musicology* 3 (June): 5–33.

———. 2012. *Song Means: Analysing and Interpreting Recorded Popular Song*. Burlington, VT: Ashgate.

Moylan, William. 2020. *Recording Analysis: How the Record Shapes the Song*. New York: Routledge.

Nobile, Drew F. 2015. "Counterpoint in Rock Music: Unpacking the 'Melodic-Harmonic Divorce.'" *Music Theory Spectrum* 37 (2): 189–203.

Osborn, Brad. 2013. "Subverting the Verse–Chorus Paradigm: Terminally Climactic Forms in Recent Rock Music." *Music Theory Spectrum* 35 (1): 23–47.

Rothman, Joshua. 2014. "The Church of U2." *The New Yorker* (September).

Samarotto, Frank. 2012. "The Trope of Expectancy/Infinity in the Music of the Beatles and Others." Paper presented at the annual meeting of the Society for Music Theory, New Orleans, LA.

Temperley, David. 2007. "The Melodic-Harmonic 'Divorce' in Rock." *Popular Music* 26 (2): 323–42.

Waters, John. 1994. *Race of Angels: Ireland and the Genesis of U2*. Belfast: Blackstaff Press.

SECTION IV

Sonic Journeys

10

ON THE STRUCTURE OF FEELING IN BOB DYLAN'S "I'VE MADE UP MY MIND TO GIVE MYSELF TO YOU" (2020)

Michael Millner and William Moylan

This chapter is supported by a webpage.

"I've Made Up My Mind to Give Myself to You" is the fourth track from Bob Dylan's 39th studio album, *Rough and Rowdy Ways* (2020). It was released the month after his 79th birthday and 58 years after his first album, *Bob Dylan* (1962). Thematically, the track exemplifies key elements of Dylan's later work since at least 1997's *Time Out of Mind*, namely the volant sweeps between opposing poles—threat and conciliation, love and disgust, the affirmation of life and the haunting by death. But other elements of "I've Made Up My Mind" are strikingly fresh and even unique in Dylan's canon, especially the obbligato passages presented by soft backing vocals and guitar accompaniment. The obbligato lines often carry a strong resemblance to the barcarolle "Belle nuit, ô nuit d'amour" from Jacques Offenbach's opera *The Tales of Hoffmann* (1880), and they delicately entwine with Dylan's vocal timing and the track's instrumentation, arrangement, and recording elements. In "I've Made Up My Mind to Give Myself to You," Dylan is not resting on his Nobel laurels.

The track is intricate and repays very close analysis, both so that we can better understand the way it does its work—especially its affect-generating work—and so that we can better place it within a cultural and political context. To these ends, we will first present a close reading of the music of the track, including an examination of Dylan's vocal timbre and performance techniques. A careful examination of the most characteristic sound qualities of the recording follows, providing an evaluation of sound staging and environments and their contributions to the track's substance and expression. The chapter continues with a close reading of the track's literary form and lyrical content. We see all of these elements as generating a complex structure of feeling, and we conclude by placing this structure of feeling into dialogue with an important formation in American popular culture—the melancholic, sentimental, melodramatic,

DOI: 10.4324/9781003089926-15

detached masculine figure who most often appears in hardboiled, noir novels and films—and has been central to Dylan's work for many years.

Musical Texture: Voice and Accompaniment Qualities and Interaction

The track is characterized by a stratified texture of layers (Covach 2018, 53–56) that vary by section. Verses have three distinct layers: the lead vocal, a foundational and anchoring ensemble, and an obbligato passage. Bridges contain two strata: the lead vocal and an expanded anchoring ensemble.

The anchoring ensemble is comprised of the drum set, double bass, and acoustic guitar; this group of sources is present throughout the track. It anchors the pulse, tonality, and the sound stage of the track, providing a consistent metric pulse and a stable pitch reference throughout. Brushes on snare drum and hi-hat establish a subdued sound and also keep time without abrupt attacks; together with a gentle bass drum emphasis they maintain the pulsing metric grid. The acoustic bass is subtle in the verses; it is more prominent in bridge sections, where it serves to widen its range and timbral balance. Cymbals, organ, and flat-picked electric rhythm guitar colored by a subtle amp tremolo are added in the bridge sections. In the bridge sections, all instruments blend into a single, larger accompanying ensemble, with each instrument assuming its place within a singular undulating texture. This accompaniment layer articulates structure between verses and bridge, establishes a strong contextual and supportive function in the bridge sections, and serves more of a foundational and context function in the verses. In the verses, this ensemble plays a supportive role to the unique obbligato of the track's verses.

Instrumentation of the obbligato is two male backing voices, finger-picked electric guitar, and organ. At various times, each of these three parts appears individually with an added tremolo effect; the effect is subtle but adds to the lilting nature of the passage. The hummed backing male voices—highly unusual for a Dylan track—provide a relaxed and swaying sound layer that functions as a foundation to the character of the verses. It is an undulating, luminous accompaniment, swaying or rocking like a gondola. The obbligato functions independently to enhance the track's character and its qualities reinforce the track's affect and illustrate its sonic context. This four-measure obbligato passage fills the introduction and three of the four phrases (the statement, restatement, and conclusion, but not the departure) of each verse. A measured tremolo in the "departure" gesture of each verse's SRDC form[1] (first at 0:42–0:51) that participates in clearly defining the underlying pulse is sometimes marimba-like, sometimes with traits of a steel drum, and at times exhibits a mandolin quality.[2] The obbligato drops out in the bridge sections, articulating the track's structure, and allowing the accompaniment of those sections greater rhythmic precision; the instruments and vocals retain an identity as an independent layer though their materials shift.

Dylan's lead vocal comprises the other layer. The part is rhythmically complex and its phrasing and metric placement highly unpredictable and introspective; the nuances of performance enhance the meaning and expression of the text—especially its dynamic interplay of intimacy and detachment.

The lead vocal is largely independent of the other two strata; rarely does it directly coordinate with the anchoring ensemble and never with the obbligato. The vocal rhythms clearly separate the lead vocal line and the vocalist's persona from the accompaniment; they emphasize the meter and synchronize with the anchoring instruments (drums and bass) only at several important moments.

These features described above cause Dylan's vocal to sound detached from the rhythmic undulation of the obbligato that allows the track to float; it is as if it is disassociated from the rhythmically floating obbligato and also from the measured pulse of the foundational accompaniment. This establishes the vocal in a different place or space, or speaking from a different time. This detachment is suspended in the bridge sections, where a character of stronger unity in the accompaniment provides more direct support to the vocal.

The vocal timbre and performance qualities of Bob Dylan's lead vocal will now be examined.[3] Dylan did not release an album of original music during the eight years prior to *Rough and Rowdy Ways*—the longest period during the artist's career. These years were filled with much activity, though, in recording covers of American standards released in the albums *Shadows in the Night* (2015) and *Fallen Angels* (2016), and *Triplicate* (2017) with their focus on Frank Sinatra covers. Seemingly brought about by his performances on these albums, Dylan's vocal style and sound in *Rough and Rowdy Ways* are remarkably transformed from that of *Tempest* (2012). In *Tempest* Dylan's vocals are rough, almost hoarse in timbre, language is often indistinct, approximate pitch abounds, melodic materials undergo little development save conforming to new lines of text—one can imagine his voice has given way to age and six decades of performance. The contrast with *Rough and Rowdy Ways* could not be more stark; here, it is as if Dylan has been gifted a new instrument; his voice is fresh and clear, is masterfully controlled, and conveys sophisticated nuance and artistry.

In "I've Made Up My Mind to Give Myself to You," the *sound character and timbral content* of Dylan's voice have highly refined clarity, precision, facility, and control. There is a sound character of the vocal line and a timbre of Dylan's voice that defies his 79 years at the time of the track's release. This is a singer who has continued to develop, is in control of his technique, and knows the years in his voice. His vocal qualities reflect the text's love song of one in later life; this is golden-hour tenderness, vulnerability, resignation, melancholy, and age-weathered sentimentalism—far from the lust of youth.

There is significant clarity to the presentation of lyrics. Phrases and individual words are articulated clearly, allowing the lyrics to be easily understood. While the performance contains manipulated word sounds and timbral transformations, the lyrics' phrases and words are always clearly audible.

Strategic use of vocal timbres and performance intensity supports the content of the lyrics. Four general and contrasting vocal qualities are present throughout the track: constrained and resonant, clear or natural, and raspy or gravelly. At times these vocal qualities and performance techniques contrast timbral qualities within the same phrase or word—as in the word "lost," which begins the second line of verse 1, which exhibits a transformation within the sustained "o" between clear and gravelly that is found often throughout Dylan's performance. The sudden appearance of a resonant voice effectively emphasizes certain phrases at strategic moments; a clear example is the resonant line "a lot a people I knew" (4:40) which follows the higher-pitched line "A lot of people gone" (4:36) that is vocally and emotionally constrained. Here the sound of the voice itself shifts between poles—from constraint to resonant openness, from detachment to attachment.

Dylan's performance contains emphasized consonant sounds, such as the "s" sound that prolongs the word "terrace" at the end of the track's very first line. Such sounds not only extend word sounds past the melody, this "s" sound is gentle but also filled with timbral detail. The sound provides a clue that Dylan's vocal contains qualities that can only be heard when someone is present within the listener's personal space; thus, in the very first line of the lyrics, we learn Dylan is intimately close. One can consider him at once both intimate and distanced; it is as if sonically the listener is there with him, though his presence within the content of his lyrics is distant as he speaks to himself, unaware of the listener.

The character of Dylan's vocal quality and manner of performance conveys tenderness and portrays a delicacy of control. There is a vulnerability and sentimentality present within the performance and the presentation of the lyrics of this love song for one of age and maturity. A romantic spirit pervades the track and is resident in Dylan's vocal—its timbre, character, rhythm, intonation, and affect.

The vocal at times sounds effortless and at times needs to be pushed and forced out. There are moments of withdrawal and detachment ("It just takes me a while to realize things" [5:23]) and of proclamation ("My heart's like a river" [5:17]), moments of ease ("Sitting on my terrace" [0:22]) and of resolution ("I've made up my mind" [0:54]). In these moments lyric and voice interact to develop the affect structure of the track, involving the oscillation between affective opposites. Further, vocal timbres blending with lyric content (as in the word "yes" within "I knew you'd say yes" [6:05]) are utilized, as well as a strategic use of text painting ("Fall of snow" [1:09]).

We recall the lead vocal embodies a unique rhythmic layer set against the obbligato and anchoring strata.

There is a sophisticated and nuanced rhythmic timing within Dylan's vocal line. As with an improvisatory flexibility of placement in the great jazz singers, it stretches and pulls the beat; it works around the pulse in unusual and creative ways. Still, the rhythms are precise, calculated, and metrically complex. These complex rhythms are periodically contrasted with the sudden articulation of the pulse, usually at the end or beginning of phrases (note pulses of "shooting star," "near or far," "Well my heart's"). These rhythmic features are illustrated within the complete rhythmic and word placement transcription of Dylan's lead vocal that appears on the accompanying webpage.[4]

From that transcription, it will be evident that each verse has a different rhythmic presentation within the performance of its recurring melodic lines. The rhythmic variations are the product of the interpretation or presentation of the text—some are subtle and some more extreme, such as delaying a portion of the line, thus creating long pauses suspending the line until at the very end of phrase a quick stream of text rushes out as in "It's just something I knew" (2:41).

Dylan's inventiveness in reshaping the melodic idea is readily evident in the track's six iterations of the refrain line. The six presentations of the refrain are all different rhythmically and in intonation. Figure 10.1 illustrates the significant variation between the six refrains, as each refrain provides a different sense of resolution or open conclusion appropriate to the unique sentiment and content of its verse.

Dylan often uses subtle deviations from equal temperament to significant expressive affect and tonal effect. Intonation is artistically crafted within his performance.

His use of pitch nuance adds tension and direction to lines and enhances the expression to the lyrics. These are evident in the six unique refrain appearances of Figure 10.1. Throughout the track, there are phrases with pitches that are

FIGURE 10.1 Six different versions of the refrain. Arrows indicate slightly flat or sharp pitches, and accidentals include quarter-tones.

slightly flat (designated by arrows) and at times in quarter-tones (notated as such) that are in clear contrast to the phrases and lines that are more precisely tuned. At times this contrast is present within a single melodic line, with a shift from the beginning of a phrase toward its conclusion—for example, it is as if there is a restraint that is preventing Dylan from reaching pitch when presenting "Been thinking it all over" that is overcome at the end with the confident "thought it all through" (0:44–0:51) fully voiced and more accurately pitched.

Further, there is a different sensibility of pitch between the verses and the bridges that also represents the interplay of affective opposites. Notably, there are verse phrases performed with an affect of restraint and slightly flat, such as in "No one ever told me" (2:33). This contrasts with bridge phrases that are more assertive or energized and very accurately pitched, such as "I'm giving myself to you, I am" (1:46). At times there is a relaxed quality that comes from singing only certain notes slightly flat, as in the first lines of verse 1. Conversely, there are portions of verses that are accurately pitched, though they tend to be the lines with higher performance intensity ("It looks at nothing here or there, nothing near or far" [2:22]), and some are performed at a lower intensity level but are not constrained, such as "If I had the wings of a snow white dove" (3:06).

To conclude, Dylan's recorded vocal performance is intricately shaped in its character, timbral content, and expressive and artistic use of pitch and rhythm. These subtle qualities are fixed within the track and are inherent qualities of its sound and its artistry. These performance-related qualities are noticeably controlled and crafted in ways that can only happen in recorded music.

The barcarolle obbligato passage is a significant presence throughout the track—from its introduction through its final refrain, it only recedes during bridge sections and the contrasting phrase of verses.

This passage carries a strong likeness in rhythm and scale-degree correspondence to the barcarolle "Belle nuit, ô nuit d'amour" from Jacques Offenbach's last opera *The Tales of Hoffman* (1880). The similarities between the two are significant. The two songs have closely aligned melodic content, have the same tempo and meter, and are in the same key.

Figure 10.2 illustrates the similarity between these materials. The central melodic content of Offenbach's barcarolle is reflected in the guitar line and in the hummed background vocals. The backing vocals contain much of the first measure of Offenbach's melody, and the guitar is a variation of the entire line. The guitar line repeats the phrase (and pitches) from the first measure while the Offenbach's melody evolves to be more elaborate. Fragments of his barcarolle appear to be present in hocket between the guitar and backing vocal lines.

A barcarolle is a traditional folksong sung by Venetian gondoliers, or a piece composed in that style. Typified by a lilting or gently rocking compound-triple meter with a relaxed pulse (as exemplified, for instance, in Mendelssohn's two Venetian boat songs, opp. 19, no. 6, and 30, no. 6), it suggests a gondolier's stroke. Barcarolles often have a character of being sentimental or reflective. Just as his early "Girl from the North Country" (1963) reflects the melody, lyrics,

FIGURE 10.2 Juxtaposition of the melody from Jacques Offenbach's "Belle nuit, ô nuit d'amour" with the obbligato guitar and backing vocal lines.

and harmonies of the English ballad "Scarborough Fair," here Dylan appropriates aspects of a musical style or perhaps of another song and incorporates the ideas uniquely into his own voice. This connection to the barcarolle shifts and enriches the cultural context or reference of the track and, at some level, connects it with another culture and another period of time; this provides a sense of not being fixed in this place and time but rather a sense of being timeless in existing in various cultures and time periods simultaneously. Supporting this association with another culture and time, the contrasting (departure) phrase of the verse without the obbligato emphasizes three bars of a D7 chord which is articulated by a finger-picked electric guitar arpeggiated passage that is at times reminiscent of a Venetian mandolin's rapid reiterations needed to sustain its sound. Thus, the obbligato provides a layer embodying a folksong from another culture and another time and weaves it within a torch song.

Structure and Harmony

The structure of "I've Made Up My Mind" contains several unique qualities that distinguish the track and support and enhance its unfolding narrative, tensional movement, and drama.

Using traditional verse and bridge sections, a unique structure unfolds, comprised of six verses and three bridges that are preceded by an introduction. These

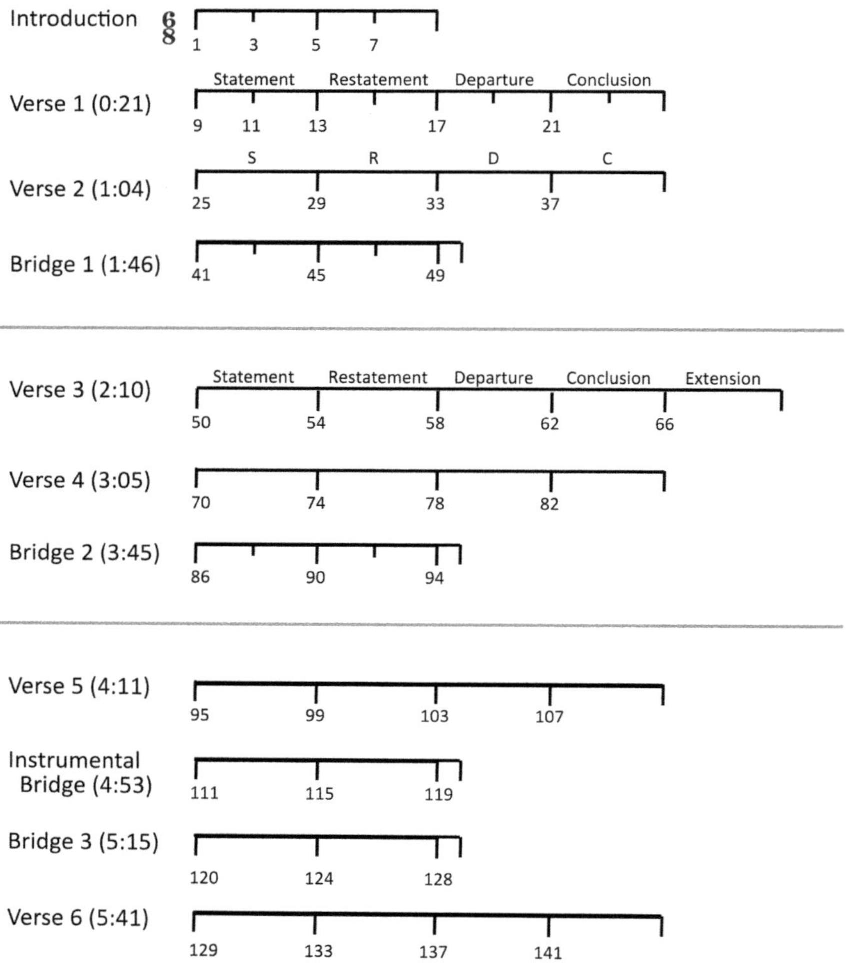

FIGURE 10.3 Structure of "I've Made Up My Mind to Give Myself to You."

sections are combined into three large-dimension groupings, each of which contains two verses and a bridge; though each grouping is also unique.

Figure 10.3 illustrates these three large-dimension groupings and the structural organization of the middle dimension.[5] Here the higher-order interrelationships between these three large-dimension divisions are evident by examining the middle dimension structure.

An eight-measure introduction precedes the first grouping and contains the obbligato accompaniment that underlies the first (statement), second (restatement), and fourth (conclusion) verse phrases.

The first large-dimension grouping presents the first two verses in a traditional sequence followed by the first bridge. This establishes the AAB structure of this large-dimension grouping.

Verses are 16 measures, divided into four-bar phrases in the sequence: aaba'—with the a' being the refrain. This is opposed to the common aabc pattern of Statement–Restatement–Departure–Conclusion (SRDC) structures (Everett 2009, 140–41). This pattern is repeated in each of the track's six verses.

Bridge sections are nine measures long and divided into two four-measure phrases, with the second phrase receiving a one-measure extension. That extra measure pauses movement in the sung bridges and allows the preceding large-dimension grouping to coalesce into a single statement; it functions much like an active fermata. This pattern and the same phrase lengths are present in each of the three bridge sections and the instrumental bridge. These phrase structures are evident in Figure 10.4.

The second large-dimension grouping follows the same verse, verse, and bridge pattern but with an extra four-measure phrase after the third verse. This interpolated phrase contains a statement of the obbligato material; this phrase suspends forward movement, perhaps to allow the protagonist time to muse or space for introspection between verses three and four. Following this added phrase, Dylan's protagonist—for the first time in the track—directs his thoughts to the subjective mood or a future possibility, as "If I had the wings of a snow white dove" begins verse four; almost all the other lines from the lyrics are statements of fact. The four-measure phrase inserted between verses 3 and 4 is the only modification to the AAB large-dimension grouping previously introduced.

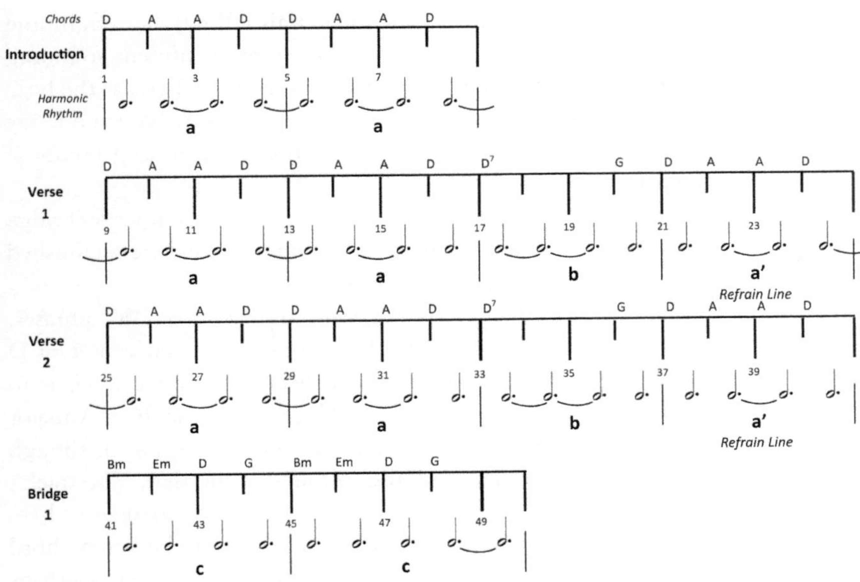

FIGURE 10.4 Chords, harmonic progression, harmonic rhythm, and phrase structure throughout the first large-dimension division of "I've Made Up My Mind to Give Myself to You."

The third large-dimension grouping has a shifted sequence of verses and bridge, and the grouping contains an added instrumental bridge after verse 5. This establishes an AB'BA sequence.

Verse 5 is followed by an instrumental bridge, which provides a new sense of energy for the track; a solo electric guitar located in the center replaces the vocal with a similar line that contributes strongly to the section's building tension and directional motion. The one-measure extension that concludes each bridge is a pause in the sung bridges, but here it builds tension and crescendos into the third sung bridge with strong percussion, bass, and accompaniment guitars all articulating each pulse, and joined by the lead vocal.

The arrival of the third sung bridge is the dramatic apex of the track—its moment of highest performance intensity[6] and dynamic levels. Here Dylan is most unrestrained as he articulates a text that reveals his feelings and the track's message; this moment is the point of arrival for all that preceded and from which all that follows tails. Dylan joins with the bass and percussion on beats five and six leading into the third bridge, adding strength and energy to his declaration, "Well my heart's like a river – a river that sings." This singer reveals himself openly at this peak moment, though it quickly shifts to a withdrawn, subdued detachment "It just takes me a while to realize things." Here, at the moment of highest drama in the track is the interplay of attachment and detachment, assertive and reserved, as affective opposites.

The relocation of the bridge, inserted between the verses, serves another important function: it allows the final verse (verse six) to end the track with its refrain. The ending is stark, no coda or added materials follow; the refrain line is sung definitively to conclude the track, with the accompaniment sounding like it is decaying in the room—or perhaps through time. Even here, as the lyric resolves its story, there is a sense of detachment and separation; as "you and I" say "yes," establishing attachment, the track quickly decays, reestablishing a sense of detachment and separation.

Figure 10.4 illustrates harmony within the first large-dimension verse/bridge grouping. The progressions, harmonic rhythm, and phrase structure established here are consistent in all other verse and bridge sections.

Harmony during the introduction is the same as the verse "a" phrases. Alternating tonic and dominant chords clearly establish the tonal center of D major. Interestingly, the harmonic rhythm from the outset of the track is in syncopation with the phrase structure and the obbligato part—with an A major chord in measures 2 and 3, and D major chord in measures 4 and 5, though the phrasing is in two-bar increments (adding to four-bar phrases). The track's sense of undulation is reinforced by the misalignment of chord changes of harmonic rhythm and the phrase structure; a pulling and waning between chord changes and its tensional motions and phrase structure is established—perhaps like the gentle wake of a gondola and arriving and receding waves against its hull. This pattern of two-measure harmonic rhythm continues until the verse's departure gesture, where the D7 chord (dominant of the subdominant) receives an emphasis of an extra measure and its resolution to G (subdominant) chord a single measure of duration; this progression and shift of harmonic rhythm provides a sense of tension and directed motion for the refrain's arrival. The refrain that concludes each verse reestablishes the harmonic progression and harmonic rhythm of previous "a" phrases.

An abrupt shift of tonality, harmonic rhythm, and harmonic progression occurs within the bridge sections. Bridges are in G major, the subdominant key—which is briefly emphasized in the third (departure) phrase of the verses. The chord progression B minor, E minor, D major, and G major is repeated to occupy the entire eight-measure bridge; these are the track's only minor chords. The harmonic rhythm is one chord per measure—a significant contrast to the two-measure chord pacing of the verses. The more complex harmonic language (iii–vi–V–I in the subdominant key) and faster, steady harmonic rhythm that is now aligned with the phrase structure provide the bridge sections with more directed motion and reinforce a more intense expression compared to the verses.

Recording Elements

Recording elements establish this track's sonic context and contribute substantively to its overall sonic qualities—especially important are holistic environment (Moylan 2020, 337) and timbral balance (ibid., 254). In the middle dimensions these also function to enhance and support the track's expression, ideas, and materials, and further define and differentiate the texture's three independent strata.

Central to the structural organization of "I've Made Up My Mind to Give Myself to You," verses and bridges maintain characteristic and essential spatial (especially sound stage and sound source host environment[7]), timbral, performance intensity, and dynamic qualities that delineate and characterize the sections, even as they undergo development in each reappearance.

The sound stage spanning the introduction and verses 1 and 2 establishes a set of spatial relationships for the voices/instruments of the track as well as the boundaries of the space within which they perform. Analogous to a primary set of a theatrical play, this sound stage provides a space of reference, and it contributes substantively to the track's spatial context and character. Sound source lateral widths and locations and their distance positions establish a basic set of relationships and qualities that subtly shift within successive verses. A contrasting sound stage (another set of spatial relationships) is present in each bridge section, that also gets subtly developed as the track unfolds.

Figure 10.5 is a sound stage diagram of lateral and distance locations of the primary instruments and vocals and identifies changes to the sound stage within bridge sections. The lead vocal is centered and the most prominent image of any sound source, as it is the source at an intimate distance position and nearest the listener. The anchoring ensemble is clustered around the center of the sound stage and spread throughout the furthest half of the personal zone;[8] all

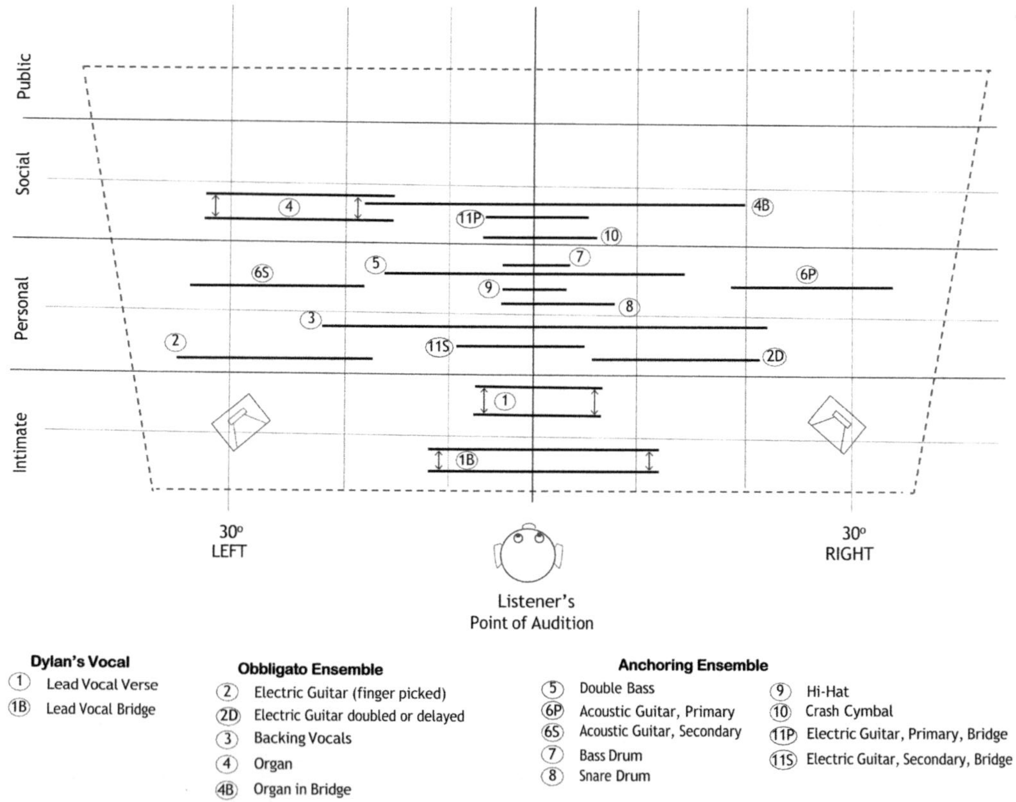

FIGURE 10.5 Sound stage representing placements of instruments and voices, including differences between verses and bridges: stereo image locations and sizes, distance location positions from the listener's centered point of audition.

instrument images have a width of less than 10° except for the double bass; the drum set instruments are distributed within the 15° span around the lateral center point of the sound stage. The exceptions are the double bass placed slightly left of center, and the acoustic guitar is split across the width of the sound stage; the right, secondary position of the acoustic guitar is often at a lower loudness level and often mirrors the primary acoustic guitar on the right side of the sound stage.

The foundational ensemble establishes a sense of stability against which the spatial fluidity of the obbligato contrasts. The obbligato parts provide a sense of motion to the verse sound stage; backing vocals and guitar have split images, where the dominant image on the left is mirrored by a softer and slightly delayed image on the right. This brings a gentle undulation and rocking to the verse sound stage. Obbligato parts are located across the width of the sound stage. Backing vocals pull to center and are wider than Dylan's lead vocal; at times, these hummed vocals pull hard to the left, and their delay on the right is substantially softer, emphasizing a gentle rocking movement. The finger-picked electric guitar is consistently located on the left and is doubled on the right, though much softer and delayed slightly, even in the verses' contrasting "b" (departure) phrase. The obbligato organ appears on the left; it is consistently the sound source furthest from the listener position and is located in the social zone. In contrast, the obbligato electric guitars and backing vocals are located in the front half of the personal zone and are the sources closest to the listener after the lead vocal.

The lead vocal and obbligato parts are subtly varied as later verses progress, with the anchoring ensemble providing consistency. The most notable changes are in verses 5 and 6, where the obbligato images become wider and more stable and seem to support the focused build to the apex of the track and the following waning toward resolution.

The anchoring ensemble retains most of its verse positioning in the bridge sections, though its substance shifts noticeably. The ensemble expands to include new flat-picked electric guitar lines and cymbals, while the finger-picked obbligato guitar and organ (contributing new material) continue into the bridge section in their same or similar positions on the sound stage. Instruments enter and exit unobtrusively at diverse lateral and distance positions while delivering short gestures often in hocket between sources and locations. Their varying sizes and placements contribute to a gentle positional shifting and rolling oscillation in the mix. This allows the bridge to acquire a gentle motion similar to that established by the obbligato parts in the verses. This effect becomes more pronounced in the instrumental and third sung bridges.

The lead vocal has the most significant physical presence of any source within the bridge sound stages, and it is wider than in the verses. It dominates the center of the sound stage not only by its width but also by it being closer to the listener than in the verses. Further, this location is nearest to the listener of any source within the track. During the line "Sometimes it takes me a while to realize things" (5:23), Dylan's voice is the closest and most intimate of any place in the track, and also at a low performance intensity level; while the previous line is in stark contrast as Dylan's performance intensity reaches the highest level of any instrument or voice in the entire track—"Well my heart's like a river, a river that sings" (5:17)—at the start of the final bridge section. This point in the track quickly contrasts the attachment of declaring to the world at a very near distance with the detachment of leaning into the closest level of proximity, only to confide with the listener something fragile and closely held.

The solo guitar of the instrumental bridge is centered, with a width slightly smaller than Dylan's lead vocal that it replaces. The guitar is more distant and softer than a typical solo and is played through the neck pickup, which emphasizes the low frequencies of the strings. The understated melody and its rather relaxed performance intensity do not exhibit the virtuosity or assertiveness often found in solos; rather, its considerable expression is woven with a gentle and simple quality that still fills the bridge with a strong presence that builds and opens to the final sung bridge.

An interesting feature of this track is how loudness and performance intensity of instruments and vocals confuse distance positions and environmental qualities. Nearly all instruments and voices perform at low loudness levels and exhibit restrained performance intensities throughout the track. Confused by the less complex waveforms of low performance levels (which generate fewer partials), the sources often provide an impression of substantial physical distance between them and the listener. In actuality, the sources are located reasonably close to the listener, as illustrated in Figure 10.5. A different sense of distance or space is, however, created by the reserved character of the performances; the withholding of energy and expression of the lines elicits a sense of affective detachment—that the performers (especially the accompanying materials) are somehow more distant in their connection to the listener.

The track's simulated environments play a significant role in shaping several of its key characteristics. The most substantive of these is the intermittent use of tremolo on obbligato instruments. The track's obbligato layer pulsates, as a tremolo is often applied to the organ part, and at times it also appears on the guitar and backing vocals. This adds to the undulation of the lateral axis of the sound stage and also provides the illusion of a peculiar space. The tremolo simulates the slap back echo of a space with highly reflective surfaces and a size that conforms to the time delay of the echoes. Here that space is rather cavernous, perhaps the street-like corridors of a Venetian canal through which a gondola might travel—reinforcing the subtle underlying impression of multiple cultural influences and different moments in time. The tremolos are often subtle, but their presence on the obbligato is significant. This effect might further extend to the verses' departure-phrase guitar arpeggios, that mirror the tremolo effect and rhythm with its rapid articulations that are often reminiscent of a Venetian mandolin or Hungarian cimbalom.

Between verse and bridge sections, Dylan's vocal vacillates between two distinctly different environments. Throughout each verse, the lead vocal is placed in an environment that exhibits a natural early time field with a slight delay before the arrival of the initial reflection, a dense reverberation with closely spaced reflections, and a substantial and noticeable decay (reverberation time). This is the most pronounced and substantive environment of any sound source in the track and has a distinct character. It provides the verse vocal with a greater sense of space that further differentiates it from the other two sound layers. Dylan's bridge environment is smaller; it has a shorter reverberation time, a more pronounced and unique early time field that suggests an unusual space, and a lower proportion of reverberant sound relative to the direct sound of Dylan's voice; this establishes the vocal's timbral detail as heightened beyond what is heard in even the most intimate conversational distances and allows Dylan to lean in closely to the listener's position.

The environment of Dylan's vocal in the bridge and its sound stage placement bring the lead vocal to be more intimately connected to the listener than in the verses, though considerable connection exists in the verses to be sure. The bridge section lyrics often address "you" or are actually speaking to "her"—they speak directly to either the listener or the object of the singer's attention. Bridges have a directness reflected in the sounds of the recording and in the vocal performance, sometimes even in the imperative voice that contrasts significantly with the unfolding verses' conversations "to her" in the singer's head. This is contrary to the common practice for verses to have a closer connection between singer and listener than a chorus or a bridge. Choruses especially assume a more detached position to the listener and speak with a more generalized voice; they often present a more universal message that extends beyond the confines of the track (Moylan 2015, 406). In this track's bridge sections, the lyric is contained within the track and often directed toward the listener.

The track's holistic environment[9] exhibits the qualities of a broad and deep space. The track's three strata occupy much of the lateral width of the sound stage, yet the holistic environment extends considerably beyond the lateral positions of sources; sound sources are diffuse across the lateral axis of the stage. Considerable depth of sound stage is the result of the low performance intensities and reverberant sound of the individual instrumental parts; these bring depth placements to coalesce in the social and public zones, though their timbral detail are often surrounding and filling the personal zone. As illustrated in Figure 10.5, the three layers are localized individually by distance placements:

- The lead vocal resides in the intimate zone where its position shifts within the boundaries identified in both the verse and bridge sections; it is the closest sound at all times,
- The obbligato layer is split between the closest half of the personal zone and the farthest source placement of the track, with the organ in the social zone, and
- The anchoring ensemble occupies the rear half of the personal zone and extends slightly into the front of the social zone.

The holistic environment emphasizes the high low-mid and full mid pitch/frequency ranges[10] of timbral balance (Moylan 2020, 95–97). This establishes a consistently balanced frequency area around the male vocals and guitar parts. The track's timbral balance itself expands and contracts around these ranges as the track progresses; this movement is especially noticeable between verses and bridges.

The holistic environment has a reverb time of slightly less-than two beats in that same high low-mid and full mid pitch/frequency range. This blends the instrumental parts and allows the subtle, unobtrusive undulation of instrument entrances and exists to sway the sound stage (in depth and width) within the gently active environment. The presence of a large space is established, though the sound sources are close to the listener but not brought to encroach upon one's sense of personal or intimate space. Significantly, the holistic environment is exposed at the end of the second bridge (4:06–4:11); there, these described qualities are readily observed in a sparse texture that contains only light cymbals and gentle but quick finger-picked electric guitar iterations.

The track's overall character of reflection, melancholy, and ease is reinforced by the holistic environment;[11] within it is a sense of timelessness and perhaps simultaneous cultures and time periods. The spaciousness of the environment can accommodate the cross-cultural connections and breadth of the protagonist's experiences of age. Despite the diversity of sounds and ideas, the holistic environment allows none to startle or challenge and resolves them all into a unified place.

Literary Form: Apostrophe and the Production of Affect

The recording and musical structures that create the track's sway between attachment and detachment—intimacy and distance—are further supported by several elements of the track's literary composition. The track, like so many of Dylan's (and so much pop lyrics generally), is structured by an apostrophe—the address of another who isn't present and perhaps isn't even a person (like in Percy Bysshe Shelley's poem "Ode to the West Wind" [1820]). Of course, it seems like Dylan addresses a lover, but, as in many of Dylan's tracks over the last 60 years, the addressed lover might well be a higher power, or perhaps a muse, or even death, as in this couplet from the penultimate verse.

> I'll see you at sunrise—I'll see you at dawn.
> I'll lay down beside you, when everyone is gone.

It's impossible to settle such slippages between lover and death, muse and god, and surely part of Dylan's genius is to keep them in play, sliding the "you" in the track from one identity to another. Indeed, the apostrophe is built on a slippage, a contradiction. The power of apostrophe is that the other, the you, isn't actually there. She/he/it is imagined, fantasized, an object of desire constructed by the singer. One is pledging one's attachment to one that is forever detached and at a remove. That experience is at the heart of all apostrophes. Literary critics have noted the way apostrophe often establishes a tension or friction between attachment and detachment, between the intimacy of imagining another person and the gap involved in imagining that other (Johnson 1986, 16–17; Berlant 2011, 25–27).

This movement between attachment and detachment seems essential to the track in a number of ways. It helps us better understand the development of the singer's persona and the production of affect in the track. Various formal elements of repetition and rhyme scheme help to develop the movement. We get to know the speaker well, and he is detached—emotionally detached from others and most things he encounters. In the opening lines, he's far away in thought, if not in body:

> I'm sitting on my terrace, lost in the stars.
> Listening to the sounds of the sad guitars.

And then he is like a shooting star—alone in cold space and looking at nothing:

> My eye is like a shooting star.
> It looks at nothing, neither near nor far.

As the singer moves through the verses, we learn more about his sense of remove. For instance, he travels the road alone—"the long road of despair"—and he "met no other traveler there." In many ways, this depiction of desolate detachment is a blues cliche, a trope of alienated masculinity that frequently turns melodramatic across the history of American literature and song. But the track seems saved from this fate by the intensity of the speaker's declarations of a desire for the opposite of detachment—an attachment and one that makes him exceptionally vulnerable. The refrain is a declaration of the deepest possible form of attachment: "I'm giving myself to you." "Attachment" is perhaps too bland a word for such a vulnerable, self-abnegating decision or desire, although in psychoanalytic discourse, it references the intensity of bonds between mother and child. As used here, attachment is the kind of relationship one might have with a person, idea, muse, or entity that produces meaning and fulfillment. And it does so in Dylan's track: the singer will be able to "preach the gospel – the gospel of love" and the singer's heart will "sing like a river." Attachment promises power, significance, and artistic creativity in "I've Made Up My Mind to Give Myself to You."

The singer approaches this attachment very cautiously. He keeps it at a distance. While the desire for the apostrophic "you" in most pop tracks is one of infatuation and sexual desire, the desire here is the opposite—it's very cerebral and cool: "Been thinking it over and I thought it all through" and "I've made up my *mind*" (italics added). This isn't a track about lustful attachment but rather the most careful attachment. The speaker isn't quite rational in his deliberations—it isn't all thought—"No one ever told me, it's just something I knew"—but whatever process he's gone through, it is cool and collected. Sentence structures, even beyond the refrain, repeat, as do words: "I saw …/I saw …/I don't think …/I made." "My eye is like …," "My heart's like …" All the repetition tends to have the effect of slowing down the emotional process at the heart of the track. This songwriter is being so careful—it's as if he wants to limit himself to a small pool of potential expressions so he can remain in control. The rhymes are also, for the most part, modest and humble. The exceptions are "I am"/"Birmingham" and "San Antone"/"life alone," which come in the first bridge and the first moment that really lilts; it is a moment where careful emotional control is replaced by something that swings up the emotional ladder but is then gathered in. A hotness gathers just beneath the surface—the singer might break out, he might sing like a river or fly like a dove—but he attempts to hold onto his detachment.

And he does, at least until the very end of the track. In the final couplet, the "you" speaks back. The answer given by the you is "yes"—to what question we do not know.

> I knew you'd say yes—I'm saying it too.
> I've made up my mind to give myself to you.

Some connection seems achieved, and the song can come to a formal close. Attachment and detachment seem reconciled, at least for a moment.

But most of the track—its lyrics, its apostrophic form, its recording elements, Dylan's voice—is unresolved in its movement between the poles of detachment and attachment, as we've seen, and this weave constitutes the track's dominant affective structure. We can call this oscillating affective structure melancholy—where melancholy is understood in Freud's terms as an unresolvable relationship to the other that moves between holding on and getting distance (see "Mourning and Melancholia" [1915, 243–58]). This irresolute melancholy is integral to Dylan's style over the last two decades, and he appropriates established conventions in American popular culture in order to further its development. A brief turn to Dylan's use of these conventions can help us understand what is at stake in what we might call Dylan's hard-boiled melancholy.

The Cultural Context of Dylan's Hard-Boiled Melancholy

Since 1997, Dylan's work has often been characterized by this particular type of melancholy in ways that are absent from the jagged modernism of the mid-1960s work or the sense of nostalgia in 1975's *Blood on Tracks* (more mourning for something lost than the ambivalence of melancholy), or the often unambivalent, fire-and-brimstone albums of the born-again period (1979–early 1980s). This affective structure is central to "Love Sick" and "Not Dark Yet" (1997), to "Sugar Baby" (2001), "Forgetful Heart" (2009), and "Soon After Midnight" (2012). It is one of the most important features of Dylan's late style.

This affective structure is nothing new in American culture; it has been one of the 20th century's most popular affective formations, although it has been reshaped in response to shifting forces. The paradoxically tender-hearted tough guy—the hard-bitten but ultimately empathetic figure—who feels intensely but keeps that feeling under restraint is a type of masculine protagonist (frequently, but not always, white) in American cultural life who threads his way through important writers and genres of the 20th century, from Ernest Hemingway to Cormac McCarthy and Walter Mosley, from late westerns (like John Wayne's melancholic ex-Confederate ranger in *The Searchers*) to noir detective novels. This endemic figure in American popular literature and film has also been put to use by popular music stars. Tom Waits, Nick Cave, and Bruce Springsteen (see Katie Kapurch and Jon Marc Smith [2019] on Springsteen's noir in *Darkness at the Edge of Town* [1978]) have all at various points in their careers called upon the hard-boiled aesthetic. Anyone who has read or watched the films of Dashiell Hammett's 1930 novel *The Maltese Falcon* or Raymond Chandler's *The Big Sleep* (1939) knows the type: the detached, nearly libertarian private investigator who stays clean by working outside the law and other entanglements but is nevertheless drawn into emotional attachments which threaten his detachment. The prototypical example is narrated at the end of *The Maltese Falcon* when Sam Spade melancholically chooses his self-reliant code of honor over his love for the novel's femme fatale, with the classic hard-boiled lines of emotional (in)expressivity: "All we've got is that maybe you love me, and maybe I love you" (214) (Dylan riffs off these very lines in "When the Night Comes Falling from the Sky" [1985]).

Dylan has clearly been attracted to these genres of masculine melancholy. As far back as his album *Empire Burlesque* (1985), he was appropriating lines from *The Maltese Falcon* and *Rear Window* (1954), and he was carped at by some critics upon the release of *Love & Theft* (2001) for lifting lines from Jun'ichi Saga's *Confession of a Yakuza* (1995), a Japanese gangster narrative (translated to English) which calls upon American notions of the hardened but not unsentimental outlaw. Dylan's turn to Sinatra in the mid-2010s reiterates the importance of noir hard-boiled detective films—a genre to which Sinatra contributed both acting and singing performances. Dylan's radio show—*Theme Time Radio Hour* (2006–2009)—is often infused with a hard-boiled aesthetic, and the 2015 video that Dylan starred in for his cover of "The Night We Called It a Day" (originally made famous by Sinatra) is fashioned after noir detective films of the 1940s. In his 2004 memoir, *Chronicles*, Dylan speaks of himself as a kind of conversion artist—"Opportunities may come along for you to convert something—something that exists into something that didn't yet" (2004, 51)—and in his late work among his favorite objects for conversion into song are the genres of masculine melancholy.

A helpful account of what is at stake in the affective structure we have been outlining is provided by Leonard Cassuto's *Hard-Boiled Sentimentality: The Secret History of American Crime Stories* (2009). Cassuto captures an aspect of many 20th-century crime novels and films that at first seems counterintuitive: the hard-boiled detective—cool, hyper-rational, cynical about love, shut off—is also a man of feeling. Cassuto even traces the hard-boiled genre to the 19th-century genre of sentimentalism: "Inside every crime story is a sentimental narrative that's trying to get out" (7). In the 19th century, popular authors like Susan Warner and Harriet Beecher Stowe imagined a society based less on reason, laws, and traditional forms of politics (structures which upheld institutions of patriarchy and slavery in the 19th century) and more on emotional attachments and affinities, which were grounded in the private individual. By focusing on sympathy and love, the 19th-century texts posited, the world could be redeemed. In Cassuto's view, the 20th-century's hard-boiled genre develops a similar view, but one that moves between the detachment of rationality and attachment of fellow feeling. Its protagonists detach from a society shaped by the state, by the capitalist markets, and by other forms of abstract and corrupting power but nevertheless imagine a connection to the community through affective affinity. The genre encourages an affective structure which is both hard-boiled rationality and intense attachment. Hard-boiled heroes are melancholics. Most of the time, as Cassuto points out, the genre's characters and plots never achieve any kind of redemption or even fully believe that redemption is possible (as did the 19th-century writers). But this kind of structure of feeling—an ambiguous and ambivalent position—seems to be all that is possible in the modern world. What we have described here as the sway in Dylan's work between attachment and detachment—what we've described, thinking of Freud as unresolved melancholy—is aligned with Cassuto's "hardboiled sentimentality." Over the past couple of decades, Dylan has gravitated toward the figure of the hardboiled-but-sentimental PI from Chandler and others as a way of developing not just a popular, commodified persona or as a way of providing a generic reference point for his songwriting but as a way of articulating this paradoxical affective structure, this paradoxical relationship with the modern world.

We might understand this affective structure in personal terms: the recognition that perhaps comes with experience that our relationships with others are often melancholic, often simultaneously characterized by attachments and detachments, intimacies, and distances. The drama in Dylan's track is the singer's struggle to accept this fact of our relations with others. But we might also understand hard-boiled melancholy in less personal terms and as a broader structure of feeling that is about our relationship with institutions and ideologies. We live in a world that values the detached rationality of markets and the impersonality of institutions in which we must participate, at the same time that we long for a feeling of communal attachment. This conflicted situation can lead to a dangerous, even life-threatening melancholy, and it is this final note that we are left with in many works of hard-boiled sentimental film and fiction. It is what makes them so noirish. But this common final note isn't the case with "I've Made Up My Mind to Give Myself to You." In the final couplet of the track, the addressed "you" says "yes," as we have seen. There is an optimism at the lyric's closure that is absent from almost all noir works. This optimism doesn't exactly suggest that attachment can win out over detachment. Rather "I've Made Up Mind to Give Myself to You" offers us the opportunity as listeners to participate in an affective structure that is ambiguous and ambivalent—that moves between very different poles—but doesn't become schizophrenic. In this position, Dylan seems to suggest, we might hold onto both poles at once.

Notes

1 Statement–Restatement–Departure–Conclusion (SRDC) structure (Everett 2009, 140–41).
2 Here Blake Mills shares his performance of the obbligato passage including the departure phrase: https://www.youtube.com/watch?v=ae77IGXOYnA (accessed January 24, 2022); https://www.instagram.com/p/CBrs_SaHwbo/ (accessed January 31, 2022).
3 Qualities of vocal performance as described in Moylan (2020, 439–41) will be explored.
4 The rhythms of Dylan's transcribed vocal line throughout the track appear as Webpage Figure 10.1. While illustrating the performed rhythms of Bob Dylan's vocal line, the transcription also includes the lyrics as they are performed, which at times contrasts with the version published on Dylan's official website.
5 The division of structure into three dimensions (large, middle, and small) is related to perspective (Moylan 2020, 54–56). "Perspective" is one of the guiding principles of the analysis framework in Moylan's *Recording Analysis: How the Record Shapes the Song.*
6 Performance intensity is the timbre/sound quality of the instrument or voice created by the loudness and expression of the performance as it is tracked. It is the dynamic level at which the sound source was performing when it was recorded. Performance intensity is comprised of the loudness, energy exerted, performance technique, and expressive qualities of the performance, that manifests as a particular timbral quality of the sound source. See Moylan (2015, 45–47, 2020, 91–94).
7 See Moylan (2020, 323–53) for discussion of sound stage and host environments.
8 See Moylan (2020, 313–20) for explanation of distance zones and the qualities of sources within each zone.
9 "The holistic environment is the space in which the track, in its entirety, resides. It is the environment of the sound stage; it contains all sound sources fused with their host environments. The holistic environment provides a sense of place for the track. … It also presents other significant attributes and connections of space and place; it establishes an aesthetic context for the track, and provides its character (including ambience and spaciousness). The holistic environment contributes substantively to the unique world of the track" (Moylan 2020, 337).
10 Pitch/frequency registers (Moylan 2020, 245–47) are specific divisions of the hearing range that were identified as related to frequency placement in the basilar membrane. The registers represent pitch/frequency ranges that can be used to identify pitches and frequencies (either at generalized or specific levels) throughout the hearing range.
11 These qualities of reflection and melancholy are reflected in the track's reference dynamic level of a moderately high mezzo-piano. "Reference dynamic level (RDL) is the overall or global intensity level of a piece of music [track] and is the reference level for evaluating dynamics for musical balance and program dynamic contour; it is the 'perceived performance intensity' of the work as a whole, conceptualized as a single entity out of time …" (Moylan 2015, 453).

References

Berlant, Lauren Gail. 2011. *Cruel Optimism.* Durham: Duke University Press.

Cassuto, Leonard. 2009. *Hard-Boiled Sentimentality: The Secret History of American Crime Stories.* New York: Columbia University Press.

Covach, John. 2018. "Analyzing Texture in Rock Music: Stratification, Coordination, Position, and Perspective." *Pop weiter denken: Neue Anstöße aus Jazz Studies, Philosophie, Musiktheorie und Geschichte, Beiträge zur Popularmusikforschung 44*, edited by Ralf von Appen and André Doehring, 53–72. Bielefeld: Transcript Verlag.

Dylan, Bob. 2004. *Chronicles*. New York: Simon & Schuster.

———. 2015. "The Night We Called It a Day." Accessed July 16, 2021. https://www.youtube.com/watch?v=iOxy_hy22CA

———. 2020a. "I've Made Up My Mind to Give Myself to You." *Rough and Rowdy Ways*. CD. New York: Columbia. 19439780982.

———. 2020b. "I've Made Up My Mind to Give Myself to You" (lyrics). The Official Bob Dylan Site. Accessed July 16, 2021. https://www.bobdylan.com/songs/ive-made-up-my-mind-to-give-myself-to-you/

———. 2020c. *Rough and Rowdy Ways*. CD. New York: Columbia.

Everett, Walter. 2009. *The Foundations of Rock: From "Blue Suede Shoes" to "Suite: Judy Blue Eyes."* New York: Oxford.

Freud, Sigmund. 1915. "Mourning and Melancholia." *The Standard Edition of the Complete Psychological Works of Sigmund Freud. Volume XIV (1914–1916): On the History of the Psycho-Analytic Movement, Papers on Metapsychology and Other Works*. 1957. Vol. XIV, 243–58. London: The Hogarth Press.

Johnson, Barbara. 1986. "Apostrophe, Animation, and Abortion." *Diacritics,* vol. 16, no. 1, pp. 29–47.

Kapurch, Katie, and Jon Marc Smith. 2019. "A Fear So Real: Film Noir's Fallen Man in Bruce Springsteen's *Darkness on the Edge of Town* and the David Lynch Oeuvre." *Interdisciplinary Literary Studies*, vol. 21, no. 1, pp. 89–108.

Mills, Blake. 2020a. "I've Made Up My Mind to Give Myself to You" guitar obbligato. Accessed January 24, 2022. https://www.youtube.com/watch?v=ae77IGXOYnA

———. 2020b. "I've Made Up My Mind to Give Myself to You" guitar obbligato. Accessed January 31, 2022. https://www.instagram.com/p/CBrs_SaHwbo/

Moylan, William. 2015. *Understanding and Crafting the Mix: The Art of Recording*, 3rd ed. New York: Focal Press.

———. 2020. *Recording Analysis: How the Record Shapes the Song*. New York: Focal Press/Routledge.

Saga, Jun'ichi. 1995. *Confessions of a Yakuza: A Life in Japan's Underworld*. 1st pbk. ed. Tokyo; New York: Kodansha.

11

"YOU NEVER GIVE ME YOUR MONEY"

The *Abbey Road* Medley (1969)

Albin Zak

Abbey Road marked the end of the Beatles' illustrious recording career. The band on the leading edge of developments in 1960s record production was on the verge of collapse and this album would be their final project together. Emerging from the shambolic "Get Back" sessions that would eventually yield the *Let It Be* album, the Beatles determined to make a farewell record worthy of their legacy. Reunited with engineer Geoff Emerick and producer George Martin, the band entered the studio with a renewed, if temporary, sense of purpose.

In keeping with the band's typical forward-looking approach, *Abbey Road* saw several technological and musical innovations. In terms of gear, the introduction of the solid-state TG12345 desk in Studio 2 brought a crisp new sound to the recording and mixing processes, as well as a new range of mixing possibilities. Engineer Geoff Emerick has claimed that, in comparison to the tube-based REDD desks the recording team had used previously, the "softer and rounder" sound of the TG12345, heard by the musicians through their headphone cues, "overtly influenced" the band's overdubbed performances (Emerick and Massey 2006, 277). How significant this, as Emerick admits, "subtle" change may have been is difficult to assess. As Ryan and Kehew observe, tube-based gear was also employed extensively during tracking (Ryan and Kehew 2006, 514). At times, the venerable REDD 5.1 desk in Studio 3 was pressed into service, and the microphones and outboard gear were the same as those used on previous albums.

But the new desk certainly had an effect on the mixes, including tracking mixes where multiple microphones were mixed down to one track (e.g., drums). With a greater selection of EQ frequencies, compression and limiting on every microphone input, and continuous panning on every channel, the desk provided Abbey Road engineers with an unprecedented ability to fine-tune individual sounds and blend them in the stereo mix. And even if sounds were first captured by tube-based microphones and preamps, and perhaps processed with a trusty

DOI: 10.4324/9781003089926-16

Fairchild or Altec compressor, all ultimately passed through the TG12345's solid-state signal path for mixdown.

The musical side of the project also saw a new kind of Beatles composition: a 16-minute through-composed track with remarkable formal coherence throughout the track's various parameters. The Beatles had flirted with the concept of large-scale design on *Sgt. Pepper's Lonely Heart Club Band*—its loose suggestion of an overarching narrative supported by the omission of gaps between tracks, including a couple of actual crossfades—and the White Album, again with a variety of hard edits and crossfades between tracks. But these were simply techniques for moving between songs in a way that would impart to the album a sense of continuous flow. The songs themselves were self-contained. Conversely, the track that consumes most of the second side of *Abbey Road* is made up of song fragments, all parts of a larger whole.

In an April 1969 interview, John Lennon described a feverish period of songwriting. "I probably average about twelve a night. Paul too," he said. "As soon as I leave here I'm going around to Paul's place and start work." He went on to say, "Paul and I are now working on a kind of song montage that we might do as one piece on one side" (Smith 1969, 3). Lennon was hinting at what would become the so-called medley on the second side of *Abbey Road*. The term "medley" (working title: "The Long One/Huge Medley") has stuck as a term of convenience, but as Thomas MacFarlane has argued, this sequence of eight songs—beginning with "You Never Give Me Your Money" and concluding with "The End"—is not in fact a medley in the conventional sense. Rather, it is an "extended form" that is "cyclic" in nature (MacFarlane 2007, 5). Indeed, motivic connections among elements of musical syntax, timbre, instrumentation, performance style, lyrics, and stereo image act as narrative devices in an inclusive thematic process. Lennon's characterization of the project as a "montage" tells us that the composers envisioned not simply a series of tunes, but rather an extended narrative composed of related musical events.

In the spirit of the comprehensive analytic methodology specified in William Moylan's *Recording Analysis: How the Record Shapes the Song* (Moylan, 2020), this essay takes a look at the interactions among the constellation of elements that shape the *Abbey Road* medley. As we shall see, the track's compositional development merges issues of tonality, timbre, musical motive, performance style and character, lyrics, arrangement, and stereo choreography through the course of an overarching narrative design.

Tonal Overview

The medley's arrangement falls into three sections. The formal divisions are suggested by brief pauses in musical momentum (songs within sections segue without pause), and by key. Table 11.1 illustrates the key(s) and length of each song.

Among the medley's central topics is a contrast between two keys: A and C. The opening song, "You Never Give Me Your Money," which has a through-composed

TABLE 11.1 Formal organization of the medley

Formal section	*Song title*	*Tonal centers*	*Length*
Section 1	"You Never Give Me Your Money"	A minor, C, A major	[4:02]
[fade]			
Section 2	"Sun King"	E, C, E	[2:26]
	"Mean Mr. Mustard"	E	[1:06]
	"Polythene Pam"	E	[1:12]
	"She Came in Through the Bathroom Window"	A	[1:58]
[pause]			
Section 3	"Golden Slumbers"	C	[1:31]
	"Carry That Weight"	C, A minor, C	[1:36]
	"The End"	A, C	[2:21]

form in four parts, presents the conceit from the outset. The song's opening phrase begins with a non-committal A minor seventh chord, kicking off a harmonic phrase that leads within four bars to the clarity of a C major tonic (0:09).[1] But no sooner is C established than the phrase moves on to a cadence in A minor two bars later (0:18). The move to C and then to A minor imbues the song's opening statement with a dualistic tonal ambivalence. The contrast, however, is a gentle one. A minor and C major, after all, share a strong family resemblance. As the song moves to C for its next two segments, the A minor opening seems to have been little more than a moody atmospheric moment.

The sharper contrast comes with the displacement of the pitch C by C♯, the third scale degree of A major, which arrives in the last part of "You Never Give Me Your Money" ("one sweet dream"). But though this concluding segment introduces C♯, its first eight bars include many chords that hark back to C major (C, D minor, G, A minor; 2:33–2:44). And it ends with a looping C–G/B–A harmonic progression that repeatedly juxtaposes C and C♯ (2:48 ff.). Though downgraded from key to chord, C lingers. Ambivalence persists, but now with two strong tonal characters testing one another.

The second large section of the medley, beginning with "Sun King," contains three songs in E, a key that turns out to have a structural association with A (I'll make general reference to songs in E or A as "A music"). As Walter Everett points out, the songs in E "work as a large dominant" arriving emphatically in A for "She Came in Through the Bathroom Window" (Everett 1999, 267). Even as the medley moves to the A side of the A–C dualism, however, "Sun King" returns to C for its 18-bar B section ("here comes …"). Thus, "Sun King," rather than a sudden tonal departure, can be heard as a transitional song slowing down the medley's move away from the C music. The residue of C also lingers in the form of a cadential chord in the songs that follow: "Mean Mr. Mustard" (0:27) and "Polythene Pam" (0:14). Then, in another transition setting up the return to C in "Golden Slumbers," the *key* of C makes a brief appearance at the end of the chorus of "She Came In through the Bathroom Window," only to be immediately usurped by A (0:48).

In "Golden Slumbers" and "Carry That Weight," as bits of "You Never Give Me Your Money" filter back into the narrative, the medley settles back in C major, which now asserts more tonal control than before. The A section of "Golden Slumbers" begins with a familiar sounding A minor seventh chord and reiterates the harmonic progression of the first four bars of "You Never Give Me Your Money," though in a more leisurely harmonic rhythm and in support of a new melody. (Each of the first three chords now takes up two bars instead of one, slowing down the phrasing.) Then, after the initial move to C ("to get back *home*"), as the harmonic momentum picks up, we return quickly to A minor, as in the earlier song. But this time A minor is only an interim destination. A new phrase extension, absent in "You Never Give Me Your Money," leads to a cadence in C ("I will sing a lullaby"), which concludes the A section leaving none of the previous tonal ambiguity. Another brief move to A minor during the song's B section ("sleep pretty darling") likewise gives way to a gentle C major cadence.

In the following song, "Carry that Weight," the bridge (0:24–1:03) presents two restatements of the melody and harmonic progression from the first part of "You Never Give Me Your Money" (one instrumental, one vocal), and once again we hear an A minor inflection. In its bold, brassy restatement of previous material, this is an important thematic moment in the medley's overall form. But though A minor now sounds more powerfully than ever, its presence is fleeting. In fact, surrounded by the song's exuberant C major outer sections, the bridge confirms an impression that A minor belongs on the C side of the A–C dualism.

More intriguing tonally is the coda for "Carry That Weight" (1:26 ff.), which restates the descending C–G/B–A progression from the coda of "You Never Give Me Your Money." Initially, this feels like a kind of denouement. We are left once again with a faint tonal ambivalence, but whereas earlier the A chord signaled the medley's departure from C major, it now seems only a fading chromatic inflection in what has become a C major landscape. As we soon learn, however, A major is not finished. In a spirited effort—via a rhythmically urgent modulating transition, with McCartney yelling "oh yeah, all right!"—it returns for "The End," where the two keys meet for a final reckoning. Lennon, McCartney, and Harrison's three-guitar cutting contest unfolds in a bluesy A major over a driving groove set up by Ringo's drum solo. Along with "Polythene Pam," this is the medley at its most propulsive. Yet the energy ultimately drives the music back to a now magisterial C major for the medley's final cadence.

Tonal Character

When analyzing the recording, it is impossible to think of musical syntax elements like keys, apart from their aural presentation, which imparts to the key a palpable character. I hear the C music variously: introspective ("you never give me …"), nostalgic ("golden slumbers …" or the rollicking 1950s-style piano accompaniment to "out of college …"), anthemic ("boy, you're going to carry that weight"), transcendent ("the love you make"). By contrast, the A music has an edgier rock and roll feel—"Polythene Pam," "Bathroom Window," "The End." We arrive in A initially (in part four of "You Never Give Me Your Money") through a wrenching transition (2:12–2:28). And when the key returns for "The End," after C major has been reestablished in "Golden Slumbers" and "Carry That Weight," it arrives through a similarly energetic transition (0:00–0:19). It is as if some special effort is required to reach the key and unleash its energy.

The keys are also associated with instrumentation. The C music features piano or brass or string sonorities. One of the markers of the C major return in "Golden Slumbers" is the prominent reappearance of the piano, which is featured in "You Never Give Me Your Money," but recedes to a supporting presence through Section Two. By contrast, the energy of the A music is heightened by the sound of electric guitars (and, in the case of "Polythene Pam," a heavily compressed acoustic). The first prominent electric guitar entrance comes in part three of "You Never Give Me Your Money" (1:32), where the music is still in C major. Its chiming quality marks a departure from the mellow low-register tone of the guitar riffs in part one (0:09, 0:17). This timbral change previews the transition to the A music. Timbre and tonality subsequently converge in the modulating passage between parts three and four (2:12–2:28), where the guitar drives the harmonic progression that moves away from C and lands us in A.

With this in mind, think again of "Sun King," which illustrates a complementary connection between tonality and sound in the shaping of musical form. Because the song's key of E is associated with A, it would seem to demand a lively setting, like "Mr. Mustard" and "Polythene Pam," the other songs in E. But "Sun King" is languid. Its dreamy texture slows down the transition to the A music. Although electric guitars feature prominently, their mellow tones recall the sounds of the riffs in part one of "You Never Give Me Your Money." The drum part de-emphasizes the backbeat and any other sharp attacks, featuring instead a slightly muffled low end with a percolating kick drum filling most of the space. Just as C major still exerts enough tonal influence (in the bridge of "Sun King") to soften the move to the A music, the instrumental texture eases the track's timbral transition. With the arrival of "Mean Mr. Mustard"—where C fades to a mere passing chord—the tempo increases, the backbeat returns doubled by a guitar chop, the bass buzzes to life with a pronounced distortion effect, the voice shakes off its drowsiness—in short, the A music, tonally and timbrally, has finally arrived in full.

Motivic Connections

As we have seen, Part Three of the medley restates harmonic, melodic, and tonal material from "You Never Give Me Your Money." There is, however, more to the recapitulatory effect. When the medley's opening harmonic phrase returns

FIGURE 11.1 Guitar figure of "You Never Give Me Your Money," transformed later in "She Came in Through the Bathroom Window."

in the A section of "Golden Slumbers," for instance, it is presented by the same instrument and in a similar rhythm: a piano playing steady eighth notes. The chord voicings are different, but the chords inhabit the same register as the previous iteration; the overall sound is familiar. Similarly, though the vocal melody and lyrics are new, the introspective tone of the singing reminds us of the medley's opening phrases. The sense of return, then, is evoked in instrumentation and performance, as well as musical syntax.

There are several such motivic amalgamations in "You Never Give Me Your Money," which are developed in the course of the medley. In terms of thematic significance, they vary from trivial to essential, but all contribute to the piece's overall cohesion. The guitar figure that drives the transition to A major (Figure 11.1), for instance, makes a fleeting reappearance in the last bar of "She Came in Through the Bathroom Window" (1:51). This recurrence seems little more than a distant echo of the earlier track. On the other hand, the guitar arpeggios in the opening song's coda (2:48 ff.), which outline the C–G/B–A chord progression and its harmonic rhythm, play a prominent recapitulatory role at the end of "Carry That Weight" (1:26 ff.), making it one of the piece's clear signs of a cyclic conception (see Figure 11.2).

These are instances of motivic repetition. There are also motivic transformations that connect sections in less obvious ways. For example, the three-chord descending fourth progression B♭–F–C in the third part of "You Never Give me your Money" ("oh that magic feeling") turns up in "Polythene Pam" transposed to D–A–E. The mood of the two, however, is dramatically different, befitting the motive's relocation from C to A music. The latter is faster and altogether more urgent, its vocal character, instrumental texture, and rhythmic drive more aggressive. And as a foundation for the guitar solo in the song's coda (0:46 ff.) it takes on a blues feeling previously only hinted at.

FIGURE 11.2 Guitar arpeggios in coda of "You Never Give Me Your Money."

FIGURE 11.3 Guitar figure from part 3 of "You Never Give Me Your Money."

This third part of "You Never Give me your Money" also introduces a prominent guitar figure: descending eight note arpeggio fragments outlining the chord progression and emphasizing the dotted quarter rhythm played by the bass (Figure 11.3).

This motive begins its development in the song's coda over the C–G/B–A bassline (Figure 11.2), reversing direction and expanding from fragment to continuous accompaniment. The syncopated feel of the eighth note that follows the dotted quarter is heightened by forcing a chord change on the "and" of the second beat, creating a sense of metric disruption. At the track's moderate tempo, this syncopated harmonic rhythm feels like three beats cutting across the duple groove.

A similar three-over-four phrasing recurs in augmentation—filling the space of a dotted half note—in the vocal melody of "Mean Mr. Mustard" (Figure 11.4).

In this case, the harmonic rhythm is not affected initially. But in the song's outro ("dirty old man," 0:59 ff.), the three-beat phrasing is compressed once again to the dotted quarter level (which is emphasized by switching the groove to 6/8), with chords changing on the dotted quarter pulse. This metric modulation also determines the tempo of the next song, as the dotted quarter becomes the half note for "Polythene Pam."

"Pam" raises the syncopated energy of the dotted quarter harmonic rhythm to a new level of intensity, but instead of chord changes on the "and" of two, they happen on the "and" of four. The figure appears first during the verse—on bass and acoustic guitar each time the chord progression moves to the tonic E—and then throughout the looping chord progression during the guitar solo.

A dotted quarter rhythm also supports the groove during the guitar solos in "The End" (0:34–1:29). Here, the dotted quarter moves back to the downbeat,

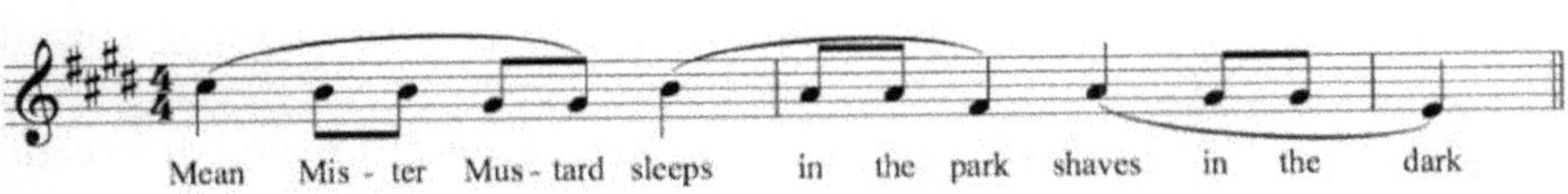

FIGURE 11.4 Vocal melody of "Mean Mr. Mustard."

played on rhythm guitars doubled by voices ("love you"). It also drops the syncopated harmonic rhythm, making it a faster version of the accompaniment for "oh, that magic feeling," a subtle link between the medley's first and last songs. After the guitar solos, the dotted quarter is repurposed for the song's vocal finale. It starts as a melodic figure for the words "*and in* (the end)" and "*love you* (take)" (1:33–1:39), which sets up the dotted quarter metric slowdown ("equal to the" [1:41–1:44], which, by the way, is similar to the one in "Mr. Mustard") leading to the medley's final cadence.

The syncopated effect of the dotted quarter recurs in many guises through the course of the medley. It is first heard in the second bar of "You Never Give Me Your Money," in the right hand of the piano (0:03), mimicking the vocal melody that follows ("mo-ney"). It moves to the 16th note in the song's second segment (1:10–1:30), leaving each phrase ending on the last 16th of a beat ("spent," "rent," "go," "sack," "back"). It highlights the C to A chord change at the end of the chorus in "Bathroom Window" (0:47, again in the form of a guitar arpeggio), and is further emphasized by the voice in that song's final gesture ("oh yeah"). Moreover, the melodies of "Golden Slumbers" and "Carry That Weight" repeatedly end phrases between beats two and three or four and one. On the one hand, these are stylistic habits. But over and above style, the importance of the syncopated figure in various contexts throughout the medley gives it a substantive thematic significance.

Vocal Character

Among the medley's dramatic devices are its vocal characters, whose tone and performance style change from one song to the next, or between sections of a single song. The dreamy choral sound of "Sun King," for instance, is followed by Lennon's droll description of Mr. Mustard, and then his heavily Scouse-accented story of Polythene Pam. The hearty group chanting of "Carry That Weight" is followed by the transcendent choral finale of "The End." In the two sections of "Golden Slumbers" McCartney uses markedly different tones—one soft and earnest, his "Yesterday" voice; the other a near growl, leaning toward his Little Richard side.

This sort of character play begins almost immediately in "You Never Give Me Your Money," setting in motion another of the medley's narrative techniques. By the song's second phrase, McCartney's vocal tone has changed. Suddenly, the dry solo voice is doubled ("funny papers"). When this opening melody repeats ("I never … number"), the voice is harmonized, ending with an ambient bloom on the final word ("break *down*"). In the song's next section ("out of college"), McCartney adopts a new, more incisive vocal character that is again dry. It is also EQ'd so as to give it an aggressive edge, the medley's earliest sign of a move away from the C music. McCartney will not return to the sweetly introspective persona of the opening phrase until the beginning of "Golden Slumbers."

Vocal tone also connects the medley with the preceding song on the album, "Because," which some consider a kind of prelude to the medley proper (Gauldin 1990; MacFarlane 2007, 39–58). The hanging D diminished chord with which "Because" ends does resolve nicely on the A minor chord that begins the medley, providing a sense of tonal continuity. But there is also a clear aural connection between the choral voices of "Because" and those of "Sun King" that feels like a thematic affinity.

All of these vocal sounds, of course, are tied to the words they express. The medley's song lyrics, though a collection of fragments of varying provenance, seem to tell a story. Because song lyrics impart a significance beyond their words' everyday meaning, widely reported anecdotes about a song's literal origins are largely beside the point. What matters is that the words support a narrative that makes "song sense," a satisfying flow of vocal sounds imparting images and characters that feel authentic, if fanciful. One might imagine that Polythene Pam is Mr. Mustard's "sister Pam" who "works in a shop." (A connection that would not occur if she were called "Shirley," as on the Esher demo version of "Mustard.") Perhaps she's the "she" who "came in through the bathroom window." And perhaps it means something that "she could steal, but she could not rob"—at least as much as the linguistic mashup in "Sun King." And maybe the wistful character in "Golden Slumbers" musing how "once there was a way" to return home is the same one who, in the same voice, sang "I break down" in "You Never Give Me Your Money." Or perhaps all of this leans too much to the literal and makes for a too-prosaic storyline. Maybe I'd rather leave things fragmented, a collage of encounters with unrelated personalities, a celebration of randomness. That works, too.

What we do know is that Mustard and Pam and the club dancer in "Bathroom Window" invite us into a whimsical, mildly deviant world framed by the medley's outer sections. And that their songs inhabit their own key and are enacted in voices with a distinct style and sonority. And that when "Golden Slumbers" begins, we recognize the sound of a familiar character returning to the story. We know all this without thinking about literal meaning. Whatever is going on with the enigmatic words, we have a clear sense that we are being carried forward.

There is one instance, however, where lyrics participate tangibly in the medley's cyclic form. When the "You Never Give Me Your Money" music returns as the bridge of "Carry That Weight," the lines begin in the same way as before—"you never"—but they have different endings. Each ending, however, follows the same rhyme scheme as in the earlier song: "situation" and "investigation" are answered five songs later by "invitation" and "celebration." Along with the musical reprise in this section, the lyrics circle back to pick up a familiar thread.

Stereo Staging

Mono, which dominated the pop market in the 1960s, had been the Beatles' primary format through most of their recording career. But with the White Album set for a stereo-only release in the US market, mixers devoted a new level of

TABLE 11.2 Stereo positions of the voice in "You Never Give Me Your Money"

Timing	*Left*	*Center-left*	*Center*	*Center-right*	*Right*
0:23–1:06			C		
1:10–1:18					R
1:21–1:29				C-R	
1:32–1:38			C		
1:40–1:47		C-L			

attention to the stereo image. *Abbey Road* took things a step further; the album would be released worldwide exclusively in stereo. With only eight tracks of tape to work with, instruments continued to be tracked in mono and mono reduction mixes were the rule (notable exceptions include the voices in "Sun King" and the drum solo in "The End"). Nevertheless, treating the mono sources like characters on a stage, the mixing team incorporated stereo choreography into the creative process.[2] The medley's stereo stage is dynamic, its characters changing position as the track progresses. Although many of the moves happen from moment to moment, at times they exhibit a more long-range unfolding. When, for example, that familiar piano part returns at the beginning of "Golden Slumbers," the attentive listener will notice that something about the sound is different. It is now on the right side, opposite of where it was in "You Never Give Me your Money." Having passed through the A music, the track has experienced a spatial realignment.

As it does with other thematic elements, "You Never Give Me Your Money" calls attention to stereo choreography early on. We are alerted to the dynamism of the stereo spectrum, as well as connections between stereo panning and phrase structure, by the track's central character: McCartney's voice. In the first part of the song, his voice sits at the center of the mix (0:23–1:06). In the second part ("out of college"), it moves to the far-right side for a phrase (1:10–1:18), then to the center-right ("any jobber") (1:21–1:29). As the song moves to its third part ("oh that magic feeling") the voice is back in the center for the first phrase (1:32–1:38), then moves to center-left (1:40–1:47), leaving room for the harmony voices ("ah") to claim the center. Table 11.2 illustrates this vocal placement analysis.

The end of the song sees all voices and instruments gradually drift to the right as the track fades, as if exiting the stage. All that remain on the left side are the sparse wind chime sounds, which prepare the space for the entrance of a soft cymbal roll and electric guitar riff on the left as the medley moves to the next song.

This new guitar riff begins as an upbeat. As it reaches the first downbeat of "Sun King," the bass and kick drum enter, also on the left. Almost immediately the image begins to split as the guitar and drums move right across the stage, leaving the bass behind. Reaching the right side—the drums far right, countering the bass on the left, the guitar center right—they stop, and the guitar hands the baton to a second guitar, which begins moving back across to the left while the first guitar and the drums remain stationary. The second guitar continues to float back and forth across the stereo field, imparting a wooziness to the track and setting the scene for the dreamy vocals that follow ("ah …"). The voices are on two tracks, allowing for a true stereo spread across the track's center, supported by a wash of guitars and organ. The sound is richly psychedelic, yet the track's foundational elements—the bass and drums—hold the amorphous texture within a clearly bounded left-right frame. (Figure 11.5 maps the spatial framing of image placement and size for the instruments that are foundational rhythmic elements as the medley unfolds from "Sun King" through "The End.")

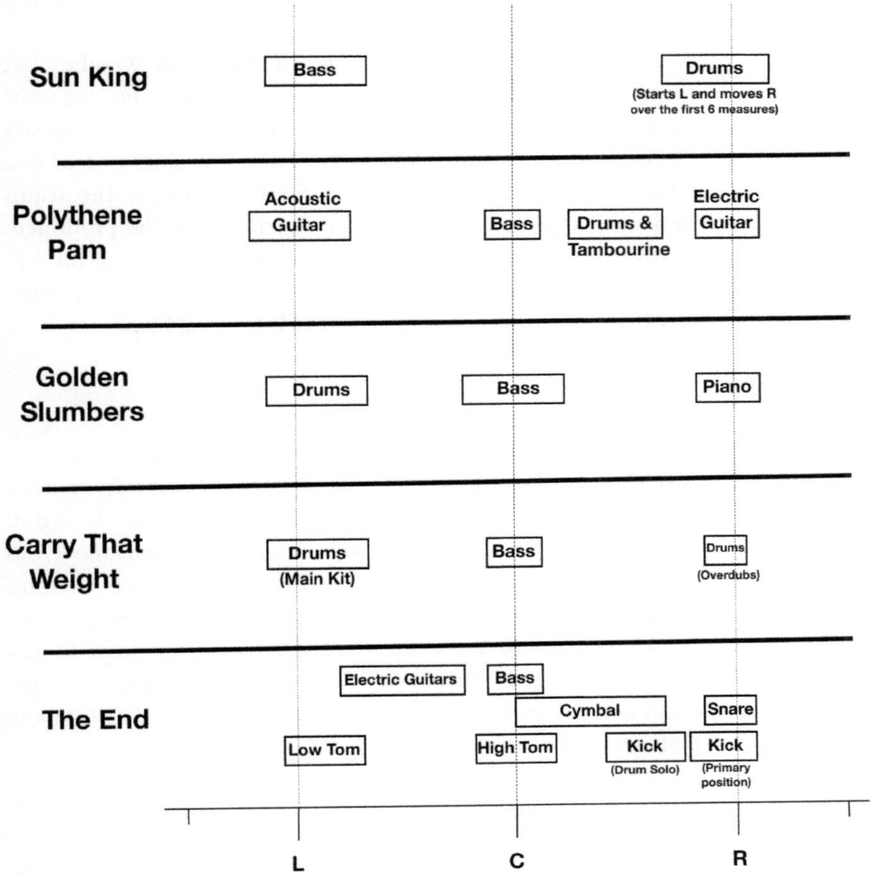

FIGURE 11.5 Stereo framing of foundational rhythmic elements.

Following the stable stereo image of "Mean Mr. Mustard," a curious effect in the intro to "Polythene Pam" again draws attention to the track's dynamic panning. The acoustic guitar chords that interrupt the "Mustard" outro ("dirty old man") claim the center of the mix, only to shoot out to the left on the last of the three chords in the riff. Following a repeat of this gesture, the guitar remains on the left, balanced on the right by an electric guitar. This quirky bit of panning is more than creative whimsy. It announces a reframing of the stereo stage, as the guitars now define the outer boundaries while the bass moves to the center and the drums to center-right. As the track picks up momentum and intensity, moving the fundamental groove instruments toward the center enhances the propulsive effect.

This stereo orientation continues through "Bathroom Window," and then, with "Golden Slumbers," things change again. The bass remains in the center for the track's duration, but the drums move for the first time to the left side of the mix. Like the accompanying change in piano position from left to right, this new drum orientation contributes to the sense that the stage has been reset.

The main drum kit remains on the left side for "Carry That Weight," but overdubbed drums on the right signal a lingering tie to the drums' previous position. At the end of the song, at the beginning of the modulating passage that sets up "The End," the main kit moves suddenly back to the right side. Except that the tom toms are no longer with the rest of the kit. For the first time, the drums are in stereo. The kick and snare are situated on the right, but the prominent toms are spread across the stereo field. This conspicuous opening up of the stereo space helps to broaden the stage and frame the piece's grand closing gesture, where voices and instruments fill the center with a rich textural blend.

Conclusion

These necessarily brief observations sufficiently illustrate some of the ways in which the thematic elements of a track—its song, arrangement, performances, and mix—are drawn together in an integrated compositional process. Using all the resources at their disposal, the Beatles recording team had spent the 1960s developing ever more sophisticated production practices. Despite their attempts to "get back" to a leaner approach in the sessions that would become *Let It Be*, studio artifice and a comprehensive conception of track creation were fundamental to crafting Beatles records. The *Abbey Road* medley is a fitting conclusion to their years of collaboration, an enduring emblem of their ambition and achievement as pioneering pop recordists.

Notes

1 For convenience, all timings refer to points within individual songs rather than the medley as a whole.
2 It should be noted that the desk's effects returns had their own pan pots, so a mono track might achieve a degree of stereo spread via its reverb return.

References

Emerick, Geoff and Howard Massey. 2006. *Here, There and Everywhere: My Life Recording the Music of the Beatles*, New York: Gotham Books.

Everett, Walter. 1999. *The Beatles as Musicians: Revolver through the Anthology*, New York: Oxford University Press.

Gauldin, Robert, 1990. "Beethoven, Tristan, and the Beatles," https://symposium.music.org/index.php/30/item/2078-beethoven-tristan-and-the-beatles (accessed June 15, 2021).

MacFarlane, Thomas. 2007. *The Beatles' Abbey Road Medley Extended Forms in Popular Music*, Lanham, MD: Scarecrow Press.

Moylan, William. 2020. *Recording Analysis: How the Record Shapes the Song*, New York: Routledge.

Ryan, Kevin and Brian Kehew. 2006. *Recording the Beatles*, Houston, TX: Curvebender Publishing.

Smith, Alan. 1969. "Beatles Music Straightforward on Next Album," *New Musical Express*, May 3.

12

LISTENING BEYOND THE RECORD: THE JUDDS' "JOHN DEERE TRACTOR" (1984)

Jocelyn R. Neal

Eight seconds into the Judds' recording of "John Deere Tractor," the ping first appears. It is in a high register, pitch B5, ringing in the back of the imagined sound stage, slightly to the left of center, close enough to draw focus, far enough away to suggest space and time between us. A slow, five-note motive emerges, its timbre full of possible meanings, references, and mystery, followed by a tell-tale, strummed glissando (0:15). The available discographies and session information offer no clues to the intrepid fan as to what instrument played that melody. And although that particular sound appears as an apparently innocuous, brief fill during the introduction, how a listener might interpret it is entirely dependent on when, and from what musical perspective, the listener encountered the record.

"John Deere Tractor" had an unusually long, slow burn on its way to becoming a modest hit: eight years passed from the time it was recorded to the time its record label released it as a single, and during that span (1983–1991), country music underwent seismic shifts in musical style and cultural meaning. From one perspective, that pinging melody places the record at the headwaters of a sonic revival: traditional acoustic instruments, folk songs acquired through oral transmission, and bluegrass and mountain music poured back into the country genre. But from another, it focuses the listener on the heavy-handed electronic processing and the studio engineering that manipulated the sounds of country-pop and sold Urban Cowboy country to an R&B-loving, mostly white, fan base. Those many interpretations coexist because sound waves themselves, captured on the record's medium, and regenerated through our speakers and headphones, invite multiple meanings and associations that our listening minds assign to them.

This analysis begins with the premise that the sounds encapsulated on this track shape its meaning in ways that extend far beyond the lyrics and music. Three aspects of the recording's reception are explored here; the first is philosophical: a

DOI: 10.4324/9781003089926-17

sense of wandering, evoked by musical components that, by convention, listeners expect to find in predictable structures, e.g., rhyme schemes, chord progressions, and vocal harmonies. Here we encounter unrhymed lyrics; a chord progression that cycles well beyond the typical harmonic realm of the genre; and intimately entwined vocal harmony. The second aspect is the effect of timbre, specifically how a listener interprets the sources that generated a particular sound, and what meaning arises from that interpretation. Finally, the third aspect of this recording's reception is how its sonic contents prompt a philosophical engagement with time: the sounds transport the listener through time as a current hit record that contains the voices of its artists when they were a decade younger. For fans, the record appeared simultaneously as both new and old, eliciting an experience that engaged fans with their own memories of listening in the past.

This analysis ultimately suggests that an attentive listening to the sonic content of a recording uncovers compelling extramusical narratives. Of particular importance to this endeavor: none of the specific sonic elements on which this analysis sits are obscure or difficult to hear, nor do they require complex methodologies to identify. The goal of this analysis is not merely to point them out, but rather to model the accumulated effect of them on the listener's experience with the track: in other words, the analysis here attempts to answer the question of how historically rich stories and self-reflexive contemplations emerge from hearing "John Deere Tractor."[1]

Origins of the Record

The mother-daughter country duo Naomi and Wynonna Judd entered the studio for the first time March 27, 1983, under the tutelage of Brent Maher, who would go on to produce all their studio albums. They cut seven tracks that Sunday, including "John Deere Tractor," a song they had learned from a bluegrass record and had been singing together for some time. The music director for that studio session was guitarist Don Potter, who had played the song with them the first day the group had met to rehearse (Judd 1993, 238). Potter served as the band leader for the session and made the arrangement of the tune in consultation with the Judds and Maher. That session was the only time the Judds recorded "John Deere Tractor" in the studio, even though it was a staple of their live performance set list. The track appeared in January, 1984, on their eponymous EP, *Wynonna & Naomi* (RCA Nashville), but was passed over by the label as a potential single. Three years later, it gained more circulation with its appearance on the European release of their third studio album, *Heartland*, retitled *Give a Little Love*. And three years after that, it finally landed on a domestic long-play album, *Love Can Build a Bridge*, with the original instrumental tracks but newly recorded vocals. The song then appeared on *Greatest Hits Volume Two* (1991) back in its 1983 version with the original vocals, where it caught a larger fanbase's attention as the first and only single from the album, peaking at No. 29 on the *Billboard* "Hot Country Singles" chart. What was particularly striking about the song's final release was that many reviews of the album made no mention of the fact that it was not a new recording.

The Sounds of a Wanderer: Lyrics

Where conventions within country music suggest predictability and stable structures, this particular recording presents sonic content that leads the listener to mystery, contradiction, and wanderings. There is an aspect of undeniable subjectivity in this claim; the basic question of "Is my experience the same as yours?" (Moylan 2020, 19) inevitably invites debate. But analysis of the musical elements explored here—rhyme scheme, intertextual references, chord progressions, and phonetics—supports this interpretation.

For the listener, the lyrics on the record prompt questions and pose mysteries. The narrative takes on an eerily personal tone as Wynonna opens with the words, "Dear Mama," with her mother Naomi's photo right next to her on the album cover, and their mother-daughter voices about to blend a mere 50 seconds later. The ending of the first line ("your girl," Example 12.1) opens up potential and expectation: within the general styles of folk, bluegrass, and country, rhyming lyrics are the norm, and as songwriting theories explain, listeners hold the end of a line of text in mind, awaiting a matching word to create closure (Moylan 2020, 111; Pattison 1991, 45). When the second line finishes without

[0:20] Dear mama, well, here's a letter from your girl Well I think my city days are done, mom And it ain't been three weeks since I *came*	*Unrhymed phrase goals of "girl" and "done."*
[0:47] And mama, I do remember what you said Say your prayers before you go to bed, child And remember city boys ain't the *same*	*Rhyme of "said/bed" appears where expected, amplifying the missing one in the first section.*
[1:10] I'm like a John Deere Tractor in a half-acre field Tryin' to plow a furrow where the soil is made of steel How I wish I was home, mom Where the blue grass is growin' And the sweet country boys don't com*plain*	*Lyrics "made of steel" often transcribed as "mighty still."*
[2:01] And mama, so much perfume I thought I'd drown And the Lord didn't seem to be nowhere around Hey, I felt just like a flower from the *vine*	*"Felt like a flower..." line's meaning ambiguous.*
[2:26] I'm like a John Deere Tractor in a half-acre field Tryin' to plow a furrow where the soil is made of steel How I wish I was home, mom Where the blue grass is growin' And the firelight shimmers and it *shines*	*"Firelight" out of context.*
[2:58] I'm like a John Deere Tractor in a half-acre field Tryin' to plow a furrow where the soil is made of steel *fade to end*	

EXAMPLE 12.1 Lyrics from The Judds' recording of "John Deere Tractor," annotated to mark rhymes, missing rhymes, and contextual ambiguities.

a matching rhyme, the listener's attention is primed to expect rhyming closure from the third line. When that, too, passes without resolution, the effect is one of an unfulfilled expectation and structural ambiguity. As the verse continues, the second section reverts to a more predictable structure and satisfies the listener's expectations: "said" and "bed" rhyme, and then the larger structural rhyme of "came/same" creates a link across the first six lines (0:47). Even with the arrival of more predictable structures, however, the lack of a rhyme in the first three lines leaves a lingering impression on the listener.

Additional segments of unmoored text emerge as the song progresses. At the end of the first chorus, Wynonna intones, "… sweet country boys don't complain." Without any context or complaining boys, the line hangs ambiguously in the song. The second verse sets a scene of women drenched in perfume, acting outside the boundaries of church-sanctioned behavior, suggesting perhaps an urban bordello (2:01). In the last line of this verse, Wynonna sings, with unambiguously clear diction, that she "felt just like a flower from the vine." The phrase seems familiar, resonating loosely with other remembered phrases from literature, but in my hearing, it leaves a lingering impression of vagueness. Furthermore, the first verse sets up expectations of six-line verses in this story, but here Wynonna only sings three lines. Finally at the end of the second chorus, Wynonna describes a night setting where the "firelight shimmers and it shines," an image that feels slightly out of context with the tractor in the field and the rest of the story. The listener's understanding of the narrative is left uncertain, unsatisfied, and incomplete.

The distinction between lyrics as text and vocalized phonemes is also blurred on this track. Notably, for each chorus, fans who post lyrics online seem to be equally likely to transcribe the second line as "mighty still" or "made of steel;" the top four results in a Google lyric search turn up two of each, with both LyricFind and www.lyrics.com showing "mighty still." In other contexts, a listener might seek an authoritative source to determine what the "right" words are, but if we instead allow the record itself to be our central text and object of study, then the sounds we hear are, in fact, the authoritative source, and from the recording, the phonemes that would differentiate between the various options are simply not perceptible. "Steel" voiced by a Kentucky native with a distinctive regional pronunciation, might well sound identical to "still," pronounced by someone from elsewhere. In terms of coherence, "mighty still" makes less sense in the text: soil that is still would not necessarily be impossible to plow, whereas steel soil would be. But on the other hand, fans who transcribe the lyrics and, presumably sing them along with the record, get a better phonetic match for Wynonna's vocals with the printed words "mighty still." With the record as the primary source, what is "right" is less clear.

The Sounds of a Wanderer: Harmony

The chord progression inscribed on the record amplifies the sense of wandering through the use of harmonic language that does not fit in the genre of the recording (Moylan 2020, 81; Neal 2019, 220–22). "John Deere Tractor's" verse

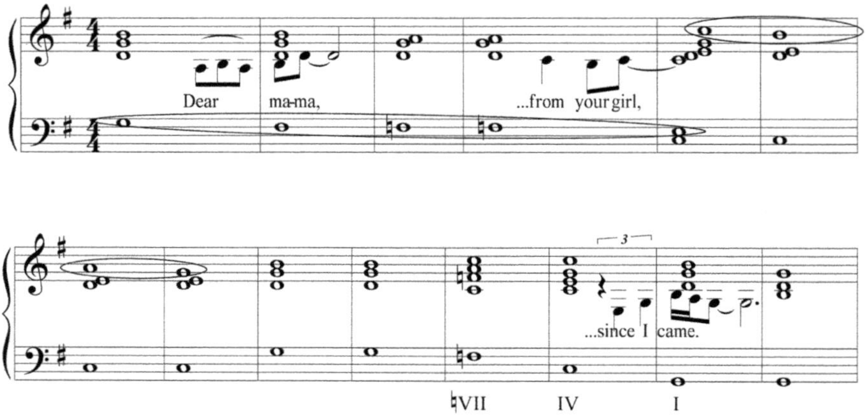

EXAMPLE 12.2 Harmonic sketch of the first half of the verse. This example is not a transcription per se but rather a representation of the harmonic structure. Cue notes for vocals are included to orient the listener. Pitch bends have been normalized. Circles show descending chromatic lines and tetrachords (0:19).

starts out in a clear G Major, then features descending chromatic lines and tetrachords, a plagal emphasis on C, and the falling-fourths progression F–C–G at the end of the phrase (Example 12.2).

As the song slides into its chorus, it begins with a dramatic whole-step drop into an F Major chord. It then repeats the falling-fourth pattern with a move to C, then launches into a circle-of-fifths progression (E–A–D–G), ending with a surprise twist that first tonicizes D (D–A–D) and then converts it to a D7 chord that denies closure and instead pushes the recording forward into the next section (Example 12.3).

EXAMPLE 12.3 Harmonic sketch of the first chorus. First and second chorus use the same progression (1:11).

EXAMPLE 12.4 Transcription of vocals and bass line, end of second chorus, showing the elegant pump-up modulation into the third chorus, which now begins on G Major rather than F Major (2:52).

At the end of the second chorus, the recording takes an unexpected journey through an incredibly smooth and sophisticated pump-up modulation to loop into another chorus. Potter, as the music director for the session, exploited the unusual chord progression at the end of each chorus that concludes on a D7 chord. At the end of the second chorus, after a brief extension that explores the F Major harmony again, that D7 chord resolves predictably into G Major. Instead of continuing with an interlude or the beginning of another verse, either of which would normally begin on a G chord, the recording launches into the chorus, now transposed a whole step up compared to its previous iterations (Example 12.4). Whereas the specific nuances of this chord progression are certainly not the most salient feature of the record for the average listener, the sense of ungrounded wandering is entirely perceptible—even by someone without the technical vocabulary to label shifting tonics, the flow of circle-of-fifth progressions, the use of flat-VII chords, and the chromatic pitches that shape the vocal lines. To any listener, the conventions of mainstream country music chord progressions are thwarted on this record.

Some of what I hear in this record are sounds that I cannot make sense of without going beyond the contents of this one track. Naomi frequently introduced "John Deere Tractor" from the concert stage as a bluegrass number, as she did in 1985 at the Hampton Coliseum concert in Virginia: "We were born and raised in the bluegrass state of Kentucky! And mighty proud of it. This is an old-timey bluegrass song, one of our favorites. It tells the story of a young girl who leaves the country for the first time …" (*Sweet Virginia*).

The Judds learned "John Deere Tractor" from the title track of a bluegrass record Larry Sparks recorded for Rebel Records in 1980. Sparks' version, which is reasonably well known in bluegrass circles, provides resolution to some of the lingering questions: for instance, Sparks opens his lyrics with "… here's a letter from you *son*/Well, I think my city days are *done*," supplying the missing rhyme that was lost when Wynonna flipped the gendered lyrics for her version. Sparks also sings a complete second verse, where Wynonna only offered half. In Sparks' version, the protagonist finds perfume-drenched women where the Lord is absent, and sings that he "fell like a flower from the vine," a phrase that here makes more contextual sense as a Biblical reference about temptation knocking the blossoms from the vine (Song of Solomon 2:15). The rest of the second verse recounts Sparks' personal encounter with a woman, who then mocks his country persona and thereby contributes to his desire to retreat from the city. With the complete second verse present, the song's narrative follows a more logical arc, and the fragments of the story left disconnected in Wynonna's version achieve coherence. The lyric ambiguities in the Judds' version become, in effect, signposts that direct the attentive listener to seek the rest of the story beyond the recording. The track's chord progression, however, is not so easily explained, as neither the Judds' nor Sparks' version evokes the conventions of bluegrass. Simply put, the harmonic language of "John Deere Tractor" is entirely different from any of the other recordings on Sparks' album.[2]

While the Judds learned the tune from Sparks' record, Sparks, in turn, found the song in a box of forgotten demo tapes tucked away at the Lemco Recording Studio in Lexington, Kentucky (Vernon 1983). The tape had no name on it, so Sparks left the song uncredited on his album.[3] The songwriter of "John Deere Tractor," was, in fact, Lawrence John Hammond (Breznikar 2014). Hammond's musical career started at Antioch College in Ohio in 1965, when he joined the Mad River band as bass player and lead singer. A few years later, the band moved to California, catching the burgeoning psychedelia scene in Berkeley and San Francisco and riding it to some success. When the band broke up, Hammond enrolled at UC Berkeley to finish his education, and returned to his folk-influenced musical roots on a few solo projects. Like several other musical veterans of the psychedelic California rock scene who had grown up on country and folk, Hammond's solo music in the 1970s took on the guise of country rock; Hammond kept the harmonic language that had been the bedrock of his psychedelic songwriting and layered it with country narratives and timbres. "John Deere Tractor" was the product of that musical recipe.

How exactly the song ended up in a box of demo tapes in Kentucky remains unclear.[4] Hammond gave the song to Jerry Corbitt (guitarist for the Youngbloods), who recorded it as a mid-tempo rock track on a solo album in 1970.[5] Hammond also made a recording, and kept the song in his set list throughout the early 1970s, yet Hammond's recording languished unreleased until 2012, only then allowing the curious listener to follow that trail all the way back to its origins. In Hammond's version, the song surges forward as an uptempo country-rock saga, where the protagonist meets the perfume-drenched girl, takes up with her, is rebuffed, and then sings in a third verse (which did not make it onto Sparks' recording) that he's heading home on the 10:30 train. Hammond's recording pushes even further into the domain of unexpected chord progressions and wandering tonics. Yet some of the lyrics' disjunct imagery finds more coherence: Hammond sings how the "fireflies shimmer and shine" in the bluegrass. Although nothing in the Judds' track explicitly recounts the song's history and origins, the sounds themselves compel the listener to seek out that story to explain the ambiguities and wandering nature of the Judds' recording. The record teases the listener with its vocal harmony as well: to a fan with some passing knowledge of the Judds as a star duo in country music, the idea that the two of them would be singing in two-part harmony seems a reasonable assumption. Yet their public personas as a duo contradict the sonic material on the recording in provocative ways. Here I offer my own account as a subjective description of that listening experience.

I was once a fan in a younger incarnation, listening to this recording on the radio, singing along at the age when one learns to sing harmony. The sonic mix of this record places Wynonna's lead vocal front and center on the perceived sound stage (Moylan 2020, 288–90), and tucks carefully blended harmony vocals into the wash of sound behind her lead, any subtleties of their placement lost in the casual listening environment of a radio or speakers in my 1980s home. With intense focus, I pull my attention to the pitch space where I expect the vocal harmony to be so I can learn the part, and I find the low vocal line sitting under Wynonna's lead, following its contour in intervals of thirds and fourths, filling out the bottom of the vocal texture. But then I refocus my listening to the pitch space above Wynonna's lead, and I hear the high harmony part, shimmering above the melody and tracing it by thirds and fourths, right where I anticipate it will be. Slight cognitive dissonance results: I expect one line of harmony, emanating from one named body (Naomi Judd), but I hear two voices sandwiching the lead vocal. Such a structure is a time-honored tradition in many country styles including bluegrass. But in the context of the Judds as a duo, the anonymous third voice provokes curiosity. Music director Don Potter explained the source of those sounds: in the studio, Wynonna overdubbed the high harmony vocals; on the road live, Potter himself sang the third line of harmony (email to author, August 4, 2021). Such an explanation satisfies the historian in me, but not the young fan listening for the two voices of the country duo whose pictures were on the album cover.

Mountain Music

Records often hold secret the techniques and instrumentation that were used to create the sounds locked in their grooves. Without video or photos of the recording sessions, we can only infer the sound sources from our general knowledge of timbres gained from past listening/viewing experience, and from whatever external sources we can unearth. Gallons of ink have been spilled during the past decades by scholars, critics, and fans speculating on exactly what the sources were for the sounds we hear on our favorite records, and one of the most intriguing aspects of a record, in my opinion, is that the sounds themselves seldom supply an indisputable answer as to what created them in the first place. Such is the case with the instrumentation on "John Deere Tractor," especially one intriguing melody in the introduction, where the interpretation of timbre relates to how a listener finds meaning in the track (Moylan 2020, 436).

By the time the Judds' fame crested in the late 1980s, two competing narratives about their music were firmly entrenched in the country genre. One focused on Wynonna's vocal timbre and technique, lauding her R&B sound and her commanding alto resonance as she growled her way through sensual vocals and blues-styled tunes. The other focused on the Kentucky roots through which a family duo had re-kindled the traditions of Appalachian traditional music in its most appealing forms. In particular, songs such as "Born to Be Blue" played into the R&B interpretation, while songs such as "Grandpa (Tell Me 'Bout the Good Old Days)" and "John Deere Tractor" amplified the traditional Appalachian roots narrative. Both narratives found support in the broader trends within country music. When the Judds first broke into country radio, country was at the height of its crossover period, still enjoying a commercial boost from the movie *Urban Cowboy* and celebrating the outrageous pop success of tracks such as Dolly Parton and Kenny Rogers' "Islands in the Stream," tellingly written by Barry, Robin, and Maurice Gibb (disco superstars the Bee Gees). But as the Judds' own career moved through the 1980s, mainstream country music shifted back toward its roots, with bluegrass, honky-tonk, and old western swing tunes resurfacing. And in that context, the instrumental opening of "John Deere Tractor" similarly tells two tales.

The Judds' public image was firmly linked to mountain music traditions by the 1990s. Many of their records featured the mountain dulcimer, played by David Schnaufer, who publicized his work on their albums quite widely, although he did not appear on their early sessions (Wright 2006). Within that narrative, the opening melody (Example 12.5) evokes an acoustic folk instrument. One might perceive the clean attacks to be a hammer rebounding from a dulcimer's strings for the first few notes (even though Schnaufer played mountain, not hammered, dulcimer). The glissando-arpeggio (0:15), however, provides a more distinct clue: autoharp. Autoharp had resurfaced in the 1980s on such albums as Dolly Parton, Linda Ronstadt, and Emmylou Harris' *Trio* (Warner Bros., 1987), and even appears in the discographic credits for one later Judds recording session. Any mention of autoharp in country music harkens back to the Carter Family and Sara Carter in particular, who played it on most of the family's now-classic recordings from the 1920s and 30s, lending even more credence to hearing an old-time, traditional element in the Judds' music.

I asked two knowledgeable music scholars, both experts in country and bluegrass, to listen to and talk with me about that five-note melody, and to speculate about its acoustic source. "My first sense with that note at 0:08 is that it sounds super synth-y. Like it almost sounds like a keyboard at first, a real 80s-ish sound," said one. "Frankly, it sounds to me like a keyboard synth of some 1980s type using a fairly cheesy 'strings' setting," commented the other. From that perspective, we

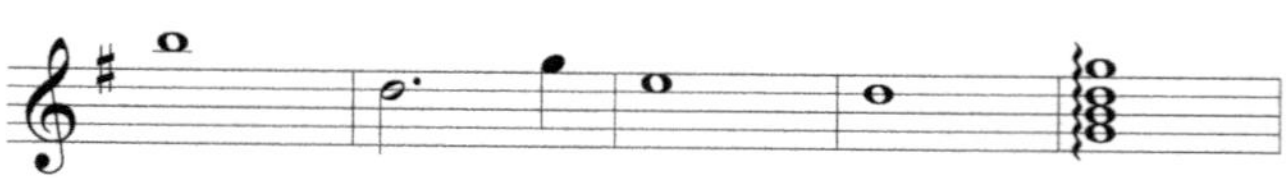

EXAMPLE 12.5 Opening "pinging" melodic motive that invites timbral interpretation (0:08).

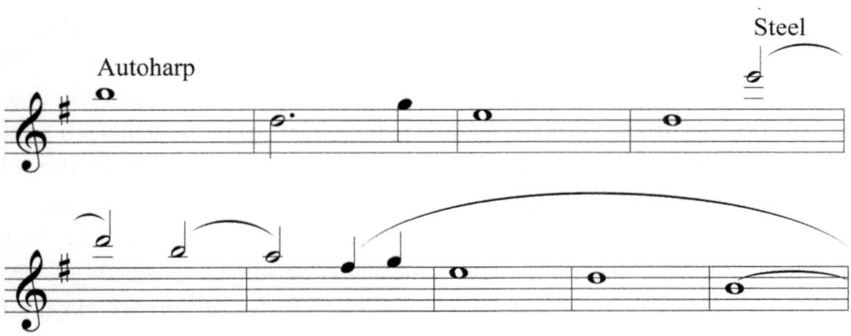

EXAMPLE 12.6 Referentially traditional timbres in the interlude (1:42).

find ourselves squarely in the midst of the 1983 mainstream country studio model: synthesized, highly processed music created in a sonic laboratory constructed to simulate acoustic instruments. That specific brand of country music has drawn the derision of many fans for decades, some of whom describe the early '80s crossover music as the worst of an inauthentic, sell-out period in country's history. It also fits the narrative of the Judds in the studio, specifically in the hands of producer Brent Maher, who cultivated Wynonna's R&B sound and pulled in as much rock energy as possible. "Wynonna Judd's often R&B-tinged vocals, consistently lucid and regularly electrifying, were a principal attraction," summarized critic Barry Mazor (2016). An autoharp was not part of that narrative.

Don Potter, the guitarist and music director for the Judds' early sessions, shared with me his recollection of making the record: The "sound you hear is me plucking an autoharp single string at a time. It's very hard to do that without bumping other strings and this was an analogue recording so no computer fixes" (email to author, July 13, 2021). Yet far from closing the book on this timbral mystery, Potter's interview invites another listen, and then another. His autoharp was recorded and processed in such a way that it plausibly sounds like a synthesized version of an autoharp. In this instance, however, behind that sonic processing is an actual autoharp, strummed as accompaniment for two Kentucky women singing a song they learned from a bluegrass record. It leaves us with a fascinating "both/and" perceptual moment as the autoharp gives way in the interlude to the sighs of a steel guitar (Example 12.6), an instrument conceived outside the traditions of mountain string-band music but which had come to signify, by the 1980s, a kind of traditionalism in country music all its own.

Running Out of Time

Many aspects of the "John Deere Tractor" track contain the sonic fingerprint of its 1983 session: processing on an autoharp that makes it (ironically) sound like a synthesized version of itself; the musical signatures of a country rock artist who had spent years in the psychedelic California scene, lyric gaps and omissions that were the product of "found" songs transmitted orally across styles and genders. But the song as a radio hit did not reach many of its fans until the early 1990s. Seldom is a record presented as "new" in mainstream commercial music that was the product of a studio session from almost a decade earlier. When such a release does occur, it creates an "out-of-time" listening experience for fans: the record contains the voices and musical expression of much younger artists, presented not as a "greatest hits" reminiscence, but instead as current music. The "out-of-time" experience was amplified even more when Naomi announced in October, 1990, that she was retiring due to health issues. The Judds embarked on a final "Farewell Tour" in 1991, just as the single for "John Deere Tractor" was released. The result was that fans heard a record made at the very start of their career, when their voices were a decade younger and their studio production styled in the early '80s, as they bid farewell to the Judds at the end of their career as a duo. And whereas the release of the record itself created the effect of time travel, its sonic contents invited an even more radical time-warp.

I listened to country radio compulsively in the 1980s, fantasizing about catching a bus to Nashville and becoming a country singer. I do not recall when I first encountered "John Deere Tractor," although I know it was sometime after the Judds' career was in full swing. Thus, my experience of the record was out-of-time chronologically with their overall output. I remember listening to the radio in decidedly inferior acoustic conditions, with what any audiophile would dub entirely deplorable equipment, and trying to make out the words well enough to sing the song myself. Having traced the song's sonic pathways to their historical sources, I now know that my best efforts to figure out the words mirrored the Judds' own efforts to decipher the words from Sparks' record. That practice matched what so many earlier generations of country musicians had done. When the Carter Family sang "Wildwood Flower" in 1928, for instance, their lyrics showed evidence of having been extracted from many layers of oral transmission, some lines ending up as nonsense syllables (Neal 2019, 42). Historians have since referred back to published 19th-century sheet music to disentangle the Carter Family's lyrics, suggesting that those published lyrics are more correct by some measure. But the lesson from these recordings is, perhaps, that the record does not contain "the song" at all, but rather merely an imprint of the musicians' re-creation of whatever sonic memory lingered from their encounter with some other imprint of it.

I find the track all the more interesting when it is not cross-referenced with published sheet music or a definitive version. For "John Deere Tractor," the recording that is arguably the original—that by songwriter Lawrence Hammond himself—was, in fact, the last to enter the chronology of this song's public history when it was finally released in 2012, long after fans had forged strong connections with the other versions. I listen again to the Judds' "John Deere Tractor," asking myself, as a fan, what moves me in this record. The answer is not the sounds that are most clear, but rather those that are the hardest to understand: ambiguous, mysterious, and wandering.

Notes

1 I am grateful to Larry Sparks, Don Potter, and Lawrence Hammond for sharing personal accounts and histories of this song and recording with me so generously, which have made this analysis possible.

2 Bluegrass, like many other genres of music, finds its songs and source material in a variety of different musical arenas, including but not limited to blues, folk songs, fiddle tunes, gospel music, Southern gospel as sung and composed in the 19th-century religious revival movements, show tunes, minstrel songs, and more. Across this wide and varied repertory that makes its way into bluegrass, there are many different harmonic languages, including the modal chord progressions that underpin many fiddle tunes, the blues progressions, and more. All I am asserting here is that there is a common-stock core that emerges in bluegrass around three-chord songs, which can be modeled by basic I–IV–V progressions. Even that common-stock core gets challenged in the 1960s when progressive bluegrass artists import songs from psychedelia, the standard "Fox on the Run" (originally by Manfred Mann) being the most commonly discussed instance. But as evidence for my assertion here, consider that the other four tracks on Larry Sparks' record that sit next to "John Deere Tractor" on the A-side of that album are all entirely three-chord songs (I–IV–V), with no modulations ("Girl at the Crossroads Bar," "Hot Sauce," "I'd Rather Be Alone," and "Great High Mountain"). It is in that context that "John Deere Tractor" stands out as fundamentally different in its underlying harmonic language.

3 When the Judds first released the song, neither they nor their record label knew who the songwriter was. The record label placed the mechanical royalties from the song in a trust. Hammond had known about Larry Sparks' recording but did not reach out to Brent Maher to claim the royalties from the Judds' version until it was released as a single (email to author, December 16, 2021).

4 I asked both Larry Sparks and Lawrence Hammond if they knew what recording had made its way into the box of tapes in the Lemco Studio in Lexington, Kentucky, from which Sparks learned the song. Hammond did not know, and Sparks could not recall (phone conversation with author, January 20, 2022), so that specific point of transmission remains, at this time, a mystery.

5 The account of the song's history prior to Larry Sparks' recording comes from Lawrence Hammond via email to author, December 16, 2021.

References

Breznikar, Klemen. 2014. "Mad River Interview with Rick Bockner and Tom Manning." *It's Psychedelic Baby Magazine.* https://www.psychedelicbabymag.com/2014/03/mad-river-interview-with-rick-bockner.html.

Judd, Naomi, with Bud Schaetzle. 1993. *Love Can Build a Bridge.* New York: Villard Books.

Mazor, Barry. 2016. "'Wynonna & The Big Noise' Review." *Wall Street Journal,* February 16.

Moylan, William. 2020. *Recording Analysis: How the Record Shapes the Song.* New York: Routledge.

Neal, Jocelyn R. 2019. *Country Music: A Cultural and Stylistic History.* 2nd ed. New York: Oxford University Press.

Pattison, Pat. 1991. *Managing Lyric Structure.* Boston, MA: Berklee Press.

Vernon, Bill. 1983. "Larry Sparks." *Bluegrass Unlimited* 17, no. 10 (April). https://www.bluegrassunlimited.com/article/larry-sparks-3/.

Wright, Jerry. 2006. "Interview with David Schnaufer," *The Grand Old Dulcimer Club.* https://www.thegrandolddulcimerclub.com/david-schnaufer.

Recordings

Hammond, Lawrence. *Presumed Lost*. Shagrat Records, 2012.
Parton, Dolly, Emmylou Harris, and Linda Ronstadt. *Trio*. Warner Bros., Nashville, 1987.
Sparks, Larry. *John Deere Tractor*. Rebel, 1980.
The Carter Family. *Wildwood Flower*. Victor Records, 1928.
The Judds. *Wynonna & Naomi*. RCA Nashville, 1984.
The Judds. *Give a Little Love*. RCA, 1987.
The Judds. *Love Can Build a Bridge*. Curb/RCA, 1990.
The Judds. *Greatest Hits Volume Two*. Curb/RCA, 1991.
The Judds. *Sweet Virginia*. Freefall Records, 2021.

SECTION V
Sampling and Reframing

13

SONIC MATERIALITY AND BOOM-BAP[1] EMBODIMENT IN CONWAY'S "BISCOTTI BISCUIT" (2018)

An Autoethnography of Recording Analysis

Michail Exarchos (aka *Stereo Mike*)

This chapter is supported by a webpage.

Introduction

Fourteen years ago, I found myself at a telling crossroads in my sample-based music journey: sweaty, on a hot summer's night in Athens, Greece, I had just finished performing a few pages of fresh lyrics from my writing pad to one of the country's most celebrated folk singers. The beat[2] I rapped to was playing back from a traveling Akai MPC1000[3] I set up on the floor, hooked up to the singer's home studio system. The verse sections featured chopped[4] instrumental elements from a remix/remaster of one of the singer's biggest 1980s hits; the choruses were largely intact—if filtered—samples from the original including vocals, and juxtaposed against layers of beat programming, synth bass, and string elements. Kneeling on the floor so I could control the basic AB structure of sequences playing out of the sampling drum machine, I tried to perform—without a microphone—at a loudness and intensity that matched the amplified beat, attempting to hit my envisioned cues and expression as intended. The last lyric of each rap verse was meant to rhythmically land upon the same word commencing the singer's sampled chorus, and I interjected some of the rests she left within the chorus phrasing with a rapped (call-and-)response line. At the end of what felt like an intense performance, the singer asked me if I needed access to the multitrack tape, or whether I would prefer her to perform the vocal parts again. She was seemingly concerned about the amount of (1980s-inspired) reverberation on the vocal elements of the sampled master. Despite my excitement at what sounded like a licensing approval, I somehow managed to relay the importance of the *sample-based aesthetic*: the source material had to *feel* sampled and recontextualized. It had to *sound* like there had been no access to anything

DOI: 10.4324/9781003089926-19

but the released stereo track: essentially, a *bootleg*—preferably sampled from vinyl (although the samples featured were actually recorded from a "greatest hits" CD the singer had gifted me in our previous meeting).[5]

The collaboration became the leading single of my first major label album,[6] and with further production on the track, I inserted a heavily manipulated introduction from the remaster, using iZotope Vinyl (a software emulation) to turn the full-range CD sample into something of an exaggerated vintage sonic signature.[7] The remainder of the album production, however, turned out to be a more adventurous journey, as the transition from an indie to a major label landscape would bring about unexpected challenges. Emboldened, perhaps, by the positive collaborative tale, or naïve to the licensing power—I thought—I was about to assume, I did not anticipate that my further (internationally sourced) sample-laden productions I was considering for release, were never going to be cleared by EMI Music Greece: I was a small—still alternative music—fish, in a rather large pop market pond. The predicament I was in mirrored a large proportion of the international beat-making community, finding themselves in a similar creative limbo when attempting to practice authentic forms of sample-based music making. Wayne Marshall explains:

> Producers working for large record labels, enjoy production budgets that permit them to license any sample they like [...] Independent and largely local artists, operate well enough under the radar to evade scrutiny or harassment and continue to sample with impunity [...] Acts with a sizeable national, if not international, following but who lack the resources of a "major label" [...] find themselves in a tight spot: to sample or not, to be real or not, to be sued or not?
>
> *Marshall 2006, 869*

Situating the Listener: Analyst Identity, Imagined Community, Background, and Method

Analyst Identity and Imagined Community

Marshall, of course, considers a "major label" context as the threshold for a licensing carte blanche, but this is a variable relative to geographical location and market size. My response to the above challenge became a lifelong pursuit of creating originally composed and recorded source content that could legitimately serve a sample-based creative process (a pursuit shared by a growing proportion of the beat-making community, famously exemplified by artists/producers such as Frank Dukes, De Le Soul, J.U.S.T.I.C.E. League, Marco Polo, and Griselda/ Shady Records collaborators Beat Butcha, Conductor Williams, Just Blaze, Alchemist, and Daringer[8]). This, in turn, necessitated a detailed examination of what the content, context, and character of recorded material that can be regarded as sample-worthy may be, bringing me face-to-face with questions about: what is it that is "phonographic" at the heart of samples; why is it that phonographic sources are the preferred raw sampling material in the practice of (sample-based) Hip-Hop; and which elements and domains contribute to what Goldberg describes as *impactful* and *gestural* in hip-hop music:

> Everything that people love and hate about rap music deployed at car stereo volume (where sound becomes *gesture* and *impact*) has to do with the *spatial* modification accomplished by hip-hop compositions. People are not necessarily referring to the voice of the rapper, which is frequently reduced to another rhythm in the track. Instead they are referring to the *exploding* kicks, the *echoing* snares, and the sometimes terrifying *sonic manipulations* of DJ scratches.
>
> *Goldberg 2004, 130, emphasis added*

Moylan's thorough analytical framework in *Recording Analysis* (2020) provides a systematic set of guidelines toward approaching such questions with rigor. Although this introduction has situated myself as both practitioner and analyst in this undertaking, my choice of track for analysis is from a different producer, to enable me to assume a somewhat removed and more objective analytical vantage. Nevertheless, I have chosen to share my identity, history, and biases, as the reader engages with how these will shape my listening-analytical process. The chosen track for this analysis is an exemplar of the sample-based aesthetic, and one with which I share a kindred stylistic and sonic worldview: "Biscotti Biscuit" is a hip-hop track from *Blakk Tape*, the fourth solo mixtape released by rap artist Conway (The Machine) on March 27, 2018. It is produced by Bozack Morris and features a sample of "Super Slow Down" (1975) by Bernard Lubat. As we explore its domains, sonic elements, context, and characteristics, I will attempt to get closer to what it is "that people love and hate about [sample-based] rap music" (Goldberg 2004, 130), with a particular focus on the less explored spatial-timbral material/ity (dimensions) at the heart of its sonic mechanics.

Background

Previous literature on Hip-Hop has overwhelmingly favored notions of "musical borrowing," which—even when inclusive of broad themes such as the intertextuality and reception of genre, the rapper's voice, and sampled sound—promote elements of the musical and/or lyrical domains (musical style, motifs, words, and rhythm) as the primary focus (see, for example, Williams 2010). On the occasions when timbre, sonic content, and the overall phonographic texture resulting from sample-based processes are considered, the majority of studies stop at surface observations. For example, Tricia Rose's otherwise foundational mapping of Hip-Hop's sonic priorities to "black intent" in *Black Noise* (1994) mostly "focuses on the conventions of orality in hip-hop rather than on the underlying sonic product that supports and sustains it" (Goldberg 2004, 130).

Schloss's exhaustive ethnographic study of Golden Era[9] practices in *Making Beats* (2014)—which accurately identifies the sonic characteristics that define the period's sample-based aesthetic—only extends as far as filtering, deployed in the service of sound-object isolation (in the recording domain); analysis of essential processes such as (swing) quantization and sample chopping revert back to rhythmic and structural/motivic ramifications. Adam Krims' Kantian-inspired "hip-hop sublime" goes as far as recognizing "timbre [...] as a crucial means of organization" and a catalyst in the listener's sense-making sonic experience; but it does not explore how the "combination of incommensurable musical layers [...] are selectively and dramatically brought into conflict with each other" (Krims 2000, 41–54). Krims' timbral sublime does echo Moylan's "gestalt percept" as a "singular impression that coalesces from its many parts" (Moylan 2020, 239). Moylan, however, considers the acoustic component parts within timbre, as will this study alongside the underlying spatial architecture. And Amanda Sewell's useful typology of hip-hop samples across a range of seminal phonographic case studies identifies their structural organization and aesthetic implications, but not the inner workings of a sample-based record's mixing architecture and textural synthesis (Sewell 2013, 26–67). In other words, while Williams (2010, 1) and others choose to transcend "narrow discourses of 'sampling,'" this examination (re)considers the importance of close readings of content within the recording domain, in order to evaluate the character and context (and uniqueness) of the chosen record.

Method

The methodology (as is already clear from the introduction), fuses an autoethnographic/neonarrative approach with aural and literary analyses, as well as ethnographic strategies (an interview with the producer of the track has been conducted in order to objectify aural findings and enrich perspectives).[10] Within the remit of accepting/celebrating that "listening is [indeed] personal" (Moylan 2020, 59–60), I will aim to connect personal observations and evaluations to wider cultural practices (of the imagined beat-making community) through—as transparently as possible—situating the nuanced, lived process, and identity of myself as the listener within the narrative of the analysis. As Ramsey states in *Race Music* (2003, 22), one of the goals of such projects is "to have readers understand something about some of the sources and grounding of [the author's] own critical voice and biases."

Analytical Listening and Structural Observations

Following the open listening engagement (described in more detail in "Section 1," the listening notes of the companion website), the attention and intention of the analytical listening process have focused on deciphering the layers/stratification making up the multi-domain texture (Moylan 2020, 198):

commencing by observing the musical and lyrical domains (defining structural aspects), through to noting and then evaluating spatio-timbral phenomena and underlying mechanics in the recording domain.

Structure and Form in the Music Domain

Figure 13.1 illustrates that "Biscotti Biscuit" is constructed around a four-measure prevailing time unit (PTU), which corresponds to the sample taken from "Super Slow Down" (between approximately 00:39 and 00:59 on the original). The sample has been pitched up by about six semitones bringing it to a tempo of 72 bpm (from approximately 49 bpm). The instrumental hip-hop production by Bozack Morris features the sped-up sample alongside an additional drum beat and sub bass element, and the layered construct is cycled around its PTU sixteen and a half times, forming an overarching structure of: introduction (IN: 2.5 PTUs), two verses of four PTUs each (16 measures per V1 and V2), two choruses of two PTUs each (8 measures per CH), and a coda (OUT) of two PTUs (another 8 measures)—at the end of which follows a sub/808 bass "tail" lasting approximately one and a half measures. The listening notes on the companion website illustrate the progressive observations of the analysis that, through repeated and attentive listening, led to composite Figure 13.1. The intention in this phase has been to observe the structure informed by the music and lyrics domains, and extract information on the instrumentation and layers comprising the arrangement.

Progressively deploying Sewell's sampling typology, the layers comprising the music production could be categorized as follows:

- The approximately 20-second sample from "Super Slow Down" could be described as an "intact" structural sample type including "every element from the source material"—in this case: drums, electric bass, electric piano (Fender Rhodes), vibraphone, and electric guitar—"looped throughout [almost the entirety of] the new track" (Sewell 2013, 36–54). The exceptions come in the form of a number of effect manipulations "performed" upon the sample, such as delay processing, cuts,[11] filtering, and tape-stops/slow-downs, which will be discussed as part of the recording domain analysis below.
- In comparing the source content to "Biscotti Biscuit," it becomes clear that the hip-hop production features an additional drum beat, which could have been programmed out of individual drum "hits" (isolated hi-hats, kick, and snare drum). This forms a complimentary sequence that supports the existing drum sounds and the pattern contained within the intact sample. The new drum beat can be categorized as a "percussion-only" structural layer containing "sampled drums … looped throughout the new track" (ibid.). The snare drum of this percussion-only layer becomes exposed at the last beat of measure 10, synchronously to the voice entering the mix to introduce Verse 1 (the snare is also manipulated with delay processing).

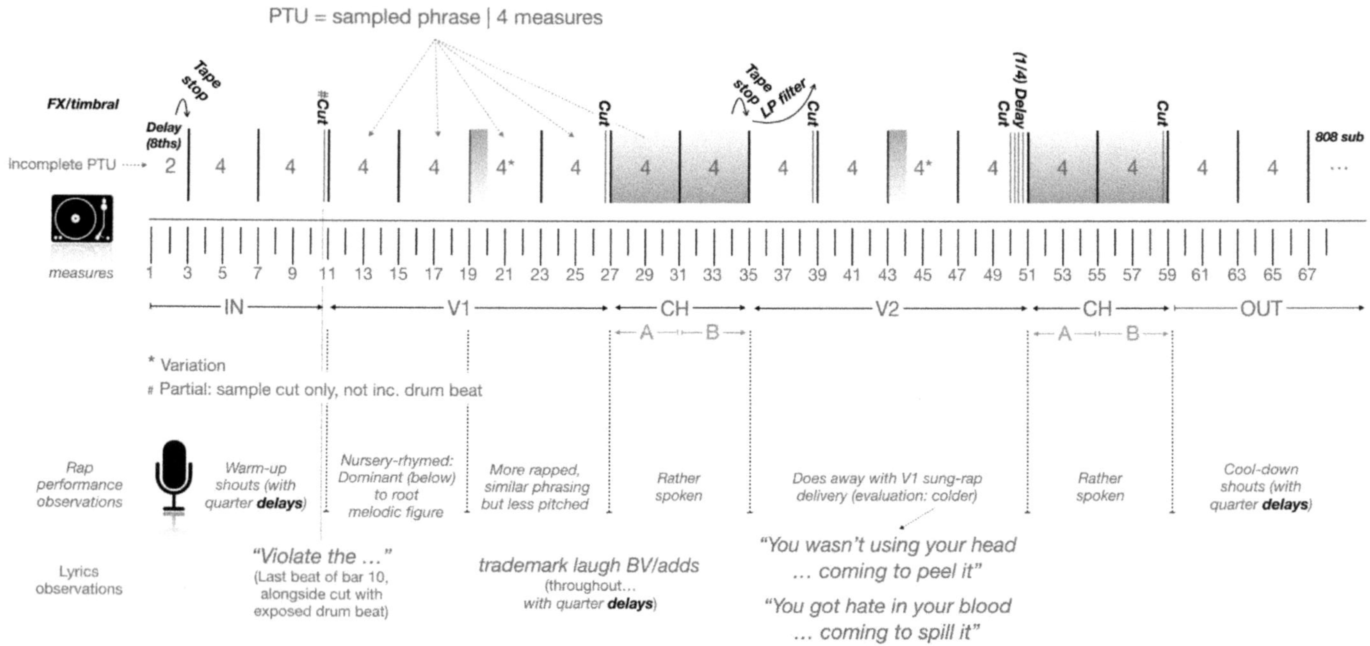

FIGURE 13.1 Schematic representation of structure and musical form of "Biscotti Biscuit" by Conway (The Machine) [beat produced by Bozack Morris], with recording domain (effect manipulation) and rap performance (lyrics domain) observations.

- As the two layers remain combined for the majority of the hip-hop track's duration, their structural type could be categorized as "aggregate," which is defined as consisting of component "layers … sampled from different sources" (ibid.).

Finally, Bozack Morris brings in an additional sub bass element (808) alongside the source bass part, which is most audible on the first and final beat of every measure throughout—this becomes exposed at the very end of the coda. A subtle two-note legato string line is blended upon the aggregate layer during the choruses (and this variation is indicated in Figure 13.1 by the graded shading in chorus sections).

Rap Performance and the Lyrics Domain[12]

The rap performance delivers the lyrical content forming six distinct sections (with further subsections identified within). The introduction features spoken delivery (during the first, incomplete PTU), a cough, and some shouts ("aah" and "yeah") during the first complete PTU, and lyrical content from the forthcoming first verse during the second complete PTU (with the phrase "bullet with your name on it …" repeated twice alongside further shouts). The first verse enters on the last beat of the introduction (measure 10, beat 4) with the words "Violate the mob"—the underlined word "mob" synchronized to the downbeat of measure 11, the beginning of the verse structure proper. The first half of the verse is delivered in a somewhat sung dominant-below-to-root melodic figure. The second half of the verse assumes a more classic monotone/rapped delivery, after its entry at just before measure 19, but with very similar rhythmic phrasing to the first half (featuring rests of about one-eighth at the last beat of every measure and allowing early entry of the lyrical content tied to the following measure). The rests left on the lead rap delivery are utilized by the backing vocals ("adds" in hip-hop parlance), which are also effected by synchronized delay processing (more on this below).

The first chorus entering at measure 27 is more spoken, delivered with reduced performance intensity, and using spaced phrases with considerable silence in between (up to half a measure of rests). Again, the spaces are filled with background—delay-processed—adds and shouts. The second verse follows a similar rap monotone to the second half of the first verse, although this time, the rhythmic phrasing is slightly busier leaving fewer rests (until the last two measures, where the "blowing thick clouds in the clouds" lyric and phrasing is borrowed from the chorus and mirrored by the similarly phrased "busting bricks down since a child" response). Off-beat (eighth) breaths become audible toward the end of the second verse (in measures 47–49, matching the increased performance intensity). The second chorus echoes the first, while the coda mirrors the intro with delay-processed shouts and spoken phrases. Figure 13.1 includes *rap performance* and *lyrics* highlights pertinent to these observations.

Effect Manipulations and the Recording Domain

The first two measures of the structure (the incomplete PTU) are constructed out of chopped eighths from the first two measures/chords of the sample and then either processed with a delay effect synchronized to eighths in the tempo of the new track or repeated using a DAW (seven times for the first sample and six times for the second) with the volume decreasing. Importantly, there is no considerable reduction of high-frequency spectral content during the repetitions. The very last repeat of the second chord's chop (measure 2, beat 4, repeat 6) is slowed down, resembling a tape or vinyl deck "stop" (a tape or record gradually slowing down to a complete stop). Conway's cough, processed with a quarter delay, repeats over the instrumental slow-down. The tape stop is indicated by the downward arrow in Figure 13.1, and a second occasion of this effect takes place on the last beat of the first chorus (beat 4, measure 34), this time over the whole "aggregate" layer.

Following the stop, the pitched-up but otherwise unaffected "intact" sample is presented within its "aggregate" layer, which includes the new "percussion-only" drum beat and sub bass elements. The four-measure PTU is repeated twice in this form (8 measures)—featuring the warm-up shouts and introductory phrases by Conway—until the last beat of measure 10, when the "intact" sample is cut (completely this time, not "stopped"), exposing the snare from the "percussion-only" layer. The snare is processed with a 16th delay effect synchronized to the track's tempo with a resonant low-pass filter over the snare repetitions—the repetitions stop just before measure 11 (and the resonant filter makes it resemble a high to low tom fill). During the introductory sections, Conway's warm-up shouts, adds, and phrases are processed with a quarter delay effect, a feedback setting allowing two to three repeats (with repetitions stopping before new phrases are performed), and a gradual reduction of high-frequency content.

Verse 1 proceeds with the full aggregate PTU repeating (almost identically, with the exception of measure 19; see below) four times (for a total of 16 measures), until another, complete cut of the aggregate layer takes place at the last beat of the verse's last measure (beat 4, measure 26). The third repeat of the PTU on measure 19 introduces a variation within the aggregate layer (splitting the verse structure half-way). This features a different sample from "Super Slow Down" on the intact layer, as well as a sixteenth delay on the first snare (second beat) of the percussion-only layer. This variation is repeated in the equivalent half-way point of the second verse (measure 43), and both measures featuring the variation are highlighted in Figure 13.1 by the gradient grey shading.

Verse 2 is introduced right after the tape stop effect that brings the first chorus to a close, and it is processed with a resonant low-pass filter gradually opening to reveal the full spectral content of the aggregate layer over a full cycle of the PTU (this is illustrated in Figure 13.1 by a longer upward arrow). A full aggregate cut at the second eighth of beat 3, at measure 38, reveals the phrase "kill it" in Conway's lead rap performance. The earlier cut is illustrated in Figure 13.1 by a vertical bar slightly further to the left than the cuts in measures 10, 26, and

58. The last measure of the second verse is constructed by processing an intact sample chop from the first quarter of the PTU's final measure using a quarter delay with a feedback setting resulting in three repetitions receding in volume (similar to the early introduction figure, but accentuating the almost isolated guitar element in the sample). However, the aggregate chop here features a variation in the percussion-only layer, created by a snare roll audible in the drum beat. The aggregate chop variation is indicated in Figure 13.1 by gradient shading on the first beat of measure 50, while three further equidistant vertical bars illustrate the quarter repeats of the intact sample layer. Finally, the second chorus ends with a full aggregate cut on the last beat of measure 58, synchronized with "now," the last word of the chorus rap performance.

Critical Listening and Staging Architecture

In this section, both the source sample and the new track are critically listened to out of a temporal context as phonographic sound objects. The emphasis of the examination begins by establishing the spatial placement/staging of (instrumental) elements within the intact sample, followed by an evaluation of how the sampled stage dimensions are then negotiated within the hip-hop production (and how this "sound world" interacts with the drum beat, sub bass, and rap elements in the new track). The approach to describing and illustrating element positions and sound stage dimensions combines notions of:

i. William Moylan's two-dimensional concept of an illusory stage where musical elements are located in front of the listener (establishing aural images with *width* on a horizontal plane, and *depth* with a distance perception); with
ii. representations of *height*, inspired by Moore and Dockwray's "sound-box" diagrams (Moore and Dockwray 2008) and David Gibson's visualizations of mix elements as 3D sonic objects (Gibson 2008). These enable the inclusion of proximate illustrations of combined pitch and spectra areas on a vertical axis.

Figure 13.2 is the result of an initial critical listening session to the 20 sampled seconds of the source track on repeat, taking some artistic license in "depicting" the section's perceived and crystallized aural image/form (Moylan 2020, 57). The figure identifies, however, Bernard Lubat's instrumentation, proximate placements, and communicates aspects of their spatio-timbral dimensions. A number of conventions are deployed to this effect: the sizes of the instruments correspond to perceptions of relative loudness (volume), the level of opacity represents proximity (depth), their vertical placement impressions of pitch and spectra, and their lateral placement stereo position; use of close, blurred shadows indicates prominent early reflections, while distinctly offset and less blurred shadows alert to more identifiable (and less diffuse) time-based effects (e.g., the tremolo effect on the Rhodes piano). More specifically, the mirroring of the keyboard

FIGURE 13.2 Schematic representation of the proximate/perceived sound stage in "Super Slow Down" (representing general placements of elements contained in the section from 00:39 to 00:59 in the original).

illustration communicates the oscillating tremolo effect that enables the electric piano to appear in two positions in quick succession (at an approximate rate of eighths), and the opaque spaces around the snare and the whole record's environment—respectively—an applied reverb effect in post-production, and the perceived recorded environment tying all instruments together. The electric bass is loud with a substantial amount of low-to-sub spectral information, and it is matched in prominence (volume and proximity) by the snare and vibraphone (least opaque and largest objects in the figure).

Using similar conventions, Figure 13.3 illustrates the notion of a *record-within-a-record*, with the full sampled stage "elevated" upwards on the pitch-spectral axis, reduced in volume relative to the new elements, but occupying similar "width." The new kick and snare drums occupy the position of the equivalent sounds on the sample, offering blended combinations with a high degree of masking (yet benefitting from the "transposed" spectral content of the original sounds), while the 808 acts as a sub bass extension to the sample's electric bass (note that the two illustrations intersect). The new kick drum shares a spectral area with the sub bass, but appears more frontal in terms of proximity (as a result of its spectral envelope). The new, notably louder, hi-hat appears on the opposite side of the stereo field relative to the original. Conway's lead rap (the larger "rapper cap") masks the lower area of the combined snare but is positioned centrally (the combined snare placement is slightly off-center), and the smaller white cap with the non-blurred, offset shadow represents the rap adds and their clearly separated quarter delays, respectively.

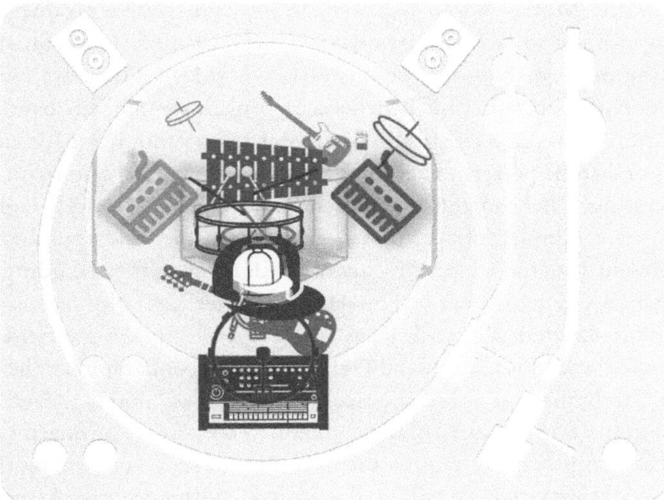

FIGURE 13.3 Schematic representation of the proximate/perceived sound stage in "Biscotti Biscuit" (representing general placements of elements contained in an aggregate PTU).

Recording Domain Evaluations and Conclusions: How Sample-Based Processes Shape the Song

The preceding structural analysis demonstrates that although, on the surface, "Biscotti Biscuit" follows a simple, repetitive music (domain) form borne out of the PTU's perpetual looping, the recording domain manipulations create a wealth of dynamic and structural difference. This phenomenon is congruent with the Black cultural practice of the "changing same" (Jones 1969, cited in Williams 2010, 30), which refers both to the recontextualization of a past idea (here, sampling a 1970s phonographic section for a 2018 reinterpretation in a different style), *and* the constant, real-time revision of that idea within the structure of the new piece. Bozack Morris' cyclical production expresses Afrological sensibilities and traditions (Lewis 2017), and the recording domain manipulations (cuts, stops, filtering, delays, and layering organization) are an expression of what Tricia Rose has identified in hip-hop practices as manifestations of "openness, ruptures, breaks and forces in motion" (Rose 1994, 83). In Morris' own words:

> I sometimes throw [the effects] in to end sequences or introduce the next part of the song, or to switch things up a bit. Sometimes I'll put them in before someone writes a song to it, but I will definitely add it after to put emphasis on what the artist is saying in parts.
>
> *Bozack Morris, Email interview with author, July 11, 2021*[13]

Furthermore, Morris' boom-bap stylizations (and relative production minimalism) pay tribute to Golden Era practices, when sample-based processes first emulated the performative innovations of hip-hop DJs. These are exemplified in the choice of materials (psychedelic soul/funk instrumental, overlayed and chopped drum break, 808 expansion of sub spectrum of source material), as well as the cuts and stops performed at key structural moments. The richness of the delay processing synchronized to different subdivisions (quarters, eighths, and sixteenths) over sampled chops, layers, exposed snares, and lead and backing vocals represent yet another historic acknowledgment in the recording domain, in this instance, paying tribute to Hip-Hop's dub heritage. Not only is this contextually multi-layered as a result (featuring many of Hip-Hop's stylistic predecessors in one track: Soul/Funk and Dub) but it also contributes to the hypnotic and viscerally rhythmical *feel* of the track (described as "noddage factor" in the listening notes; see "Section 1" of the companion website). The quarter delays on Conway's backings literally sync to the beat subdivisions, reinforcing them, and echoing his cold narrative, threats, and warnings, with a feeling of eerie, uneasy timelessness extending beyond the physical duration of the track.

At the heart of the track's narrative, the driving force is a well-selected sample, which notably becomes the PTU, and the foundation for further creative building blocks. Morris attests:

> I really don't try to take away from what's already there with the sample. I'll hear something that catches my attention and in my head I'll come up with a rough sketch of where I think I can take it and build around it from there. I'll add things and layer different sounds on top. In the case of "Biscotti Biscuit," the sample already had a bassline in it, but I tucked an 808 and played it alongside the original bassline just to give it more depth.
>
> *ibid.*

It can be argued that effective samplists possess the unique ability to recognize and select the ideal sound objects and sections from phonographic ephemera, in a way, solidifying (crystallizing) what is the best fit for their envisioned revision. "Super Slow Down" features many more psychedelic, even humorous, sections prior to Morris' 20-second selection, courtesy of whole-tone descending figures on the Rhodes and vibraphone, and an unstable major/minor tonality implied by the constantly variated chord voicings. Yet, the four-chord sequence that becomes Morris' PTU (00:39–00:59 in the original), settles into a darker, somber, and more contemplative harmonic progression, echoed only once more to close the track (at 01:38 in the original). This selection is crucial to "Biscotti Biscuit," as the transposition lightens the spectral information, but retains the harmonic implications of the stable minor tonality that sets the character and mood of the new track. The PTU variation inserted in the middle of both verses "to switch things up a bit" (ibid.) is another manifestation of the "changing same." The sub bass addition is both stylistically typical as a means to reinforce the "woofer" frequencies (a technique frequented by early architects of the subgenre such as Marley Marl and Hank Shockley), and timbrally functional in solidifying the record, filling up the spectral gap left by the upward transposition of the sample (and therefore the electric bass).

The disparate phonographic elements brought together for "Biscotti Biscuit" benefit both from Morris' painstaking "matching" strategies, but also his choice of stylistically appropriate tools. The listening notes on the companion website showcase an early impression suggesting that the beat had been made using an Akai MPC. The variables that triggered this tacit prediction were specifically the feel of the programmed groove, and the sonic character of the combined output.[14]

The combined texture of the 1975 master sampled here communicates yet another pertinent and unique characteristic to the hip-hop production. Not only are the 1970s a phonographically fertile and frequently chosen sampling ground (see Demers 2003), but sample-based Hip-Hop is defined by its interstylistic sonics and incorporation of vintage recording elements within its contemporary utterances. Written in 2019:

> The end result may sound nostalgic because of its obsession with the past, but it is really a manifestation of an inherent genetic trait that defines its very function and aesthetic. The sample-based condition has occupied such a large ratio of hip-hop outputs and such a large portion of its 44-year-long lifetime that it has elevated and celebrated the morphing, synthesis and interaction of old and new music to the forefront of its status quo and modus operandi.
>
> *Exarchos 2019b, 445*

The careful juxtaposition, layering, exponential staging (placement of record-within-record), and interaction between past and contemporary layers Morris performs, consciously showcase and celebrate the retro sonic within the flux of the new. The characteristic coloration of the 1975 master as a result of era-specific record production processes, communicates a sense of "elsewhen" preserved within the timbral gestalt, which Pickering defines as "the temporal distance brought about by recorded music," adding that: "Musical repeatability means that we are able to hear music from various previous periods and identify them, even on a decade-by-decade basis, by their *characteristic musical sounds*" (Pickering 2012, 25–26, emphasis added). In a more material sense, the sonic attributes infused by the vintage signal flows, practices, and recording and mixing media are manifested in the sampled spectra (often by softened timbral components due to analogue coloration); physically communicating staging "distance" beyond semiotics.

Returning to Figure 13.3, the sampled stage and perceived recording environment are placed on average in the "social zone" of distance, with some sounds occupying its close range (due to heightened timbral components exaggerated by the upwards transposition—particularly the vibraphone and Rhodes), while others appear even further, in the "public zone"—for example, the original snare-plus-reverberation sonic object still perceivable beneath the left-positioned new snare, as well as the electric guitar "wail" (Moylan 2020, 314). Conversely, Morris' new sounds, feature spectral envelopes that place—particularly the low end of—the beat within the "intimate"/"body space" zone, supporting the unwritten boom-bap condition for rhythmic embodiment (ibid.). Crucially, Morris is aware of the subtlety of proximity and impact versus volume, and deploys the staging awareness as both listener and maker:

> I really like the way old records were mixed back in the 1970s. The way they made chunky drums hitting hard without being on top of everything in the mix. I try to do the same thing. I don't think I can get in there fully with modern mixing equipment, but I try to make the drums sound like they were originally part of the OG [original] sample [...] I go through so many drum breaks to find the right sounding drum for the sample. Sometimes it takes hours. Then I tweak the drums and tap 'em out in the MPC until it sounds right. I try to think of how an actual drummer would manoeuvre around the music and I want it to sound like I'm catching the groove the same way.
>
> *Bozack Morris, Email interview with author, July 11, 2021*

Concluding Remarks

What has been exciting and informative about engaging with this ARM framework is that the analysis has deciphered some of the first, open listening engagement impressions (see companion website to this chapter). Yet, it has also offered plausible explanations for the perceived effects, feelings, and aural images, enriching and contextualizing the causal relationship, through an evaluation of the forces found within the sonic content and production phenomena residing inside the strata of the record. Somehow, after immeasurable open, deep, attentive, analytical, and critical listening sessions to both "Super Slow Down" and "Biscotti Biscuit" not only have these tracks not lost their entertainment value and magic; on the contrary, they have opened up new dimensions for their enjoyment and the appreciation of their infinitely revealing tapestry of sonic materials, as they continue to shape the song narratives with every listen.

Furthermore, the interaction between the ARM framework's (phenomenologically-inspired) multi-stage approach and the auto/ethnographic strategies deployed in this analysis, has led to an iterative theoretical-practical symbiosis. Tacit/practice-based knowledge has shaped the initial listening observations upon another producer's track, and—in turn—the content/context/character observations that followed, informed the semi-structured redaction of Bozack Morris' interview.[15] The resulting gradual (and often hermeneutic) interpretations enabled a variational deciphering[16] of the phonographic phenomena at the heart of Hip-Hop's source objects, as well as an evaluation of the staging mechanics in sample-based music making (the sample-based processes that shape the song). Therefore, the findings from the multi-method examination of the song's recording domain contain the potential to inspire sample-based creativity with an intrinsic (and contextual) understanding of the sonic materials and their (architectural) relationships within the hip-hop mix. As such, the analytical cycle empowers not only future sample-based beat making but potentially also the construction of fertile new source objects powering its evolution.

Notes

1 See companion website "Section 3—note 1" for a list of terms, definitions, and conventions relating to stylistic terminology used in this chapter. Boom-bap and Hip-Hop appear capitalized in the chapter when referring to the genres themselves as nouns and in lowercase when used as stylistic adjectives preceding other nouns.
2 As above: "Section 3—note 2."
3 The Akai MPC1000, released in 2005, is one of the more portable sampling drum machines in the company's series of Music Production Controllers (originally, Music Production Centres), which were first introduced in 1988 with the MPC60.
4 Loren Kajikawa (2015, 164) defines *chopping* as "the process of dividing a digital sample into any number of smaller parts and rearranging them to create a new pattern."
5 Joseph Schloss (2014, 72–78) explains that "the idea of sampling as an aesthetic ideal may appear jarring to individuals trained in other musical traditions, but it absolutely exemplifies the approach of most hip-hop producers"; adding that "this preference is not for the act of sampling, but for the sound of sampling: It is a matter of aesthetics."
6 Stereo Mike. 2007. *XLi3h*. EMI. 50999 513333 2 4.
7 Additional layers included a performance by DJ Sparky T (Greek DMC champion), using early Serato technology to scratch elements from the original master over the new beat.
8 Westside Gunn's *Who Made The Sunshine* (2020) is a case in point, as the artist proudly announced: "I wanna thank @daringer @beatbutcha_soi @conductorwilliams @justblaze @alanthechemist for the production of the album IT HAS NO SAMPLES!!!!!!!!" (@westsidegunn, October 14, 2020).
9 Neil Kulkarni describes it as "an era in which sampling hit a dizzying new depth of layered complexity and innovation" and "a sublime 10-year period from 1988 to 1998 in which hip hop was artistically more free than it had ever been before" (Kulkarni 2015, 78).
10 See companion website "Section 2" for the full interview.
11 See companion website "Section 3—note 4."
12 Although the focus of this chapter is on the sonic-domain implications of the rap performance and its production, readers/listeners should be forewarned of the explicit content in Conway (The Machine)'s lyrics, which is typical of Griselda Records' output. There is a personal note about the lyrics domain in "Section 1" of the companion website (specifically, in the "Second, analytical listen: hip-hop production" sub-section) addressing my relationship with rap performance as sonic domain from a non-native listener perspective.
13 I am indebted to Bozack Morris for his informative and enthusiastic involvement in the interview exchange, sharing personal insights on process, vision, and sample-based aesthetics, which have hugely enriched this analysis.
14 I have covered elsewhere how the rhythmic/swing and sonic variables determined by MPCs' operating script and hardware specifications create *affordances* mapped to characteristic music and recording domain stylizations—see Exarchos (2019a).

15 See, for example, the evolving listening journey in "Section 1" of the companion website, offering progression from open listening through to early interpretations, analytical, and repetitive/attentive listening, which led to the formation of the Bozack Morris interview in "Section 2."

16 An active "probing activity" phenomenologists describe as "*variational method*" (Ihde 2012, 23, original emphasis).

References

Demers, Joanna. 2003. "Sampling the 1970s in Hip-Hop." *Popular Music* 22, no. 1: 41–56.

Exarchos, Michail. 2019a. "Boom Bap Ex Machina: Hip-Hop Aesthetics and the Akai MPC." In *Perspectives on Music Production: Producing Music*, edited by Russ Hepworth-Sawyer, Jay Hodgson, and Mark Marrington, 32–51. New York: Routledge.

———. 2019b. "(Re)Engineering the Cultural Object: Sonic Pasts in Hip-Hop's Future." In *Innovation in Music: Performance, Production, Technology, and Business*, edited by Russ Hepworth-Sawyer, Jay Hodgson, Justin Paterson, and Rob Toulson, 437–54. New York: Routledge.

Gibson, David. 2008. *The Art of Mixing: A Visual Guide to Recording, Engineering, and Production*, 2nd ed. Boston, MA: Course Technology.

Goldberg, David Albert Mhadi. 2004. "The Scratch Is Hip-Hop: Appropriating the Phonographic Medium." In *Appropriating Technology: Vernacular Science and Social Power*, edited by Ron Eglash, Jennifer L. Croissant, Giovanna Di Chiro, and Rayvon Fouché, 107–44. Minneapolis, MN: University of Minnesota Press.

Ihde, Don. 2012. *Experimental Phenomenology: Multistabilities*, 2nd ed. Albany, NY: State University of New York.

Jones, LeRoi. 1969. "The Changing Same (R&B and New Black Music)." In *Black Music*, 180–211. London: MacGibbon and Kee.

Kajikawa, Loren. 2015. *Sounding Race in Rap Songs*. Berkeley, CA: University of California Press.

Krims, Adam. 2000. *Rap Music and the Poetics of Identity*. Cambridge: Cambridge University Press.

Kulkarni, Neil. 2015. *The Periodic Table of Hip Hop*. London: Ebury Press.

Lewis, George E. 2017. "Improvised Music after 1950: Afrological and Eurological Perspectives." In *Audio Culture: Readings in Modern Music*, 2nd ed., edited by Christoph Cox, and Daniel Warner, 385–98. New York and London: Bloomsbury Academic.

Marshall, Wayne. 2006. "Giving up Hip-hop's Firstborn: A Quest for the Real after the Death of Sampling." *Callaloo* 29, no. 3: 868–92.

Moore, Allan F., and Ruth Dockwray. 2008. "The Establishment of the Virtual Performance Space in Rock." *Twentieth-century Music* 5, no. 2 (September): 219–41. https://doi.org/10.1017/S1478572209990065.

Moylan, William. 2020. *Recording Analysis: How the Record Shapes the Song*. New York and London: Routledge, Focal Press.

Pickering, Michael. 2012. "Sonic Horizons: Phonograph Aesthetics and the Experience of Time." In *Time, Media and Modernity*, edited by Emily Keightley, 25–44. London: Palgrave Macmillan.

Ramsey, Guthrie P., Jr. 2003. *Race Music: Black Cultures from Bebop to Hip-Hop (Music of the African Diaspora)*. Berkeley, CA: University of California Press.

Rose, Tricia. 1994. *Black Noise: Rap Music and Black Culture in Contemporary America*. Middletown, CT: Wesleyan University Press.

Schloss, Joseph G. 2014. *Making Beats: The Art of Sampled-Based Hip-Hop*, 2nd ed. Middletown, CT: Wesleyan University Press.

Sewell, Amanda. 2013. "A Typology of Sampling in Hip-Hop." PhD diss., Indiana University.

Gunn, Westside (@westsidegunn). 2020. "FLYGOD." Instagram photo, October 14, 2020. https://www.instagram.com/westsidegunn/.

Williams, Justin A. 2010. "Musical Borrowing in Hip-Hop Music: Theoretical Frameworks and Case Studies." PhD diss., University of Nottingham.

Discography

Conway (The Machine). 2018. "Biscotti Biscuit." *Blakk Tape*. Griselda Records, Fashion Rebels.

Gunn, Westside. 2020. *Who Made The Sunshine.* Shady Records.

Lubat, Bernard. 1975. "Super Slow Down." *Vibrations—Volume 2*. Tele Music. TM 3051.

Mike, Stereo. 2007. *XLi3h*. EMI. 50999 513333 2 4.

14

"IT AIN'T BUT ONE KIND OF BLUES"

Kid Koala's Bluesy Embrace of the Fragmented ("1 bit Blues," 2012)

Ragnhild Brøvig

Introduction

In 2012, the Canadian turntablist, multimedia artist, and music producer Kid Koala (aka Eric San) released his fourth studio album, *12 bit Blues* (Ninja Tune), which consists of heavily manipulated and juxtaposed samples from old blues records. It is the opening track of this album, "1 bit Blues," that I have chosen to analyze for this chapter. For my analytical framework, I will use William Moylan's (2020) three-step process. In the "observation stage," I will describe the track's time signature, tempo, and structure; the sounds present in the track; and these sounds' characteristics. In this chapter, these descriptions will not be presented sequentially but instead intermingled with what Moylan calls the "evaluation stage," wherein I will position these sounds in relation to the common stylistic features of the blues. This vantage point will, during the "conclusion stage," lead me to a discussion of the recording paradigms and expectations that, for me, frame this track's compelling peculiarity.

The blues genre arose and evolved in the late 1800s and early 1900s, and the earliest blues recordings are from the 1920s. Several of them came out of the Mississippi Delta and featured very influential singer-guitarists, so the Delta gained a reputation as the geographical origin of the blues despite the fact that there were also important blues pioneers working elsewhere, including Mamie Smith and Bessie Smith on the East Coast.[1] Drawing upon a black oral music tradition among enslaved people and their descendants in the American South (Zheng, 2013), as well as laying the foundation for many succeeding popular music genres, including jazz, R'n'B, and rock and roll, the blues is often considered an American "roots music."

Kid Koala is not a blues musician but is instead known for his eclectic selection of samples, his virtuosic turntable techniques, and the intriguingly playful

DOI: 10.4324/9781003089926-20

and bizarre—and, to my ears, pleasant and interesting—music that results from this approach. In contrast to his other recordings, *12 bit Blues* was primarily created using the E-mu SP-1200 drum machine and sampler. In a *Spin* interview, he describes his fascination with this machine from the time he first got involved with DJing in high school. The machine had already become iconic thanks to its frequent use, including its prominence in the hip-hop scene during the late 1980s and early 1990s, when producers took advantage of its extended ten-second sampling time to sample truly durable sections of existing music instead of only short sounds (such as drum hits), as had been the case before. Kid Koala notes that "1 bit Blues" was the first song he created after having bought this machine:

> I made a few chops in the first four pads. I didn't bother learning the sequencer; instead I just started [to] roll tape, and decided to just punch it out and that, essentially, became the methodology for all of these tracks … I just decided for ease's sake—and also it was kind of more fun too—to just do it on the fly. The whole idea that if you stop pushing buttons the music will stop, the whole danger there. Like I can't go make a coffee or something. I had to keep playing, I literally had to perform using these pads.
>
> *Weingarten, 2012*

Spin journalist Christopher Weingarten describes the album as "The turntablist's version of going acoustic" (Weingarten 2012).[2] But it is not only Kid Koala's approach to sampling that initiates a dialogue between two quite distinct aesthetic paradigms, but also the sonic result of that approach, which is part of why I find this track so engaging. The focus of this chapter, then, is on this particular dialogue, including the ways in which this track reimagines the blues by embracing the aesthetic potential of the temporally fragmented.

Kid Koala's "1 bit Blues"

So why does Kid Koala characterize this album, and this specific track, as blues? While there are several subgenres of blues, they all share certain characteristics, and "1 bit Blues" does so as well.[3] The instrumentation of blues varies—its earliest styles (including the Delta blues) consisted primarily of voice and guitar, whereas the style that emerged during the middle of the 20th century (the so-called urban blues) often expanded the vocal accompaniment to an ensemble typically consisting of drums, double or electric bass guitar, acoustic or electric guitar, and sometimes also piano, harmonica, organs, and saxophone. In "1 bit Blues," then, we hear piano, drums, bass guitar, electric guitar, a Farfisa organ, and various vocals. There is thus nothing particularly unconventional about the instrumentation of "1 bit Blues" in relation to the blues tradition save for the scratching and synth sounds,[4] as well as the fact that the instruments and vocals derive from samples. The time signature of the blues is usually 4/4, and slow blues often feature triplet eighth-note subdivisions of each beat, lending these songs a 6/8 or 12/8 feel. Similarly, "1 bit Blues" is in the time signature of 4/4 with a subdivision of eighth-note triplets articulated most notably by the drums and the bass guitar (as such, it could also be heard as having the time signature of 6/8 or 12/8). The tempo of "1 bit Blues" (78.5–106.9 BPM) is well within the range of the blues tradition, though it is inconsistent. This inconsistency of tempo is partly the result of sampling from live recordings, which seldom hold to a consistent tempo, and partly the result of Kid Koala's live arrangement of the samples—when samples are pushed in and out by hand, the last beat is often shortened or prolonged by a number of milliseconds, resulting in measures with slightly uneven durations. For example, relatively few of the 17 measures in the first verse have the same BPM—the BPM of the first three measures of each cycle varies between 78.6 and 79.3, whereas the BPM of the last few measures varies between 78.5 and 106.9, which, at its extreme, represents a deviation of 794.9 milliseconds from the verse's otherwise average tempo of 78.9 (this deviation, in turn, equals the duration of a quarter note at a tempo of 79 BPM). Although sampling is not typical for blues, this unevenness in tempo is quite typical since the blues is mainly a live genre.

Blues is often associated with a 12-, 8-, or 16-bar structure characterized by repeated instrumental- and vocal phrases. "1 bit Blues" can be interpreted as having the following structure: Introduction—Verse 1 (A)—Chorus 1 (B)—Pre-Chorus (C)—Chorus 2 (B)—Bridge (D)—Verse 2 (A)—Chorus 3—Coda (see Figure 14.1).[5] While this is not one of the typical structures of the blues, it is, like the blues, based on the reappearance of patterns in the instruments and vocals. Blues is typically oriented around the tonic, subdominant, and dominant chords of the key signature as well as the pentatonic or hexatonic scale (the pentatonic scale consists of five tones per octaves—1, 3, 4, 5, 7—and the hexatonic scale is the pentatonic plus a flattened fifth). "1 bit Blues" primarily centers itself upon one chord alone, but this is quite common in blues as well—famous examples of one-chord blues songs include Howlin' Wolf's "Smokestack Lightning," Muddy Waters' "I'm a Man," and Willie Dixon's "Spoonful." The dominating (E) minor

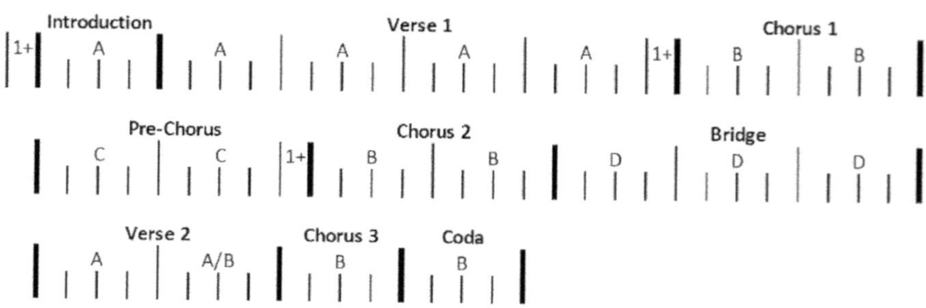

FIGURE 14.1 The structure of "1 bit Blues" (*12 bit Blues*, 2012, Ninja Tune) by Kid Koala.

seventh chord of "1 bit Blues" is a common blues chord, and both the vocals of the chorus and the Farfisa organ that characterizes the pre-chorus are presenting a melodic line in the E-minor pentatonic scale. Furthermore, a deep, fuzzy synth bass implies a typically bluesy harmonic progression during the later choruses, pre-chorus, and bridge (its licks are based upon the root tones of the tonic, subdominant, and dominant chords).

Although samples are not common in blues as such, the appropriation of prior songs is. The samples here are furthermore from old blues recordings, of course, which contribute both the signature blues sound and a concrete indexical referent to the blues. This is especially true for the spoken vocal of the first verse, which is a sample from a concert recording (by CBS Studio in New York) from 1968 by legendary blues singer and guitarist Son House (aka Edward James House Jr.). Son House was, together with Blind Lemon Jefferson, Robert Johnson, Charley Patton, and Howlin' Wolf, among others, a Delta blues pioneer. The sample is from his introduction to his performance of "Ain't but One Kind of Blues,"[6] so his delivery lacks any melody, rhyming, or repetitive rhythmic patterns (though Son House often speaks quite rhythmically and does so here as well). Kid Koala does not manipulate the tempo or rhythm of the vocal delivery (other than adding echo to certain words) but does, in the first verse, pitch it down by about five semitones. On the one hand, this distorts the delivery by making it sound unnaturally low, thus twisting the blues style into something else. On the other hand, this pitched-down twist on the vocal presents an exaggerated version of the typically low-pitched, harsh voices of these early blues singers, thus emphasizing the track's link to the genre. Moreover, it also functions as a comment on the raw emotional expression associated with blues vocalists, as the pitch-shifting hyperbolizes the perceived depression and agony conveyed by Son House's vocal performance. The association of the track with the blues and its typical conveyance of deep-seated emotions is further emphasized by Kid Koala's loyalty to Son House's lyrical message (though he does leave out some parts of the original speech). For example, during the third verse, the excerpt includes Son House explaining that the *true* blues is informed by a certain feeling—the kind of feeling you get when you are deceived by your loved one; the kind that will make you do anything.

The vocals of the chorus differ from those in the verse in several ways. We again hear a male voice but not that of Son House; this voice is not pitch-shifted but instead quite natural; the vocal part is melodic, consisting of three words sung with syncopation and melisma along the E-minor pentatonic scale; it fills one measure only and alternates (in a call-and-response) with a measure of siren-like sliding portamento synth; and it is supported by a female background vocal. This vocal also seems to be from an old blues record, replete as it is with a powerful and expressive delivery, syncopations, and melisma—all characteristic features of the blues. The pre-chorus is characterized by a third scratched-in male vocal that again speaks in an unnaturally low (that is, pitched-down) voice, and in the bridge, as well as in the song's intro, a fourth vocal sample appears, stating simply: "The kid is in rare form tonight." The last two words of this vocal line are repeated numerous times at varying rhythmic intervals and positions within the mix (often panned from side to side), which is partly the result of playful experimentation with tape-delay feedback. In the second part of the bridge, two heavily manipulated vocal samples engage in a call-and-response dialogue. While the technological manipulations of these vocals might well seem out of place in a conventional blues context, the original soulful performances boast strong associations with the blues. So does the call-and-response pattern that appears between the lead vocal and the portamento synth in the chorus, and between the two manipulated vocals in the bridge, which several scholars have traced to the genre's roots in the "shouts and hollers" of field slaves (see, for example, Oliver 1998).

The repetition of short riffs and hook lines that appears throughout the track, both in the instrumental- and vocal parts, is a common feature of the blues as well, although their sampled and looped nature in "1 bit Blues" is obviously idiosyncratic. Indeed, and partly due to its origins before the development of recording, blues is instead traditionally associated with live performances—that is, with musicians situated in the same room at the same time playing a spatiotemporally coherent performance. While Kid Koala's sampling and looping are not particularly bluesy, though, his *approach* to these techniques clearly is. By performing this arrangement of samples live and with improvisation, he invests in the blues values of process and risk, and the blues artist's faith in the "spontaneity of the long take" (Eisenberg 2005, 122).

The timbre of the various instrumental sounds is quite low in sound quality and includes vinyl crackle, tape hiss, and a limited frequency range, the result of being sampled from vinyl records of old recordings. Moreover, the SP-1200 features only 12-bit sample depth (to which the album title alludes), which results in a lower sound quality than today's recording standard of 24 bit. Joseph Auner (2000) reminds us that the contemporary use of old and outmoded sounds and machines is related to how listeners and musicians generate meaning by locating themselves in a tradition while also transforming it. That is, old and outmoded sounds and machines may reinforce a sense of authenticity by communicating a certain time and place. Stan Link adds that noise has a certain tactility to it as well: "Radio static becomes the feel of a tuning knob. The crackle of dust becomes the vinyl itself. The hum of tubes evokes their warm temperature. A stylus dropping carries the weight of a tone arm … Noise, then, reinforces the authenticity of listening even as it destroys that of the reproduction" (2001, 38–39). It also may allow for an almost "tangible experience of time" (39), which may itself reinforce the sense of authenticity derived from hearing sounds from the past. However, when situated in a context of the contemporary, these outmoded sounds and machines can also make us even more aware of the present, which adds further complexity to the experience.[7] Though the track's individual instrument sounds are mostly taken from different samples, they are spatially arranged in such a way that they are readily pictured as occurring in the same room or "sound stage" (Moylan

2020, chapter 8), both in terms of ambiance and acoustic "placement" within the mix or "sound box" (Moore 2012, 29–44), thus sonically complying with the spatial coherence of the blues.

While all of these aforementioned features explain why Kid Koala characterizes this track as blues, the many elements in "1 bit Blues" that are not bluesy at all, and that therefore stand out as incongruous amid the track's gestures toward conventional blues style, are also relevant in terms of how we conceive the overall track. That is, "1 bit Blues" embraces temporal fragmentation while still evoking strong associations with the spatiotemporally coherent performance- and recording paradigm, thus creating a captivating tension between the two. Most noteworthy is Kid Koala's extensive and exposed use of samples, as well as the sound generated by the recording medium itself, which testifies to a recording paradigm quite different from that of the conventional blues. The latter is located in his integration and aesthetic use of scratching and looped sounds throughout the track—sounds that are not at all common in blues. There are also occasional appearances of "signal dropouts,"[8] that position Kid Koala in a digital context. For example, at three points during the verse (the transitions between measures 3 and 4, 7 and 8, and 11 and 12), all the instrument sounds fall away. This signal dropout lasts only for the duration of a single triplet eighth note, but it grabs one's attention nevertheless, since the resulting silence represents a pronounced break from the relentless crackling background noise of the rest of the track.

Except for those sounds that are unique to the given recording technology itself, the track consists of nothing but samples. For example, the verse, as well as the introduction, consists of four instrumental samples (piano, drums, bass guitar, and electric guitar) and a vocal sample. Most of these samples last for one measure that is repeated, but Kid Koala does not consistently "play" them in their entire form—these respective iterations could, as such, be heard as different samples, but their shared sound and content suggest otherwise. The sampled piano presents the same chord—an Em7, voiced with chord tone 7 as the highest—on beats 1, 2, and 3, then rounds out the phrase on beat 4 with an ornamental trill. This sample is repeated in its entirety throughout the verse save for the last measure of each cycle, when its first half is played twice instead. The drum sample likewise has two parts, each lasting for one measure. The first part consists of a bass drum articulating the downbeat and its triplet eighth-note pickup, a snare marking the back beat, and a ride cymbal pattern alternating between shuffled-triplet 8th and 16th notes, supplying a prominent swing feel to the track's groove. The second part, which concludes each cycle, is more active as it builds to the next cycle, and each iteration varies somewhat depending on Kid Koala's treatment of the sample in question.

The practice of sampling can represent a kind of historically and culturally conditioned engagement with a musical genre's roots or an act of homage to earlier artists. Kid Koala's album, with its extensive use of samples from old blues records and enthusiastic embrace of typical blues characteristics, can certainly be interpreted as an homage record. Yet it is very much of the present in its exaggerated, almost subversive, approach to those samples and characteristics. As such, it could be situated within the larger subcultural "retro movement" that Elisabeth Guffey describes as half ironic, half longing, and entirely "unconcerned with the sanctity of tradition" (2006, 10–11). This retro movement can further be described as playful and humorous, as it simultaneously seeks to evoke *and* to subvert otherwise predictable associations. In this unconventional example of the blues, in fact, Kid Koala starts a dialogue between two quite distinctive recording paradigms, which is foreshadowed by the track's introductory call to arms: "The kid is in rare form tonight."

A Dialogue between Two Recording Paradigms

When we listen to "1 bit Blues," we are unlikely to pay much attention to all the details described above, yet they contribute to our general impression of the track. And this impression, in turn, shapes our affective response to it. I find this track aesthetically pleasing but also interesting and somewhat strange and funny. While such responses are often difficult to understand and explain, my fascination with this track starts with the ways in which it twists the blues, even as it comments on the very different recording paradigms within which it operates so successfully.

In addition to recreating sounds, the recording medium also always partakes in the final result or "primary text" (Gracyk 1996, 21). Yet, it can still be made to behave in distinctive ways—what I have elsewhere described as recording paradigms that emphasize either a transparent or an opaque mode of production (Brøvig-Hanssen 2013, 2018). In blues, as in jazz, folk, and classical, the recording medium has traditionally served to document or construct a spatiotemporally coherent sonic performance. As Timothy Day (2000, 23–34), Jonathan Sterne (2003, 219, 235), and Colin Symes (2004, 54–59), among others, have pointed out, the recording medium is neither neutral nor objective in this regard, but perceptual transparency is generally preferred to opacity. In his discussion of jazz, Evan Eisenberg understands this rather conservative recording paradigm as an act of resistance to that which would disrupt the recording's purported faithfulness to a pre-existing performance: "as improvisers they believed, even more passionately than old-fashioned classical musicians, in the spontaneity of the long take" (2005, 122). Today, recordings in which subtle editing is found to be perfectly acceptable likely outnumber strictly conservative recordings in which nearly all forms of technological manipulation have been rejected. Some listeners nevertheless persist in their view of recording technology as a transmitting medium whose main function is to archive musical performances. Others happily sacrifice veracity for perfection but still treat the recording medium as a window onto an *imaginable* continuous and spatiotemporally coherent performance, even if it has been constructed for the recording alone. The blues adheres to this conservative recording paradigm and its privileging of the performance and the recording medium's recreational, curational, and polishing capacities over its explicitly creative capacity.

In the alternative (opaque) recording paradigm, of course, the fabrication of the event is exposed, including the music's spatiotemporal fragmentation and juxtapositions. Here, "space and time [become] equivalent forces to be worked with like any other material substance of music," Thom Holmes writes (2002, 78).[9] The recording medium takes an active part in the creative process, in terms of not only coloring or otherwise shaping sounds but also *generating* unique sounds that carry meanings of their own. For example, in Kid Koala's music, the scratching, cut-off sounds, abrupt transitions between sound sequences, signal dropouts, exposed "quotation marks" of samples, and pitched-down and stuttering vocals all draw attention to the temporally fragmented nature of the track and the technological mediation involved in the production of the song.

While this active, opaque mode is very common in today's music (and has been so for several decades), it is not common in all musical contexts, including blues. As Andrew Goodwin observed when discussing the emergence of sampling, recording paradigms informed by "the old ideologies and aesthetics are still on the menu" (1990, 272). We remain, in other words, quite context sensitive with regard to how the recording medium is used. The "laying bare" of techniques and artifice, or the insertion of self-referential narrative disruptions, has interested the Russian formalists, including Viktor Shklovsky, in particular. To Shklovsky, such "riddles" question the common assumption that pleasure in art derives from seeing something new. Instead, he argues, pleasure in art stems from seeing something familiar *defamiliarized* (1989)—that is, when the habitual and consensus view of an object, including its collective associations and discursive entanglements, is disrupted or presented with an unexpected twist. At the moment of this encounter, we think in parallel, with the familiar and the defamiliarized simultaneously present in our minds. And it is this "double meaning," he argues, that compels pleasure, as well as attention, narrative impetus, and interest. In contrast to the aesthetic experience of immersion in a recording, such opaque mediation may draw the listener *out* of this absorption and into an insistent and gratifying metalevel regarding the recording.

Sounds generated by the recording medium itself carry with them the peculiar quality of operating at the musical metalevel, or what is referred to in theater and film as "breaking the fourth wall"—that is, when the narrator or the medium that narrates takes part in the narrative, thus fulfilling a dual function. For example, signal dropouts take away the music but also constitute it; scratching manipulates the music while also being part of it; looping and cutups repeat and/or disrupt the music while also creating a peculiar musical rhythm of their own. This duality does not make these sounds less musical, but their ambiguous nature as at once musical sounds *and* sounds that disrupt the music can trigger a pleasant feeling of bewilderment or uncanniness. In "1 bit Blues," this feeling of a metalevel—of some sounds straddling the interior and exterior of the music—is especially profound when set against an otherwise powerful evocation of traditional blues, which, again, adheres to a paradigm where the recording medium is generally in the attentional background. When these two paradigms engage in such an explicit dialogue, it changes our perceptions of both. Defamiliarization makes "the stone stony" (1990, 5–6), Shklovsky waxes metaphorical, and in the same way, recording technology's doubled function (to mediate and be that which is mediated) and its respective aesthetic paradigms are brought into the perceptive and reflective foreground through their contrasts.

Related to the notion of a dialogue between two recording paradigms is how "1 bit Blues" deflects life toward the mechanical and, conversely, the mechanical toward life. Without going into the complexities that Henri Bergson locates in the phenomenon of laughter, I must nevertheless mention his notion that laughter often results when life is to some extent deprived of its "vital impetus" and instead takes on the logics of the mechanical (Bergson 1980). To Bergson, "life" represents the ever-changing and spontaneous, whereas the "mechanical" represents rigidity, predictability, and repetitiveness. According to him, repetition in the form of the duplicate belongs to the mechanical, since life is by definition unrepeatable, and thus, he argues, humans often find it funny when life takes on this mechanical quality. In "1 bit Blues," the vitality of the blues—as located in its privileging of the live performance, spontaneity, and improvisation—is entangled in the sampler's mechanical tendencies: here, the instrumental sounds are interrupted after only one measure or less, and the repetitions of riffs are, in contrast to blues vamping, *entirely* similar. Conversely, the machine is made to subscribe to certain human or vital tendencies as well, including Kid Koala's live arrangement of his blues samples and his faith in the long take and resultant emphasis on spontaneous improvisation, process and risk, and rawness and honesty. Here, the blues is invested with qualities associated with the machine, while the machine is invested with qualities associated with the human/blues. The two do not merge but instead flourish in a playful, humorous, and compelling tension.

When Bergson states that laughter derives from "something mechanical encrusted upon the living" (Bergson 1980, 84), he does not reserve the mechanical for machines alone but also applies it to humans' habitual ways of understanding and behaving. Laughter may thus, he argues, awaken us from our confident and settled ways of living. As such, "1 bit Blues" could be interpreted as an unspoken but implied critique of habitual rigidity in the ways in which we conceive genres. This is *almost* made explicit through the self-reflective irony that Kid Koala evokes during the first verse, which starts with Son House insisting that blues is not a plaything, before complaining that the "youngsters today take anything and make the blues out of it." He then proclaims: "It ain't but one kind of blues!"—a statement thrown into relief by a signal dropout. As such, the track as a whole functions as a dialogue between Son House and Kid Koala (or between respective musical paradigms), wherein one declares, "It ain't but one kind of blues," and the other responds, "Oh really?"

Notes

1 David Brackett problematizes the common narrative of a male, authentic, virile folk blues as a definite iteration, arguing that prominent early commercial recordings of blues, and recordings of female blues musicians, in particular, were equally consequential to the genre (he uses "Crazy Blues" [1920] by Mamie Smith and Her Jazz Hounds as an example of both; Brackett 2016, 71). Brackett also questions the common assumption that the blues is an exclusively African American tradition by pointing to the heterogeneity of both blues performers and audiences throughout history. Nevertheless, he maintains that blues music has historically signified African American–ness to the greater public and can thus be described as a black genre, even if it came about in a context of racial interaction (2016, 76–78).
2 Kid Koala points out that, whereas most of the track was made "on the fly"—that is, recorded while pushing the machine's buttons to play and mute samples—some elements, including the scratching, were added in a DAW sequencer program afterward.
3 For a lengthy discussion of the blues and some of its typical stylistic characteristics, see, for example, Keil (1992), Oliver (1998), and Wald (2010).
4 "Hard Time Killing Floor" (2000) by R. L. Burnside is another example of a blues track that features scratching.
5 "Verses" are here interpreted as the sections with the voice of Son House, and "choruses" as the sections with the vocal hook. The pre-chorus and bridge are neither clear verses nor clear choruses.
6 This recorded performance, including his introduction to it, is also featured on the motion picture soundtrack of *Black Snake Moan* (2007) by Craig Brewer.
7 For a discussion of how the combination of outmoded and contemporary sounds, or of noise and silence, may draw attention to the present, see Brøvig-Hanssen and Danielsen (2016, chapter 4), Danielsen and Maasø (2009, 132–35), and Théberge (1997, 205–6).
8 Signal dropout, as a concept and an aesthetic effect, is addressed in Auner (2000), Brøvig-Hanssen and Danielsen (2016), and Danielsen and Maasø (2009).
9 Holmes is here discussing the renewed interest in the recording medium's opacity following the introduction of magnetic tape.

References

Auner, Joseph. 2000. "Making Old Machines Speak: Images of Technology in Recent Music." *Echo: A Music-Centered Journal* 2/2. Accessed December 20, 2021. http://www.echo.ucla.edu/Volume2-Issue2/auner/auner.pdf.

Bergson, Henri. 1980 [1900]. "Laughter." In *Comedy*, edited by Wylie Sypher, 61–190. Baltimore, MD: John Hopkins University Press.

Brackett, David. 2016. *Categorizing Sound: Genre and Twentieth-Century Popular Music*. Berkeley, CA: University of California Press.

Brøvig-Hanssen, Ragnhild. 2013. "The Magnetic Tape Recorder: Recording Aesthetics in the New Era of Schizophonia." In *Material Culture and Electronic Sound*, edited by Frode Weium and Tim Boon, 131–57. Washington, DC: Smithsonian Institution Scholarly Press/Rowman and Littlefield Publishers.

Brøvig-Hanssen, Ragnhild. 2018. "Listening *To* or *Through* Technology: Opaque and Transparent Mediation in Popular Music." In *Critical Approaches to the Production of Music and Sound*, edited by Samantha Bennett and Eliot Bates, 195–210. New York/London: Bloomsbury Academic.

Brøvig-Hanssen, Ragnhild, and Anne Danielsen. 2016. *Digital Signatures: The Impact of Digitization on Popular Music Sound*. Cambridge, MA: MIT Press.

Danielsen, Anne, and Arnt Maasø. 2009. "Mediating Music: Materiality and Silence in Madonna's 'Don't Tell Me.'" *Popular Music* 28/2: 127–42.

Day, Timothy. 2000. *A Century of Recorded Music: Listening to Musical History*. New Haven, CT: Yale University Press.

Eisenberg, Evan. 2005. *The Recording Angel: Music, Records and Culture from Aristotle to Zappa*. New Haven, CT: Yale University Press.

Goodwin, Andrew. 1990. "Sample and Hold: Pop Music in the Age of Digital Reproduction." In *On Record: Rock, Pop, and the Written Word*, edited by Simon Frith and Andrew Goodwin, 258–73. London: Routledge.

Gracyk, Theodore. 1996. *Rhythm and Noise: An Aesthetics of Rock*. Durham, NC: Duke University Press.

Guffey, Elizabeth E. 2006. *Retro: The Culture of Revival*. London: Reaction.

Holmes, Thom. 2002. *Electronic and Experimental Music*. New York: Routledge.

Keil, Charles. 1992. *Urban Blues*. Chicago: University of Chicago Press.

Link, Stan. 2001. "The Work of Reproduction in the Mechanical Aging of an Art: Listening to Noise." *Computer Music Journal* 25/1: 34–47.

Moore, Alan F. 2012. *Song Means: Analysing and Interpreting Recorded Popular Song*. Farnham/Burlington: Ashgate.

Moylan, William. 2020. *Recording Analysis: How the Record Shapes the Song*. New York: Routledge.

Oliver, Paul. 1998. *The Story of the Blues*. Boston, MA: Northeastern University Press.

Shklovsky, Viktor. 1989. "Art as Technique." In *The Critical Tradition: Classic Texts and Contemporary Trends*, edited by David H. Richter, 774–84. New York: St. Martin's Press.

Sterne, Jonathan. 2003. *The Audible Past: Cultural Origins of Sound Reproduction*. Durham, NC: Duke University Press.

Symes, Colin. 2004. *Setting the Record Straight: A Material History of Classical Recording*. Middletown, CT: Wesleyan University Press.

Théberge, Paul. 1997. *Any Sound You Can Imagine: Making Music/Consuming Technology*. Hanover, NH: University Press of New England.

Wald, Elijah. 2010. *The Blues: A Very Short Introduction*. Oxford: Oxford University Press.

Weingarten, Christopher. 2012. "Kid Koala Breaks Down Scratch-Pop Masterwork '12 Bit Blues': Full Album Stream." *Spin*, September 12, 2012. Accessed December 20, 2021. https://www.spin.com/2012/09/kid-koala-breaks-down-scratch-pop-masterwork-12-bit-blues-full-album-stream/.

Zheng, Jianqing. 2013. "The Robert Johnson 'Blues Trail': A Photo Essay." *Southern Quarterly* 50/3: 69–78.

15

"THREE AND A HALF MINUTES OF ATTITUDE"

Vocal Delivery, Groove, and Production in Azealia Banks' "212" (2014)

Anne Danielsen and Thea Sørli Paulsrud

"212" is the first single from Azealia Banks' debut album *Broke with Expensive Taste* (2014).[1,2] The song is co-written and produced by Lazy Jay, an electro-house duo associated with the Dutch house scene of the early 2010s who first released its beat under the title "Float My Boat" (2009).[3] The song immediately became a hit and received enthusiastic reviews in the music press; according to Pitchfork's reviewer, for example, Banks has "the unpredictable vocal range of a former theater geek" and "clicks between characters and styles casually, effortlessly. No seams."[4] Similarly, a *Guardian* writer commented on the quality of Banks' performance as well as the composite character of the song: "It's basically three different songs fused together to create a startling three and a half minutes of attitude."[5]

The aim of our analysis is to unpack the secrets to this recording's success by focusing in particular on the track's groove, production, vocal delivery, and overall stylistic fusion of rap, reggaetón, pop, and house. Following a brief introduction to the history of the song and the artist, we will look closely at its groove and production before addressing the impact of the vocal delivery in the context of its influences from battle rap. Ultimately, we will explore the ways in which the musical form of "212" negotiates among the various conventions of the musical styles involved.

The Beat, the Record, and the Artist

The beat of "212," originally named "Float My Boat" by its producers, the Belgian electro-house duo Lazy Jay (Jef and Toon Martens), was already a hit on the Dutch club scene (Lyons 2019; McNally 2016) when Banks picked it up. It is built on a reggaetón rhythm that is joined to production techniques typical of electronic dance music.

DOI: 10.4324/9781003089926-21

The same reggaetón rhythm also becomes a characteristic and prominent aesthetic feature of "212" in turn. Reggaetón can be traced back to the musical underground in 1980s and 1990s Puerto Rico, and a sound marked by cultural exchange during its migration around the Caribbean (Rivera-Rideau 2015). Defined by its mixture of sounds, vocal styles, and other stylistic elements from genres such as dancehall, hip-hop, and Trinidad soca, reggaetón can be characterized as the musical form that unites Puerto Rican and other African-based music traditions (Rivera-Rideau 2015; Rivera-Rideau and Torres-Leschnik 2019). During its initial travels in the Caribbean, it acquired its characteristic beat from a rhythm known as "Dem Bow," itself from a track released by dancehall legend Shabba Ranks in 1991 and produced by the Jamaican producer duo Steely and Clevie (Marshall 2008). This actual rhythm, however, was first used by the same producer duo in an instrumental track called "Fish Market" from 1989. In October 2021, interestingly, Steely and Clevie filed a copyright infringement lawsuit against numerous reggaetón acts for misappropriating the rhythm and using other original elements from "Fish Market" without permission.[6] Warner Music, Sony/ATV Music Publishing, and Universal Music Latin Entertainment are among those entities being sued, and the case involves songs including the superhit "Despacito" from 2017. The lawsuit is a compelling, relatively recent example of the conflicts of interest that can arise when a subcultural musical style is adopted into the "mainstream."

According to Samponaro (2009), it was during reggaetón's Puerto Rican phase that "ties to the US mainland and the corresponding interest of the music industry" paved the way for its crossover to the US pop, Latino, and hip-hop charts in the 2000s (2009, 491), and the year 2004 is often acknowledged as the time when reggaetón arrived in the US commercial music industry (Rivera 2014; Rivera-Rideau 2015). Daddy Yankee's "Gasolina" and N.O.R.E's "Oye Mi Canto" are noteworthy examples of it that circulated in the mainstream media at the time. By the time of the release of "212" in 2011, that is, the act of combining Afro-Latina/o musical practices and elements with pop and hip-hop was far from unusual (Rivera 2014; Rivera-Rideau and Torres-Leschnik 2019). Latin crossover acts such as Ricky Martin, Jennifer Lopez, and Marc Anthony had been attracting enormous Latina/o and non-Latina/o mainstream audiences for more than a decade. Such crossover acts have, of course, fostered their own debates. Before it began to meet with commercial success, the Puerto Rican underground music scene was censored, and in 1995 actually banned, amid accusations that it was closely related to violence and other illegal activity (Rivera 2014). As reggaetón found more acceptance in Puerto Rico in the late 1990s and early 2000s, then spread to the United States and Europe, that is, well beyond the "predominantly non-white communities where it originated, reggaetón's aesthetics began to shift toward what some critics have described as a 'whitening' of the music" (Rivera-Rideau and Torres-Leschnik 2019, 7). At the same time, especially as it achieved global stature, reggaetón was portrayed in the non-Latina/o mainstream media as an urban, exotic "new" musical genre associated with hip-hop (Rivera 2014; Rivera-Rideau

2015).[7] As Rivera-Rideau (2015, 12) argues, this labeling in the media can be accused of reproducing Latina/o stereotypes, but at the same time, "these links are also critical to reggaetón's African diasporic aesthetics."

Lazy Jay's "Float My Boat" could be seen as an example of the "whitening" of reggaetón, as it is indeed a crossover act joining typical mainstream EDM production techniques, such as filter-sweep buildups, with its reggaetón rhythmic pattern. In keeping with club music's traditional focus on pleasure and dancing (see, for example, Reynolds 2012), the duo probably appropriated the beat simply as a "groovy" musical ingredient with dance appeal, never associating it with reggaetón's fraught history. While the musical forms that preceded the reggaetón rhythm were "intertwined with symbols of blackness" (Marshall 2008, 132), the musical style's close link to black identity politics clearly weakened as it moved into the mainstream in the new millennium. Marshall writes: "Hence, whereas the genre had previously been aligned explicitly with a cultural politics grounded in racial community, with increasing commercialization and an expanded listener base, its vistas and pistas (i.e., the underlying tracks propelling reggaeton songs) seemed to shift accordingly" (Marshall 2008, 132). The electro-house beat of "212" was thus the product of multiple rounds of musical fusion even before it was joined to Azealia Banks' rap vocals.

"212" represents an obvious peak in Azealia Banks' career, which, despite this track's huge success,[8] has been turbulent.[9] She started to rap and write songs at the age of 17 and created a MySpace account in 2008 (Tannenbaum 2014) that is often thought to represent the kickoff of her music career. Through MySpace, Banks connected with various A&Rs and record labels, leading to a development deal with XL Records and Richard Russell (Baron 2012). After working with several producers and managers, none of whom were able to fulfill either Banks' or Russell's vision, the young rapper was let out of the agreement with XL in 2010. Banks reached out to other managers who had shown interest in her MySpace account, but they all rejected her songs, including "212." Banks later recalled that her situation at the time was miserable both emotionally and financially (Baron 2012; Insanul 2012; Lyons 2019). Broke and tired, Banks left for Montréal, where the footage for the "212" video was shot.

When the video for "212" was finally edited and released as a free download, the tables turned for Banks' music career. Banks' rap performance, airy melodic solo, and punchy shouts added hip-hop attitude and melodic vocals to Lazy Jay's electro-house beat. Combining hip-hop with electronic dance music styles was, however, not a *new* approach at the time. During the late 1980s, the merging of house and hip-hop, also called hip house, was already common in European and American club music. The trend returned in the late 2000s, and Banks' seemingly effortless fusion of hip-hop and the electro-house beat brought a new form of hip house (sometimes called electro hop) to the charts in the early 2010s.

As "212" gained its momentum, Azealia Banks was recast as a rising star and soon signed with the record label Interscope. However, in the years that followed, she struggled to maintain her spotlight in music and instead found infamy on Twitter through her feuds with rappers and collaborators such as Nicki Minaj, Iggy Azalea, Pharrell Williams, and Baauer. A rather slow and difficult work process eventually concluded with Interscope releasing her from her deal with the rights to all the songs she had recorded to date. The album *Broke with Expensive Taste* (2014) finally came out on her own independent label, Prospect Park, with, according to Pitchfork, "very little fanfare"—it was announced with a tweet (Jenkins 2014; Kayuha 2014). The album did garner critical praise, though several reviewers remarked upon its striking musical eclecticism. Pitchfork, for example, wrote that it was "all over the place, with tracks new and old and production that runs the gamut from Ariel Pink to Lone to AraabMuzik, and it functions as a sort of anthology, complete with flashes of brilliance" (Jenkins 2014). According to Pitchfork's reviewer, the album reflected the cultural blend and freedom of New York City (212 is the area code of Manhattan), where the characteristic sounds and scenes of the different parts of the metropolis had gradually merged into one.

In the following, we will take a closer look at certain aspects of the groove, production, vocal delivery, and form of the song. For this analysis, we listened to the Spotify versions of the original single version of "212" as well as Lazy Jay's original beat "Float My Boat" (both in AAC formats) on the speakers of an Apple MacBook Pro laptop,[10] as well as on Genelec 8020D active studio monitors. That is, we based our experience of the song on a format and playback system (an AAC file in Spotify played back on a laptop computer) that would commonly be used to listen to a popular track like "212" via a streaming service.[11] Then, for comparison's sake, we also listened to it on the studio monitors.[12]

Groove and Production in "Float My Boat" and "212"

"212" is an up-tempo track (126 bpm) in 4/4 meter wherein the beat—that is, all of the musical parts save for the vocals—seems to be identical with "Float My Boat" (see Figure 15.1).

The beat is built on a reggaetón rhythm originally created with an electronic drum machine and starts with a kick drum, snare drum, and two tom-toms playing the basic reggaetón rhythm pattern, with tom-toms supplying the 3 + 3 + 2 "tresillo" rhythm and the kick and snare supplying the four quarter notes. The reggaetón pattern shares this balance between the 4/4 meter and the cross-rhythmic tendency of the tresillo pattern with many other dance music styles, such as funk (Danielsen 2006), EDM (Butler 2006; Zeiner-Henriksen 2010), and clave-based Afro-Cuban music (Bøhler 2021; Peñalosa 2009; Stover 2009). The timing of the second stroke of the tresillo pattern (the syncopation before beat 2) is slightly late (by approximately 30 milliseconds) in comparison to the 4/4 layer, which gives the pattern its characteristic swing. On top of this, a rhythmically repeating noise on 1-and adds a "push" toward the second beat in each measure. Following the opening section (which is reduced to two times four bars in "212"; see comparison in Figure 15.1), an energizing hi-hat layer with a sixteenth-oriented subdivision enters, intensifying the groove pattern and

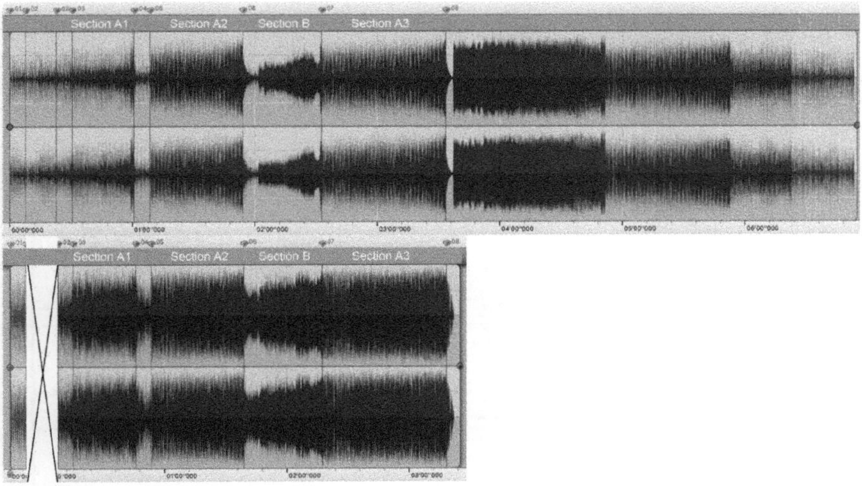

FIGURE 15.1 Comparison of waveform representations of "Float My Boat" (top) and "212 (feat. Lazy Jay)" (bottom). Material from "Float My Boat" used in "212" includes the following: two different eight-bar sections taken from the opening = intro, section A1 = verse 1, section A2 = verse 2, section B = bridge, section A3 = chorus + verse 3 + chorus.

contributing to its danceability. Hints of a periodization of the groove appear in an accented percussive rhythmic event at the 4-and position in bars 2, 4, and so on. In the middle of the first section of "Float My Boat"—that is, the beginning of the first verse of "212"—a counterrhythmic offbeat-oriented snare drum pattern intensifies the groove still further (for an outline of all the rhythmic patterns in sections A1 and A2, see Figure 15.2). This effect is soon combined with a long filter-sweep buildup before the section ends with a virtuosic snare drum break and a filter-closing effect (more on the use of filters below).

In the next section (A2)—that is, the second rapped verse of "212"—a prominent trumpet figure joins the groove. The pitch of this figure is made to descend in a portamento-like triplet fashion every eighth time it appears, effectively organizing the flow into periods of 32 bars (see Figure 15.3). In this section, further percussive emphasis on 1-and is introduced in the second period of 32 bars prior to the hi-hat layer that appears together with the filter sweep buildup (in the third period of 32 bars). The counterrhythmic snare drum pattern is not present in this section, probably to leave space for the trumpet hook.

As mentioned, each section/verse has a long buildup toward the end. In the first verse, the buildup begins with a relatively lengthy increase in intensity as several noises move upward in the frequency register. On top of this, Banks starts to repeat a particular phrase until the upward-moving sounds, or "uplifters" (Solberg 2014), give way and the listener hears only vocal echoes of the repeated phrase (for example, "I guess that c*nt getting eaten"). These echoes gradually

FIGURE 15.2 Basic groove in "212." Kick/snare and toms play the reggaetón rhythm.

fade out via the closure of a low-pass filter, which effectively removes everything but the basic reggaetón rhythm from the track's sound (see Figure 15.4). In the second verse, the same hook happens over the final eight bars (that is, the last 4 + 4 added bars), but the upward-moving noises carry right through to the start of the bridge, even as the echoes of the repeated concluding phrase ("I'm a ruin' you, c*nt") gradually fade out once again.

The use of "uplifters" and closing low-pass filters in the hooks that end each verse clearly situates the "Float My Boat" beat in the world of electronic dance music (EDM); such effects are very seldom heard in African American or Latino

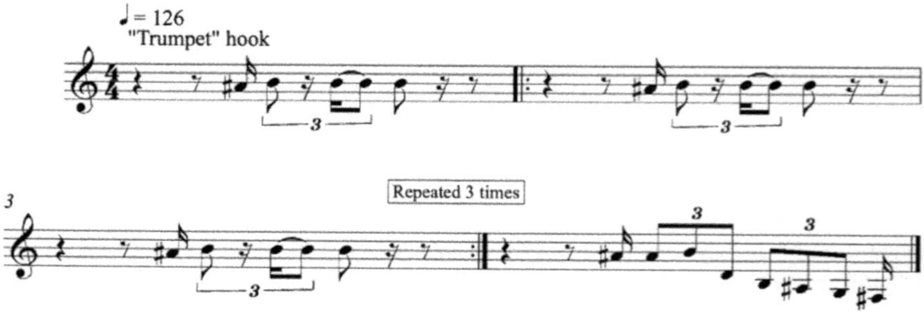

FIGURE 15.3 Trumpet hook in "212."

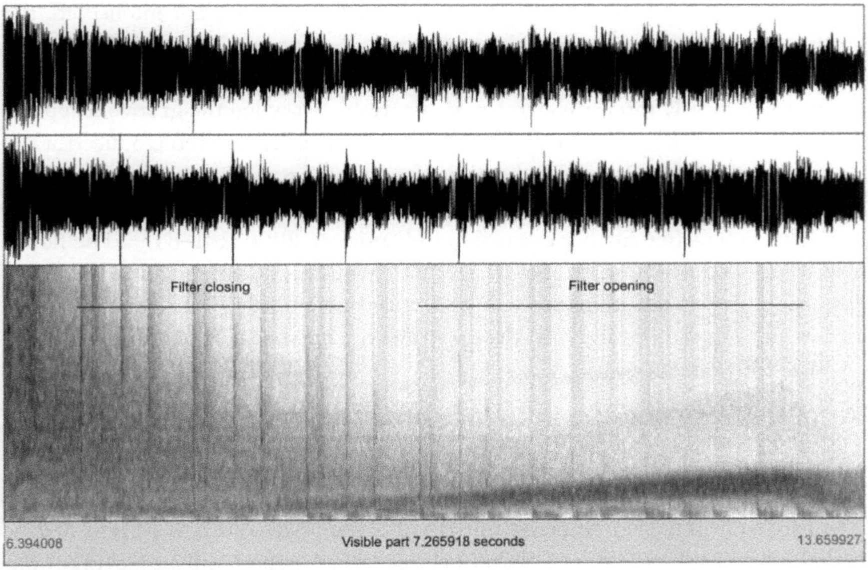

FIGURE 15.4 Waveform (upper panel) and spectrogram (lower panel, x-axis = time, y-axis = 0–12,000 Hz, window size 5 milliseconds, made in Praat, v. 6.1) of filter sweep of hook in transition from second verse to bridge, 1:38–1:45. Note closing low-pass filter followed by noise-sound filter opening ("uplifter").

rap or pop. These buildups prepare for the crucial ecstatic point in an EDM groove when rhythmic layers that have been absent from the sound for some time return to start a new section through the so-called drop.

According to Zeiner-Henriksen (2010), filtering effects reach back to the ways in which DJs in the 1970s experimented with the use of equalizers in New York clubs. The removal (and gradual or sudden return) of either low or high frequencies has since become a standard part of a DJ's repertoire. With the arrival of hard-disc recording techniques and digital effects processing units in the 1990s, such effects also became available at the production stage (Zeiner-Henriksen 2010, 237). The most common filter effect used in house music production is that found in "212," which, according to Zeiner-Henriksen, involves the removal of high frequencies with a low-pass filter. It is often called the "underwater" effect or "the neighbor's stereo through the wall" trick. It is equally common to follow this effect with a buildup wherein the low-pass filter reopens (note that this is in the opposite order from what we hear in "Float My Boat"): "On production equipment these filters are often combined with a resonance or Q-point setting that boosts the frequency area in which the filter starts its attenuation. A typical buildup, then, might involve a gradual opening of a low-pass filter, often combined with a boost in the various relevant frequency areas" (Zeiner-Henriksen 2010, 237).

According to Solberg, buildups in EDM tend to imply that the section in question is "headed towards 'something'" (Solberg 2014). In "212," on the contrary, the buildup (or hook) ends in "nothing." The final beats are not intensified, the uplifting noise drops away, and the rap vocal is "suffocated," leaving the listener with the very thin texture of the basic reggaetón beat alone. This moment of nothingness clearly intensifies the sense of a fresh start when the groove reappears and is more in line with the way in which rhythmic layers are reintroduced at the end of a break in hip-hop. This break is not directed toward a goal but cuts back to the (new) start and serves to bring the underlying layers of the song's rhythm into better focus (see further discussion in Danielsen 2006, 181–84).[13] The hooks toward the end of the main sections of "212" thus bring influences from different groove-based traditions into contact with one another, as the track features aspects of both the classic EDM drop and the rap breakbeat.

Vocal Delivery and the Influence of Battle Rap

Several aspects of the lyrics, Banks' overall performance attitude, and the forceful vocal delivery can be linked to the contemporary battle rap culture of her era. The battle rap tradition emerged initially in 1970s New York as a performative verbal alternative to physical fights between rival gangs, only to evolve into what Bacon (2018) describes as visual and theatrical exchanges. In 1999, battle rap even came to the screen, first on TV and later via the Internet and YouTube, and its visual aesthetics (featuring close-up footage of facial expressions) became as consequential as its musical style (Bacon 2018). The best prepared battle raps served as showcases for impressive lyrical skills, the forceful assertion of identity and authenticity, and often very entertaining humor, despite often being quite rough and tension laden—after all, the rappers are, as Bacon (2018, 5/16) puts it, "attacking each other with words instead of punches and kicks." Below, we will trace these characteristics in Banks' *musical* performance in "212," but we will limit our acknowledgment of the *video* in this regard to the fact that it largely consists of close-up footage of Banks' face and mouth—again, an important part of rap battle iconography at that time.

Banks has no opponent rapper as such in "212," but she is most definitely attacking other rappers through her words. The overall message is simply that Azealia Banks is "bad" meaning good—so bad that she can compete with any rapper, male or female, who might try to measure up to her. Several verified annotations from the website Genius support this straightforward interpretation.[14] The line "You could see I been that b*tch since the Pamper" is, for example, verified by Banks to mean that she has been the "baddest" since she was born. The expression "eat your food" in the line "I'm a eat your food up boo" is slang for defeating someone—in this case, "boo," which means brother. In the bridge, Banks becomes a bit more reflective, talking to herself about being patient and not wasting what she has achieved, but the chorus brings back the in-your-face attitude; Banks assures her imitators that their careers are over when she comes out and that this game has been hers from the start.

According to Tricia Rose, "rappers' rhymes are clearly influenced by, if not a direct outgrowth of, the African-American toast tradition" (Rose 1994, 86). This tradition includes toasts like the traditional story of the Signifying Monkey, an essential aspect of which is bragging about one's own achievements and ridiculing those of others (see, for example, Gates 1988). It is in this context of signifying and "playing the dozens"[15] that "the art of the dis" (Johnson 2012, 21) typical of battle rapping must be understood. The performed verses in "212" clearly display aspects of this "art of the dis" in their playfulness, exaggeration, and insistent, determined delivery. The vocal sound is as rough as the verbiage, coming across as slightly distorted or overdriven as well as somewhat low-fi. The vocal sound, likely achieved by recording "in the red" (Rose 1994, 74–75) or effects processing or both, amplifies the intensity of the lyrics and delivery and gives the overall impression that Banks is so worked up and/or angry that she is rapping too close to the microphone for it to reproduce the sound cleanly. This impression is particularly vivid via the MacBook Pro laptop speakers, which tend to allow the vocals to cut through the mix. The Genelec studio monitors, on the other hand, have a flat frequency response, so this impression is less pronounced.

The flow and delivery also add to this impression independent of their function as vehicles for the words. In the first verse, the flow is similar in each line, with a lot of offbeat phrasing and end rhymes on beat 4 that align the musical bars and text lines in a convergent metrical structure (Oddekalv 2022). The flow's timing is tight, and Banks' insistent diction reinforces the underlying beats. Verse 2 is also based on repeated flow characteristics, but now Banks stretches her lines across barlines to generate a *divergent* metrical structure. She also introduces a more speechlike delivery while still rhythmically emphasizing the offbeats.

Banks' use of a battle rap aesthetic can be interpreted as an attempt to rearticulate the figure of the black female rapper. Whether this is heard as subversive in this context probably depends upon the position and cultural background of the listener. Accordingly, Hansen and Hawkins, in an analysis of Banks' song "Chasing Time" from 2014, comment on the ways in which Banks straddles the line between accommodating and opposing norms in hip-hop and pop. In their interpretation, however, "the disruptive aspects of her performance pose a challenge to gendered and racialized representations of women in hip-hop" (2018, 158). Likewise, McNally (2016) sees "212" as an attempt to pave the way for a more active and aggressive place for female black artists in the contemporary rap scene, confronting and challenging the white gaze and the sexualized fascination with black cultural forms. Applying this lens, then, the content of "212" comes forward as a mix of aggression but also humor.

Merging Musical Forms: Between Song and Groove

Interestingly, "212" lacks the formal structure of a traditional pop song. To either side of its single bridge section, it consists of a continuous stream of groove and rapped lyrics. There is no hint of a melody or chords as such, and the formal logic

TABLE 15.1 Schematic outline of the musical form of "212" (song) and "Float My Boat" (beat)

Song section	*Section of beat*	*Start time*	*Duration (bars)*	*Vocal style*	*Vocal register*	*Groove*
Intro	A1	0:00	8			Basic (reggaeton groove)
Verse 1	A1	0:15	16 (+4)	Declamatory rap	Mid	Basic
Verse 2	A2	0:53	24 (+4)	"Storytelling"	Mid	Basic with trumpet hook
Bridge	B	1:47	16	Song (legato)	Mid	Legato groove
Chorus	A3	2:17	8	Shouted rap	High	Basic with trumpet hook
Verse 3	A3	2:33	16	"Storytelling"	Mid	Basic with trumpet hook
Chorus	A3	3:02	8 (+4)	Shouted rap	High	Basic with trumpet hook

of a groove clearly supplants the logic of the pop song.[16] This is probably partly a consequence of "Float My Boat" being the starting point of "212." Given that the form (with the exception of the intro) and the instrumentation of "212" are identical to the first half of "Float My Boat," Banks probably recorded the vocals on top of a mixed-down stereo version of Lazy Jay's beat. The vocals and the beat were then mixed down and mastered in turn (comparing the waveforms [see Figure 15.1], this process probably entailed both compression and an overall increase in loudness).

Closer inspection, however, reveals some indication of traditional song sections (see Table 15.1) in the vocal textural adjustments made from one section to the next. Banks performs the first verse (A1) in the mid-register in an insistent, declamatory style and the second verse (A2) in the mid-register in a more relaxed, storytelling style, as first mentioned above. The second verse also introduces the song's signature rhythmic "trumpet hook" (see Figure 15.3), also as mentioned above. The bridge, however, transforms the track into a pop song for a time. Here, Banks sings the original melodic synth line of "Float My Boat" in a legato manner while the reggaetón rhythm has disappeared altogether. Once again, the section ends with a filter effect, but this time it is a gradually closing *high-pass* filter applied to the synth sound that builds in intensity to the arrival of the very energetic chorus. Contrary to what has become the norm when rap and song styles are combined, the chorus following the bridge is not sung but rapped with increasing intensity until Banks is shouting her lyrics in a higher register. The lower register, in fact, is almost completely absent when the song's drop arrives and the groove/rapped vocal reappears, to great effect. The basic beat of the chorus is like the groove pattern of sections/verses A1 and A2, but the lyrics repeat, which contributes to the section's chorus-like feel.

The groove continues into the third verse, which resembles the second (A2) but is shorter (only two times eight bars). Moreover, as it transitions to the second chorus, there is a sort of thematic contraction of the previously established verse ending: the vocal echoes overlap with the beginning of the chorus so as to underpin the chorus's opening bars. That is, the echoes of the last sentence of the third verse ("I'm a ruin you, c★nt") sound in the background even as Banks shouts the first phrases of the chorus in the foreground. This solution is probably also a consequence of the fact that section A3, which contains chorus + verse 3 + chorus in "212," is one long stretch without breaks in "Float My Boat" and was, as previously mentioned, probably used as is in "212." Apart from this, the second chorus features the same lyrics as the first, then concludes the track with another relatively lengthy buildup of upward-moving noises and the vocal echoes that have to this point concluded all the verses. The groove stops on beat 1 of the fourth eight-bar segment of the chorus while its final phrase ("This sh★t been mine, mine!") repeats by itself into a quick fadeout.

Conclusion

As noted above, "212" combines musical elements from different musical styles including reggaetón, electro-house, rap, and pop. Its basic groove is a reggaetón/Latino-inspired electro-house beat, the pop influence is represented by its hints of a traditional pop song format as well as the melodic bridge, and Banks' distinctive rapped vocals point to African American hip-hop. The track distinguishes itself as a markedly successful hybrid of these contrasting musical genres, forms, and vocal styles. Its musical cohesion amid all these influences is impressive but perhaps not entirely unexpected, given that the reggaetón beat has appeared successfully in both electro-house and hip-hop; the groove-directed form suits both genres; and features of the typical pop song format are often added to make a groove-based track appeal to a wider audience and enhance its commercial success (e.g., Danielsen 2012).

As to the style mix that is characteristic of the beat, convergences between electronic dance music and hip-hop have appeared in several sub-styles of hip-hop, trap, and dance music over the last decade. Moreover, as pointed out in the introduction, reggaetón had by the time of the release of "Float My Boat" reached the popular mainstream in the United States as well as the club scene in Europe. All of this, together with the fact that a groove-directed form is basic to both genres, might explain why the reggaetón rhythms and the electronic dance music–influenced production, despite their origins in very different cultural worlds, come across as a seamless musical fusion via the "Float My Boat" beat.

With the addition of Azealia Banks' rapped vocals, as well, the track clearly transcends its electronic dance music context. The vocals profoundly impact the track's stylistic affiliations, which testify to the quality and impact of Banks' rapping. While Banks' reputation as a rebel prior to "212" also comes into play here, her battle rap attitude and trappings situate this track firmly within hip-hop. The lyrics are full of insults and exaggerations, and the delivery is assertive and alternates between an insistent speechlike style and high-register shouting, or, in the words of the Pitchfork reviewer, "from sassy and chatty during the buildups to cartoon rage as the synths rear up around her at the song's end."[17]

We also directed attention to the signature hybridity of "212." At one level, the track has a typical groove-based form. This absence of a song structure in the traditional sense is common to house, reggaetón, and hip-hop and could, as stated above, be a factor in the seamless fusion of Lazy Jay's electro-house beat with Banks' rapping. However, "212" is also a *song* whose format might be best summarized as AABCAC, where A is the verse, B is the bridge, and C is the chorus, which bears structural similarities to both the traditional ABABCB format of pop and the AABA format of Tin Pan Alley. These hints of traditional song formats, together with its sung bridge, clearly contribute a pop aspect to the track. The mix of musical formats in "212" is probably also key to opening up the song for use in different contexts by different audiences. Importantly, however, the triggers to hearing the track as a traditional song are entirely in line with a groove-directed aesthetic, meaning that they are established principally through the aforementioned textural adjustments in the vocal part: the changes in tone, intensity, and style of the vocals between sections signal the difference between verse and chorus in an effective way but do not alter the abiding groove.

In sum, Banks' expressive range and seamless fusion of singing and rapping, her negotiations between confirming and protesting hip-hop norms, the style mix inherent in both Lazy Jay's original beat and her own rendition, and the mix of the musical forms typical of song and groove-directed repertoires all allow for many different interpretations and uses of "212" and represent key to these three and half minutes of attitude becoming such a success.

Notes

1 Thanks to Nils Nadeau and the editors for constructive feedback on earlier versions of the text. This work was partially supported by the Research Council of Norway through its Centers of Excellence scheme, Project No. 262762 and the TIME (Timing and Sound in Musical Microrhythm) project, Grant No. 249817.

2 The song was first released on YouTube as "212, ft. Lazy Jay" in 2011 with a video shot in Montréal; it immediately went viral and, as of today, boasts more than 216 million views on YouTube and more than 155 million plays on Spotify. See https://www.youtube.com/watch?v=i3Jv9fNPjgk and https://open.spotify.com/track/16EMONl2vH3rt9f4ehTG8g?si=553dab2508254405.

3 See https://open.spotify.com/track/6IRPB3rtIhHBjHd701QwrO?si=2f193e2b0de447cf.

4 See https://pitchfork.com/reviews/tracks/12612-212/. "Theater geek" refers to Banks' background from music theatre.

5 See https://www.theguardian.com/music/musicblog/2011/sep/27/azealia-banks-212.

6 See https://news.law/warner-music-sony-atv-and-other-major-labels-face-a-lawsuit-from-jamaican-producers-over-despacito/.

7 The year 2004 is often acknowledged as the time when reggaetón arrived in the US commercial music industry (Rivera 2014; Rivera-Rideau 2015). Daddy Yankee's "Gasolina" and N.O.R.E's "Oye Mi Canto" are noteworthy examples of it that circulated in the mainstream media at the time.

8 "212" was named song of the year by the *Guardian*, and the track reached the *Billboard* Top 10 in 2012 (McNally, 2016; Tannenbaum, 2014). It was later included in a *Billboard* list of songs that defined the decade, and *Rolling Stone* also included it in its list of the 100 best songs of the 2010s (https://www.rollingstone.com/music/music-lists/the-100-best-songs-of-the-2010s-917532/azealia-banks-212-single-917591/).

9 Her life story is also harsh. In interviews, she has revealed that she had a difficult childhood, losing her father at the age of two and growing up with a mother whom she describes as abusive (Baron 2012).

10 The MacBook Pro was a 2019 model with Quad-Core Intel Core i7 processor, 2.8 GHz speed, and 8MB RAM.

11 Streaming in combination with laptop speakers, earplugs, or headphones is the most common means of listening to popular music today (Nyre 2015; Spilker 2018; Wikström 2020). AAC is one of three file formats supported by Spotify (https://artists.spotify.com/en/help/article/audio-file-formats).

12 Genelec studio monitors are known not to color the sound. For documentation of the frequency response of Genelec 8020D studio monitors, see https://www.genelec.com/8020d#section-technical-specifications. As to the other variables of the playback system, which include the loudness level of playback, the listening room, and the loudspeaker placement (Moylan 2020, 223–26), we would note the following: the loudness varied from soft to loud (within comfortable limits); the wooden room was approximately 22 square meters with an oblique roof; the studio monitors were placed one meter apart with the listener positioned in the sweet spot (see Moylan 2020, 227); and the distance to the laptop was approximately 50 centimeters.

13 The dynamic structure of a rap breakbeat also differs from the EDM buildup/drop in that its beginning is not gradual but rather a so-called cut (Snead 1984), whereby several musical voices—especially those that have previously attracted the most attention—are taken out to allow distinct rhythmic layers of the repeated basic unit to continue "on their own" for a while.

14 See https://genius.com/Azealia-banks-212-lyrics.

15 "The dozens" is an African American verbal ritual involving two competitors insulting each other, typically by talking negatively about the other person's mother via "outlandish, highly exaggerated, often sexually loaded, humorous 'insults'" (Smitherman 1994, 99). The objective is to outtalk one's competitor and get the most laughs from the group watching the game.

16 Normally, a standard pop/rock tune has a hierarchy of sequences that order the repeated basic units (groove pattern) into larger groups of 4, 8, 16, and so on, to eventually make up verses and choruses. A groove form lacks this hierarchical organization of the repeated basic units. Here, sections are instead created via the insertion of some form of break (drop, cut, etc.). For a detailed discussion of the differences between various hybrid arrangements of song and groove forms, see Danielsen (2006, chapter 9).

17 See https://pitchfork.com/features/lists-and-guides/8726-the-top-100-tracks-of-2011/?page=10.

References

Bacon, Eli T. 2018. "Between Live Performance and Mediated Narrative: Contemporary Rap Battle Culture in Context." In *The Oxford Handbook of Hip Hop Music*, edited by Justin D. Burton and Jason Lee Oaks. https://doi.org/10.1093/oxfordhb/9780190281090.013.36.

Baron, Zach. 2012. "The Making of Azealia Banks." *Spin*, 28 (5), 58–110.

Bøhler, Kjetil Klette. 2021. "The Political Force of Musical Actants: Grooves, Pleasures, and Politics in Havana D'Primera's 'Pasaporte' Live in Havana." *Twentieth-Century Music*, 18 (2), 185–222. https://doi.org/10.1017/S1478572220000614.

Butler, Mark. 2006. *Unlocking the Groove*. Bloomington, IN: Indiana University Press.

Danielsen, Anne. 2006. *Presence and Pleasure: The Funk Grooves of James Brown and Parliament*. Middletown, CT: Wesleyan University Press.

Danielsen, Anne. 2012. "The Sound of Crossover: Micro-Rhythm and Sonic Pleasure in Michael Jackson's 'Don't Stop 'Til You Get Enough.'" *Popular Music and Society*, 35 (2), 151–68. https://doi.org/10.1080/03007766.2011.616298.

Gates, Henry Louis, Jr. 1988. *The Signifying Monkey: A Theory of African-American Literary Criticism*. New York: Oxford University Press.

Hansen, Kai Arne, and Stan Hawkins. 2018. "Azealia Banks 'Chasing Time,' Erotics, and Body Politics." *Popular Music*, 37 (2), 157–74. https://doi.org/10.1017/S0261143018000053.

Insanul, Ahmed. 2012. "Who Is Azealia Banks?" *Complex*, February 13, 2012. https://www.complex.com/music/2012/02/who-is-azealia-banks/.

Jenkins, Craig. 2014. "Azealia Banks: Broke with Expensive Taste." *Pitchfork*. November 11, 2014. https://pitchfork.com/reviews/albums/19998-azealia-banks-broke-with-expensive-taste/.

Johnson, Lynne d. 2012. "The Art of the Dis: Hip-Hop's Battle Royale." In *Etiquette: Reflections on Contemporary Comportment*, edited by Ron Scapp and Brian Seitz, 17–32. Albany, NY: State University of New York Press.

Kayuha, Sam. 2014. "Album Review: Azealia Banks's Debut Good Enough." *The Lantern*, November 11, 2014. https://www.thelantern.com/2014/11/album-review-azealia-banks-debut-good-enough/.

Lyons, Patrick. 2019. "Songs That Defined the Decade: Azealia Banks Feat. Lazy Jay's '212.'" *Billboard*, November 21, 2019. https://www.billboard.com/articles/columns/hip-hop/8543860/azealia-banks-212-songs-that-defined-the-decade.

Marshall, Wayne. 2008. "Dem Bow, Dembow, Dembo: Translation and Transnation in Reggaeton." *Lied und populäre Kultur/Song and Popular Culture*, 53, 131–51.

McNally, James. 2016. "Azealia Banks's "212": Black Female Identity and the White Gaze in Contemporary Hip-Hop." *Journal of the Society for American Music*, 10 (1), 54–81. https://doi.org/10.1017/S1752196315000541.

Moylan, William. 2020. *Recording Analysis: How the Record Shapes the Song*. New York: Routledge.

Nyre, Lars. 2015. "Urban Headphone Listening and the Situational Fit of Music, Radio and Podcasting." *Journal of Radio & Audio Media*, 22 (2), 279–298. https://doi.org/10.1080/19376529.2015.1083155.

Oddekalv, Kjell Andreas. 2022. "What Makes the Shit Dope? The Techniques and Analysis of Rap Flows." PhD thesis. Department of Musicology. University of Oslo.

Peñalosa, David. 2009. *The Clave Matrix: Afro-Cuban Rhythm; Its Principles and Origins*. Redway, CA: Bembe Books.

Reynolds, Simon. 2012. *Energy Flash: A Journey through Rave Music and Dance Culture*. Reprint edition. London: Soft Skull.

Rivera, Michelle Marie. 2014. "Hate It or Love It: Global Crossover of Reggaetón Music in The Digital Age." PhD dissertation, University of Illinois at Urbana-Champaign. http://hdl.handle.net/2142/49834.

Rivera-Rideau, Petra, R. 2015. *Remixing reggaetón: The Cultural Politics of Race in Puerto Rico*. Durham, UK: Duke University Press.

Rivera-Rideau, Petra R., and Jericko Torres-Leschnik. 2019 . "'The Colors and Flavors of My Puerto Rico': Mapping "Despacito"'s Crossovers." *Journal of Popular Music Studies*, 31 (1), 87–108. https://doi.org/10.1525/jpms.2019.311009.

Rose, Tricia. 1994. *Black Noise: Rap Music and Black Culture in Contemporary America*. Middletown, CT: Wesleyan University Press.

Samponaro, Philip. 2009. "'Oye mi canto' [Listen to my song]: The History and Politics of Reggaetón." *Popular Music and Society*, 32 (4), 489–506. https://doi.org/10.1080/03007760802218046.

Smitherman, Geneva. 1994. *Black Talk: Words and Phrases from the Hood to the Amen Corner.* Boston, MA: Houghton Mifflin.

Snead, James A. 1984. "Repetition as a Figure of Black Culture." In *Black Literature and Literary Theory*, edited by Henry Louis Gates Jr., 59–80. New York: Routledge.

Solberg, Ragnhild Torvanger. 2014. "'Waiting for the Bass to Drop': Correlations between Intense Emotional Experiences and Production Techniques in Build-up and Drop Sections of Electronic Dance Music." *Dancecult*, 6 (1), 61–82. https://doi.org/10.12801/1947-5403.2014.06.01.04.

Spilker, H. S. (2018). *Digital Music Distribution: The Sociology of Online Music Streams.* Abingdon, Oxon: Routledge.

Stover, Chris. 2009. "A Theory of Flexible Rhythmic Spaces." PhD dissertation, University of Washington.

Tannenbaum, Rob. 2014. "How Azealia Banks's *Broke with Expensive Taste* Finally Got Released." *New York Magazine*, November 14. https://www.vulture.com/2014/11/how-azealia-banks-finally-made-her-album-debut.html#_ga=2.88843160.248795159.1629800982-1977309091.1629800982.

Wikström, P. (2020). *The Music Industry: Music in the Cloud* (2. Utg). Cambridge, UK: Polity Press.

Zeiner-Henriksen, Hans T. 2010. "The 'PoumTchak' Pattern: Correspondences between Rhythm, Sound, and Movement in Electronic Dance Music." PhD dissertation, University of Oslo.

Discography

Banks, Azealia. 2014. *Broke With Expensive Taste.* Azealia Banks Records, Prospect Park.

Lazy Jay. 2009. *Float My Boat/Disco Inferno/Oracle.* Big & Dirty.

SECTION VI

Deconstructing the Mix and Production Process

SECTIO

Decon

and Pr

16

CAN YOU HEAR THE THUNDER?

The Tech-Processual Construction of Environmental and Emotional Situ in Ghost's "Cirice" (2015)

Samantha Bennett

Introduction

This chapter focuses on the tech-processual construction of environmental and emotional situ in the Grammy award-winning track "Cirice,"[1] the lead single taken from the third album, *Meliora* (2015),[2] by Swedish band Ghost. Fronted by vocalist Papa Emeritus III, the six-piece feature five instrumentalists referred to only as "A Group of Nameless Ghouls"; their anonymity reinforced by mouthless masks and black, "clergy" inspired uniforms. Such semiotics, to include the band's use of satanic iconography, are consistent with US and European black, death and doom metal, although the musical tropes featured in "Cirice" are not as easily categorized. While "Cirice" features drums, bass guitar, and multiple, multi-tracked guitars, it also features prominent acoustic guitars, synthesized strings, organ, timpani drums, and piano. "Cirice" was produced by recordist Klas Åhlund,[3] known primarily for his work with solo female pop artists, including Kylie Minogue, Katy Perry, and Madonna. This choice of producer, in conjunction with leading mix engineer Andy Wallace,[4] goes some way in explaining the success of both "Cirice" and *Meliora* in the commercial mainstream.

That "Cirice" garnered much critical acclaim, awards, and commercial success is, however, of secondary concern here. What makes this track a particularly rich text for analysis lies in its production elements, not to mention that the central tenets of "loudness and distortion" in such music tend to "resist analysis" (Lucas, 2016). Lyrically, the track—on its own, as well as in the broader context of *Meliora*—may be read as a critique of organized religion, its doctrines, and societal power. The lyrics focus on singer Papa Emeritus III "sermonizing" and questioning a single protagonist denoted throughout by terms including "you," "our," and "us." The sonic [re]construction of multiple environmental and emotional situ creates commensurate sonic environments befitting of the lyrical content.

DOI: 10.4324/9781003089926-23

Drawing on established phonomusicological methodologies (Cottrell, 2010; Moore, 2012; Bennett, 2018; Moylan, 2020), I apply my own tech-processual methodology to the track, since I consider matters of technological and processual potential to be important, prominent, and understudied components in the creation of these sonic environments. By technological, I mean "systems featured and routinely applied to a … recording and production workflow" (Bennett, 2018, 1) and by process, I mean "recording techniques and gestures implemented by a recordist in order to capture, process and record sound" (Bennett, 2018, 2). In other words, I acknowledge recordist agency in identifying the presence of these processual occurrences; while "happy accidents" occur, tech-processual attributes manifested in any recorded text are most often intentional decisions.

In this analysis, I refer to what Moylan has termed the overall "sound stage" (2020, 294), that being the holistic environment in which instruments are perceived in terms of stereo width and depth. I also reference Moylan's "host environment" (2020, 293), that is, the sonically discernible space individual processed instruments, or pairs or groups of processed instruments, occupy relative to the overall sound stage. I agree with Moylan's assertion that spatial attributes have the potential to provide "vivid real-life context" (2020, 291) for a track and that nowhere is ecological perception "more profoundly in evidence" (2020, 287). With that in mind, I recognize "Cirice" as a 21st-century music production featuring constructed early music aesthetics, including the tech-processual construction of cathedral situ (Álvarez-Morales et al., 2014) and religious sermon. Additionally, drawing heavily on meteorological tropes found throughout Western classicism (Aplin and Williams 2011), the song's signature timpani *leitmotif* represents the 'thunder' as lyrically narrated. In this detailed analysis, I will refer to the sonic construction of cathedral and storm environments. Furthermore, I unpack the tech-processual construction of prominent emotional themes, specifically, connection, unity, and deception using the Latin "situ," meaning site or position. Emotion is of particular interest in this genre (Yavuz, 2017) and more can be revealed by analyzing how production processes can construct such expressions.

Method

In applying my tech-processual methodology to "Cirice", I also draw upon broader phonomusicological studies that recognize how a track's overall "staging" (Moylan, 2020, 294) and/or "shape" (Moore, 2012, 29) is influenced by the recording and production process. Indeed, there is no shortage of such methodologies in phonomusicology; since the turn of the millennium, plenty of scholarship has presented ways of analyzing recorded texts in varying degrees of detail. For some scholars, this means retaining focus on musical elements, including pitch, melody, and harmony, temporal matters including tempo and time signature, and lyrical content. For phonomusicologists, however, matters of technological and processual potential are of equal, if not more, importance to an overall track in that they are particularly constitutive of the recorded (as opposed to live, performed) text.

Plenty of popular music analysis and sound scholarship envisions the sound "stage" along temporal, spatial, frequency, and depth axes. Cuboid representations of sound mixes are one way of presenting such information. For Gibson (2008), genre and instrumentation are the most significant aspects of mix "imaging"; mixes may be represented as colorful, three-dimensional visual diagrams. Dockwray and Moore's "sound-box" (2010), again a multidimensional cuboid representation of sound positioning, focuses on the representation of the performance space in psychedelic recordings at the turn of the 1970s and recognizes mixes as "clustered," "triangular," "diagonal," and "dynamic." Brøvig and Danielsen (2013) note that listeners perceive a "virtual space" that may be "natural" in its representation of an existing space, and/or "surreal," in that the space may feature a number of "subordinate" spaces within the main virtual space. While these methods work for holistic analysis, detailed information on matters of spatial positioning, presence of time-based and dynamics-based signal processing, volume, and frequency are often missing; such attributes play a significant part in staging and shaping a track.

I do not attempt to recreate a sound "stage" as a parallel to live performance. On the contrary, I treat the recorded track as a separate entity altogether; sound mixes are treated as entirely constructed sonic environments featuring components that may or may not resemble "real world" environments.

In executing a tech-processual analysis (Bennett, 2015, 2016, 2018), I first research the recordists, workplaces, and technologies involved in the recording process. Once ascertained, I then ensure I am listening to a stereo lossless file in a professionally acoustically treated mix room.[5] In this instance, I iteratively analyze the CD version of the track, firstly with track time locations noted against arrangement and number of bars and beats within track sections. This creates a useful framework for accurately pinpointing significant sections and aspects of the mix. Further analysis is undertaken to ascertain matters of orchestration (instruments) and multi-tracking (placement of instruments on tracks and presence of overdubs). Notes are taken against appropriate time locations and further listening is undertaken with a detailed focus on frequency (overall balance and individual instrument ranges), spatial positioning, dynamics (large and small scale), and time-based signal processing (to be discussed below). These latter three elements are of particular interest in how they texturally "shape" (Moore, 2012) the track. Once complete, specific sections are iteratively analyzed again, noting broader matters including lyrics, vocal delivery, and the holistic track. Once this data set is complete, it is drawn upon to elucidate meaning in the track. At this stage, I draw clear and significant connections between tech-processual events in the data and how such events serve to construct environmental situ—the church and the storm—and emotional situ—connection, unity, and deception. All these tech-processual events are summarized against the track arrangement in Table 16.1.

TABLE 16.1 "Cirice" structure, form, orchestration, and position in stereo field. Alternating patterns of 4, 5, and 8 bar phrases form the body of the track

Time	*Section*	*Bars*	*Beats*	*Orchestration*
0:00	Intro	4	4+4+4+5	Acoustic guitar (L); Acoustic guitar (R); Ride cymbal (CR); Keyboard (CCR & CCL), Bass (C), Keyboard (plucked instrument) (CR)
0:22		4	4+4+4+5	Kick drum (C); Ride cymbal (CR); Bass (C); Strings (bar 1+2—*legato* / bar 3+4—*détaché*) (CCR); Acoustic guitar (L); Acoustic guitar (R); Electric guitar (w/fuzz) (L); Electric guitar (w/fuzz) (R); Keyboard (plucked instrument) (CR)
0:45		5	4+4+4+5+1	Kick drum (C); open hi-hat (CCL); [light] crash cymbal (C); Electric guitar 1 (arpeggios) (CR); Electric guitar 2 (arpeggios) (CL); Electric guitar 3 (C); Strings (*détaché*) (C)
1:09	Pre Verse 1	4	4+4+4+4	Kick drum (C); Snare drum (C); Open hi-hat (CCL); [light] crash cymbal (C); OH (L & R); Bass (C); Electric guitar 1 (chords) (R); Electric guitar 2 (chords) (CL); Timpani (C)
1:20	Verse 1	8	4+4+4+4+4+4+4+4	Kick drum (C); Snare drum (C); Open hi-hat (CCL); Crash cymbal 1 (C); Crash cymbal 2 (CR); OH (L & R); Bass (C); Electric guitar 1 (chords) (R); Electric guitar 2 (chords) (CL); Lead vocal (C); Backing vocal (C); Timpani (C)
1:41		4	4+4+4+4	Kick drum (C); Snare drum (C); Open hi-hat (CCL); Crash cymbal 1 (C); Crash cymbal 2 (CR); OH (L & R); Bass (C); Electric guitar 1 (chords) (R); Electric guitar 2 (chords) (CL); Timpani (C)
1:52		8	4+4+4+4+4+4+4+4	Kick drum (C); Snare drum (C); Open hi-hat (CCL); Crash cymbal 1 (C); Crash cymbal 2 (CR); OH (L & R); Bass (C); Electric guitar 1 (chords) (R); Electric guitar 2 (chords) (CL); Lead vocal (C); Backing vocal (C); Timpani (C)
2:13	Chorus Premonition	5 5	4+4+2+4+4 4+4+2+4+4	Piano (CL & CR); Bass (C); Lead Vocal (C); Backing vocal (C); Harmony BVox 1 (R); Harmony BVox 2 (L); Timpani (C)
2:37	Pre Verse 2	4	4+4+4+4	Kick drum (C); Snare drum (C); Open hi-hat (CCL); Crash cymbal 1 (C); Crash cymbal 2 (CR); OH (L & R); Bass (C); Electric guitar 1 (chords) (R); Electric guitar 2 (chords) (CL); Electric guitar 3 (slide) (CCR & CCL & C); Timpani (C)
2:48	Verse 2	8 4	4+4+4+4+4+4+4+4 4+4+4+4	Kick drum (C); Snare drum (C); Open hi-hat (CCL); Crash cymbal 1 (C); Crash cymbal 2 (CR); OH (L & R); Bass (C); Electric guitar 1 (chords) (R); Electric guitar 2 (chords) (CL); Lead vocal (C); Backing vocal (C); Timpani (C)

(Continued)

TABLE 16.1 "Cirice" structure, form, orchestration, and position in stereo field. Alternating patterns of 4, 5, and 8 bar phrases form the body of the track (*Continued*)

Time	*Section*	*Bars*	*Beats*	*Orchestration*
03:20	Break 2	4	4+4+4+4	Bass (sustained note/ hum) (C); Lead vocal (C); Backing vocal (C); Reverse cymbal (CCR); Reverse cymbal reverb (R)
03:30	Chorus 1	5 5	4+4+2+4+4 4+4+2+4+4	Kick drum (C); Snare drum (C); Ride cymbal (R); Crash cymbal 1 (C); Crash cymbal 2 (CR); OH (L & R); Bass (C); Electric guitar 1 (arpeggios) (R); Electric guitar 2 (arpeggio accents) (CL); Lead vocal (C); Backing vocal (C); Timpani (C)
03:55	Pre Solo	3	4+4+2	Electric guitar 1 (CR); Electric guitar 2 (CL); Kick drum (C); Snare drum (C); Bass (C)
04:01	Guitar Solo	5 5	4+4+2+4+4 4+4+2+4+4	Lead guitar (solo) (CR & CL); Kick drum (C); Snare drum (C); Ride cymbal (CR); Crash cymbal 1 (CL); Crash cymbal 2 (CR); OH (L & R); Bass (C); Electric guitar 1 (chord accents) (CL); Electric guitar 2 (CR); Keyboard (CR); Timpani (C)
04:25	Key Solo	5	4+4+4+4+4	Keyboard solo (C); Kick drum (C); Open hi-hat (CCL); Crash cymbal 1 (C); Crash cymbal 2 (CR); Rack toms (CR & C & CL); OH (L & R); Bass (C); Electric guitar 1 (chords) (R); Electric guitar 2 (chords) (CL); Electric guitar (slide)
04:39	Break 3	4	4+4+4+4	Organ (CL & CR); Ride cymbal (CCR); Distortion tail (CCR); Lead vocal (C); Backing vocal (C); Reverse cymbal (CCR); Reverse cymbal reverb (R)
04:49	Chorus 2	5 5	4+4+2+4+4 4+4+2+4+4	Kick drum (C); Snare drum (C); Ride cymbal (R); Crash cymbal 1 (CCL); Crash cymbal 2 (CR); Rack toms (CR & C & CL); OH (L & R); Bass (C); Electric guitar 1 (arpeggios) (R); Electric guitar 2 (arpeggio accents) (CL); Lead vocal (C); Backing vocal (C); Timpani (C)
05:13	Chorus 3	5 5	4+4+2+4+4 4+4+2+4+4	Kick drum (C); Snare drum (C); Ride cymbal (R); Crash cymbal 1 (C); Crash cymbal 2 (CR); Rack toms (CR & C & CL); OH (L & R); Bass (C); Electric guitar 1 (arpeggios) (R); Electric guitar 2 (arpeggio accents) (CL); Electric guitar 3 (slide) (CL); Lead vocal (bar 1–5 ad libs / bar 6–10 lead vocal) (C); Backing vocal (C); Harmony BVox 1 (R); Harmony BVox 2 (L); Timpani (C)
05:37	Outro	2	4+4	Kick drum (C); Snare drum (C); Rack tom (CR); Closed hi-hat (CCL); Bass (C); Piano (CR); Organ (C); Distortion tails (C & R)
05:43 06:02	Fade out	—	—	Keyboard (sustained chord); (CCL & CCR); Organ (C); Distortion tails (C & R)

Key: R—Right (45°); L—Left (45°); CR—Center Right (20°); CL—Center Left (20°); CCR—Center Center Right (10°); CCL—Center Center Left (10°).

"Cirice," the Sonic Environment and Time-Based Signal Processing

Moore describes the sonic environment as featuring three components: textural, harmonic, and narrative (2012, 190). In analyzing "Cirice," I am focused on how matters of tech-processual potential, which are often textural in nature, act as "interventionist" environments, "amplifying what [the lyric] signifies, or even … enacting the lyric" (2012, 191). In addition to the song's specific lyrics, I am interested in how these interventionist environments elucidate its overarching themes, mimic real-world environments implicated in the track, as well as allude to nuanced emotions. I refer to the overall sound stage in terms of width, stereo position of instruments from left (L) to right (R), and, in terms of depth—foreground, middle-ground, and background. I also refer to Moylan's "host environments" (2020, 293) when referring to specific instruments, pairs, and/or groups of instruments and their resulting, sonic identities. Plenty of signal processing is present on Papa Emeritus III lead vocal. This is used to varying effect throughout "Cirice" and, as such, Hall's (1990) four "proxemic zones"—intimate, personal, social, and public—and Moore (2012, 187) and Moylan's (2020, 308) subsequent application of these to popular music analysis–are useful, although at times limited, in denoting the position of the listener in relation to Papa Emeritus III's sermonizing persona.

In an earlier article, I singled out the ways time-based signal processing in particular "creates, alters, and manipulates" (Bennett, 2016, 5–6) sound:

Time-based signal processors can be applied in order to:

- prolong and/or extend a sound in terms of its duration;
- control its density by means of manipulating diffusion;
- contain or release a sound;
- repeat it any given number of times; and
- position it in a space alternative to that in which it was originally recorded. (2016, 5)

While this chapter focuses on broader elements of the sound stage and "host environments," it does so with a particular focus on how time-based signal processors construct such environments, broadening the width and deepening the depth of the overall sound stage, as well as "fabricating space" (Doyle, 2005) in host environments. Through its duration of more than six minutes, "Cirice" traverses a number of these host environments, through what I call environmental and emotional situ, which will form the analytical focus of this chapter. Applications of time-based signal processing, along with the proximity of instruments to the foreground and background of the sound stage, situate the track in different spaces, from *inside* a church, indicated by the foregrounding of a lone, reverberant voice and organ, to *outside* in the midst of a storm. An unusually wide dynamic range[6] is particularly effective and, coupled with instrumental positioning at the extremities of the sound stage, is designed to maximize the space(s) in which instruments are positioned. Lyrical references to thunder, in both its literal sense as heard in a storm and metaphorically as internalized emotion, are reinforced with a heavily processed timpani *leitmotif*, which varies in proximity to indicate distance(s) from the two protagonists referred to in the lyrics, for example, "I feel your presence amongst us" and "now there is nothing between us." At times, the timpani are positioned underneath or in conjunction with a kick drum; both instruments feature a significant low-frequency equalization boost, thus filling out the range <100 Hz. "Cirice" also features multiple, temporal manipulations. These include:

- the positioning of reverse cymbals from the far background to the near foreground of the stage, which I interpret as an intentional decision on the part of the recordist to denote anticipation and suspense, as well as to denote proximity;
- a progressive rock-inspired chorus in five-bar sections featuring alternating four and two beats; and,
- mid-verse micro-silences.

Furthermore, two key sections of particular interest in this chapter—03:22–03:32 and 04:40–04.50—abandon a lead protagonist in a desolate space; the relative absence of instrumentation occurs simultaneously with lyrical references to isolation.

Environmental Situ

Two environmental situ in "Cirice" musically and tech-processually contrast the inside of a church, and a storm, occurring outside. In the forthcoming sections, we will learn how matters including frequency manipulation, dynamic range, time-based signal processing, and other textural elements are used in both macro and micro ways to sonically construct environments befitting of "Cirice" and its overarching themes.

(Inside) The Church

Linguistic etymology places the word *cirice*, meaning "church," as common to both Old English and Old High German. As Wieland suggested, "Cirice" was "... used in the Lyon area in the Southern Gaul, and spread from there first to Latin-speaking Trier in the 4th Century and from Trier into the Germanic speaking world as a loanword" (2017, 197) The track title alone suggests the *entire* sound stage is the church, but the ways in which the musical and tech-processual aesthetics play out suggest otherwise. In his book *Echo and Reverb*, Peter Doyle recognized The Basilicas of the early Middle Ages operated acoustically in two key ways. "Drawing on the ancient traditions of the "speaking" of secular power, the cathedral acoustics served to amplify and give substance and import to the utterances of the priest, both ritual and sermonizing" (2005, 44). In "Cirice," we hear this very reconstruction of cathedral acoustics play out with Papa Emeritus III taking on the persona of a (likely) satanic priest. This is especially prominent in a double five-bar chorus "premonition" occurring between 02.14 and 02.37, constructed to resemble a church situ.

A sudden dynamic and textural change occurs with a 12 dB drop in overall volume, leaving the bass, a piano, a lead vocal and backing vocals, and timpani accents exposed. Each of these four instruments occupies a separate host environment in the overall sound stage, yet all are in some way related to the church situ. The three timpani accents leading into the section at 02:13, in the middle of the section at 02:25, and leading out of the section at 02:37 denote thunder and will be explored further in the next section. To the background of the sound stage, a stereo piano is placed at 30° to the left and the right. The piano is processed with a pre-delay and a medium room reverb with short decay time of around 1–1.5 seconds. The reverb is glossy and features significant high-end frequencies, with plenty of early reflections and little diffusion. This could only be a digital reverb—and very likely a plugin considering the unnatural yet highly stylized way each section of the reverb has been treated; for example, there is little to no audible reverberant tail on the piano (although this is also because the piano reverb is absorbed in the reverberant wash of the vocal reverb). In contrast to the vocal, the piano reverb is light and airy and complements the delicate performance. As such, the piano creates an intimate accompaniment to the vocals, yet with Hall's proxemics in mind, the piano is *not* positioned in the "intimate space." Rather, the intimate host environment of the piano is positioned to the rear of the sound stage in a way that appears out of reach. In the context of the overall sound stage, the piano's host environment is "so near, but yet so far."

A single-note bass phrase on each beat of the bar is positioned to the very foreground of the sound stage, almost level with the vocal, and centered in the stereo field. In stark contrast to the piano and vocals, the bass is heavily compressed, with no dynamic variation, and dry with only a very short reverb. Since most of the phrase is played higher up the fret board, there are little to no low-end frequencies present. Its role appears to provide contrast as well as to maintain the temporal flow; positioning a dry and lower mid-range heavy bass line against heavily processed piano, vocals, and timpani anchors this chorus "premonition." This creates a dramatic, cinematic effect: the bass anchors the listener as a front-and-center witness to the multiple host environments in this sermonizing chorus.

The treatment of the vocals in this section is perhaps the most significant in the track; it is here we recognize Papa Emeritus III as positioned inside the church, hence "Cirice." A lead vocal is positioned to the foreground of the sound stage and centered in the stereo field. A backing vocal with identical lyrics but sung an octave lower is also centered, yet positioned to the very back of the sound stage. This backing vocal only appears on half of each line, "breaking in your heart" and "scars inside you," leaving the "I" lyrics on the lead vocal exposed.

Two harmonizing vocal lines are positioned to the far right and far left of the stereo field beyond the piano and in the middle to the background of the sound stage. These vocals broaden both the width and depth significantly, filling out the space the guitars and drums occupied earlier in the track. The lead vocal is treated with a lengthy pre-delay allowing for the vast, cavernous, stereo cathedral reverb to swell around the entire sound stage without crowding the vocal. This reverb is highly diffuse, heavy on mid-range frequencies, and with higher frequencies cut (no doubt to create space for the glossy piano room reverb), yet the decay time, at around the two-second mark, is much shorter than a natural cathedral reverb would be. One reason for this is to make room for the timpani accents, which will be discussed in the next section. In contrast, the low octave backing vocal and harmonizing backing vocals feature the same cathedral reverb but with longer decay times, reducing their intelligibility (Bennett, 2016). The position of these backing vocals to the middle and background of the sound stage adds to the reverberant wash *without* intruding on the lead vocal and also creates a choral effect. This reverberant wash, which also absorbs the piano reverb, successfully emulates the inside of the church. As Fernandez et al. suggested in their technical paper on *Church Acoustics*, "In the Romanesque, plain stone-masonry and barrel vaults were the common elements that built up a cave acoustics given by such high reverberation times that intelligibility was reduced and focalizations were produced" (2002, 1).

(Outside) The Storm

In an extensive study of meteorological phenomena in Western classical orchestral music, Aplin and Williams stated, "The hint of a distant storm from a few timpani rolls can be as evocative as the crepuscular waves portrayed by Constable" (2011, 300). Despite excluding popular music–and all musics including words–from their study, there are key parallels to be drawn between the representation of storms in classicism and those constructed in Ghost's "Cirice," which draws on classical orchestration with its timpani, organ, and piano. I also consider how the timpani work to support "Cirice" macro theme–a critique of organized religion.

The timpani, performed as an eighth note motif and straddling the last half of beat four and beat one of the selected bars, appears twenty-four times across the middle four minutes of the track. The timpani is set in three alternating positions in the sound stage: to the rear of the sound stage and positioned underneath the kick drum (five occurrences); in the middle of the sound stage, prominent and used to accent the first beat of particular bars (twelve occurrences); and, to the very foreground of the sound stage, as the most prominent instrument in the mix and featuring a much lengthier reverb (seven occurrences). The organization of timpani in "Cirice" may be referenced in Table 16.2.

Used throughout "Cirice" to analogize thunder, the timpani serve multiple purposes. In the earlier parts of the track, the first four timpani occurrences are kept relatively dry with little reverb; only the natural decay time of the large

TABLE 16.2 Timpani *leitmotif* occurrences in "Cirice" and position in sound stage

Time	*Position in sound stage*
01:09	Background—underneath kick drum
01:20	Background—underneath kick drum
01:30	Background—underneath kick drum
01:36	Background—underneath kick drum
01:41	Middle ground—prominent, first beat of bar
01:52	Middle ground—prominent, first beat of bar
02:02	Middle ground—prominent, first beat of bar
02:13	Foreground—prominent, lengthy reverb
02:25	Foreground—prominent, lengthy reverb
02:37	Middle ground—prominent, first beat of bar
02:48	Background—underneath kick drum
02:58	Middle ground—prominent, first beat of bar
03:09	Middle ground—prominent, first beat of bar
03:20	Foreground—prominent, lengthy reverb
03:30	Foreground—prominent, lengthy reverb
03:42	Foreground—prominent, lengthy reverb
03:54	Foreground—prominent, lengthy reverb
04:02	Middle ground—prominent, first beat of bar
04:13	Middle ground—prominent, first beat of bar
04:25	Middle ground—prominent, first beat of bar
04:49	Middle ground—prominent, first beat of bar
05:01	Foreground—prominent, lengthy reverb
05:13	Middle ground—prominent, first beat of bar
05:25	Middle ground—prominent, first beat of bar

drums is heard seeping out from underneath the kick drum to the background of the sound stage. Here, the timpani are buried underneath a busy mix, simultaneously analogizing the distant "thunder" *and* a heartbeat in that the motif is always an eighth note. Indeed, the words "thunder" and "heart" appear in the same vocal lyric hook, multiple times throughout the track.

The timpani are positioned prominently, in the middle ground of the sound stage on 12 occasions. Here, the timpani are brought in front of the drum kit and level with the guitars. These occurrences are processed differently from those earlier in the track; a mid-range equalization boost separates the timpani out from the drums altogether and there is an audible, large hall reverb present.

Returning to section 02:13–02:37, here, the timpani accents are at their most foregrounded in the sound stage and work in tandem with the vocals. These accents, occurring three times in this section, are treated with a rich, low-end heavy reverb with a lengthy decay time of more than three seconds. That the decay time is so long denotes the *absence* of early reflections in a room, hall, or cathedral. The reverb is stylized, yet natural in sound. The lack of early reflections in the timpani—and their compressed, emphasized, and lengthy reverberant tails—therefore positions them as being *outside* the church and creates another "so-near-yet-so-far" effect. In other words, the retention of the natural timpani decay time and the lack of early reflections in the reverb mean that this particular instrument occurrence could not be situated in an indoor environment. These tech-processual events serve two distinct purposes: to amplify the vocal host environment as *inside* the church and to contrast this position with the timpani host environment as being *outside* the church in the midst of a storm. This also demonstrates the limitations of Hall's proxemics in analyzing heavily processed instruments in sound stages; foregrounded instruments, apparently in an "intimate" or "personal" space, with extreme reverb decay times, particularly those heavy in low-end frequencies that travel further, can simultaneously be perceived as being further away in a "public" or outside space.

Critically, the timpani placements aid in articulating the macro theme of "Cirice," that is, critiquing organized religion. At two points in the track, Papa Emeritus III poses questions. At 01:32, "Can you hear the rumble?," there is no timpani on the obvious beat that follows at 01:36. At 01:41 he repeats the question. This time, the timpani suddenly appears in the middle of the sound stage, prominent and present. The same technique is applied later on: At 02:03, on "Can you hear the rumble?," there is no timpani response on the following beat at 02:08. Again, at 03:00 after the question, "Can you hear the thunder?," there is no timpani at 03:04. Yet, after the question is repeated, the timpani appear at 03:09. Here, Papa Emeritus III is asking the protagonist to listen for something that at first, quite clearly, is not there.

The timpani varies in its position in the sound stage throughout, but its position to the background of the mix in the first half of the track is no coincidence. The final time the timpani are placed to the background of the sound stage occurs at 02:48; the timpani are prominent from then on. Here, the preceding lyrical questions and premonitions give way to allegiance—"you and I see eye-to-eye"—at 02:54. At this stage, the timpani appear to take on a third meaning, that of a deity; the protagonist now hears the thunder.

Emotional Situ

Further segments in "Cirice" musically and tech-processually construct dichotomous emotive expressions of connection and unity on the one hand and deception, isolation, and loss on the other. Here, I am focused on how the use of dynamic range is exploited, particularly at the start of the track. Furthermore, I will outline how a series of tech-processual events including spatial positioning, frequency content, and time-based signal processing work to construct emotional situ befitting of "Cirice" and its macro and micro themes.

Deception

Critical to the tech-processual construction of this track is the retention of wide dynamic range. Let's consider the introduction between 00:00 and 00:45.

Unusual for 21st-century commercial music of any genre, "Cirice" begins hovering around −24 dBFS, before a huge jump on two kick drum hits at 00:23 to just under −6 dBFS before settling back at −15 dBFS. These extreme dynamics, coupled with significant low-end equalization boost on the kick drum, lull the listener into a quiet introduction before the sudden volume leap. Since the introduction is more than twenty seconds in length, listeners might be deceived in turning up the volume on a playback device before being blasted with the instrumentation. Furthermore, the song's whole-step-down tuning (the guitars are tuned to DGCFAD), minor key and ominously spiraling, Hermannian triplet acoustic guitar lines lures the listener in before a sudden, aggressive jump in volume. The spatial construct of the introductory "call and response" acoustic guitar lines—these are positioned to the extreme left and center right of the sound stage—adds to this construct of deception. As we listen, our attention is initially fully drawn to the guitar from one direction, before an instrument appears from the opposite side of the spectrum while a lone ride cymbal cuts through the center. Furthermore, a keyboard motif appears at 00:07, only just audible, centered and furthest to the background of the sound stage of any instrument in the entire track. This keyboard motif features a lengthy reverb and foreshadows a prominent keyboard appearing later in the track. Here, perhaps to self-referential effect, the quiet, reverberant keyboard analogizes a ghost. This introduction immediately sets out a vast sound stage; the extremes of spatial width, depth, and dynamic range are laid down like a sonic gauntlet within the first 25 seconds of the track. The introduction is also designed to disorientate: where is the sound coming from? As the acoustic guitars spatially exchange twice, the sudden jump in volume—kick drum, bass, and electric guitars—is, by contrast, centered in the stereo field. Additionally, the sharp attack of what a Nameless Ghoul stated were "synthetic and simulated" (Hartmann, 2015) strings and foregrounded plucking mimics a harpsichord. This is a track that, in featuring myriad religious signifiers, also pays its proverbial dues to early European music.

Isolation/Loss

"Cirice" portrays the power and control of religious authority (the persona of Papa Emeritus III) over a single protagonist and to emphasize this, tech-processual events work to construct emotional situ of isolation and loss. As the track progresses, we hear musical and lyrical references, first to the singling out of the protagonist and later, the coercion, persuasion, and manipulation of this protagonist. By the middle of the track, lyrics such as "you and I see eye to eye" at 02:54, "nothing between us" at 03:08 and "our merge is eternal" at 03:16 portray a strong emotional connection and unity between the two. However, between 03.20 and 03.30 an alternative scenario is presented—one where the deity is absent from the protagonist's life—via a dramatic break. Similar to the introduction, we hear a sudden drop in level from around −4 dBFS to −12 dB as the lead vocal is exposed. Unlike the "church" section, at this point the lead vocal is isolated with no backing vocals. Here, the lead vocal is set against a compressed and sustained bass note tail that holds for the duration of the section. The lyrics present two questions: firstly, on "Can't you see that you're lost?" the level drops to −24 dBFS simultaneous to the introduction of a lengthy reversed cymbal reverberant tail at the far background of the sound stage, before rising to −12 dBFS on the second question "Can't you see that you're lost without me?" Over the course of this second question, which lasts 5 seconds in the song, the reverse cymbal increases in volume until brought level with the vocal, abruptly ending on "lost," leaving the words "without me" dramatically exposed. These two words are comparatively dry, double-tracked, and between 6 and 12 dB louder relative to the preceding line. Furthermore, the time-based signal processing featured on the lead vocal is wound back simultaneous to the reverse cymbal. Through the trajectory of this question, therefore, we hear a heavily reverberant, quiet, and isolated emotional situ at the beginning, swiftly followed by an intensely projected, loud "reality" by the end. A similar section, again analogizing isolation and loss, appears at 04:39–04:49. This time, the voice is accompanied by a synthesized organ and the ride cymbal, positioned in the middle ground of the sound stage and off to the center right. The ride cymbal strongly alludes to the track's introduction; the reality of the now-united protagonists is shadowed by deception. In previous papers I have recognized the ways in which the rapid, alternate manual spatial positioning of instruments has been used to analogize madness or the experiential aspects of thoughts inside a protagonist's head. Here, it is not the spatial aspects of the voice, but the dynamic-based and time-based signal processing on both the lead vocal and cymbal combined to tech-processually construct both isolation and manipulation.

Connection/Unity

Earlier sections of "Cirice" tech-processually construct dichotomous inside/outside environmental situ. Toward the end of the track, the aforementioned sections denoting loss and isolation are contrasted with a tightly cohesive mix denoting connection and unity. The track is at its loudest at just under 0 dBFS and features no dynamic variation until fading out at 05:43. Additionally, the stylized time-based signal processing of previous sections is gone; instruments now feature tight, short room reverbs. The stereo width remains broad, although it lacks the extremities of earlier in the track. Here, the way in which the holistic sound stage has narrowed in terms of width, shortened in terms of depth, and heightened in terms of sustained level is in stark contrast to the extreme "sonic gauntlet" of the opening 25-second introduction. The final chorus beginning at 05:13 features a lead vocal "lead in" ad-lib positioned to the very foreground of the sound stage. At this point, two separate, harmonizing vocals performing the identical chorus line appear to the center left and center right and to the

TABLE 16.3 "Cirice"—summary of environmental and emotional situ and key tech-processual events

Time	*Section*	*Situ*	*Events*
02:14	Chorus	"The Church"	Sudden drop in dynamic range; timpani occurrences denote (outside) thunder; stereo piano to rear of sound stage; piano time-based signal processing features pre-delay, reverb with short decay time with plenty of high-frequency content. Bass heavily compressed, foregrounded in sound stage. Lead vocal foregrounded in sound stage with lengthy pre-delay, large, stereo "cathedral"-style reverb with approx. 2s decay time; low octave backing vocal (half lines) to rear of sound stage; harmonizing backing vocals with lengthy cathedral reverb positioned to left and right.
Throughout	All	"The Storm"	Timpani set in three alternating positions in the sound stage. Set 1—to the rear, positioned underneath the kick drum (five occurrences), relatively dry with natural decay time Set 2—in the middle, prominent and used to accent the first beat of particular bars (12 occurrences), level with guitars, mid-range EQ boost with lengthy hall reverb Set 3—to the very foreground, as the most prominent instrument in the mix and featuring a much lengthier reverb (seven occurrences), rich, low-end heavy reverb, lengthy decay time (3 s+).
00:00–00:45	Intro	"deception"	Sudden boost in dynamic range at 00:23; low-frequency boost on kick drum; whole step guitar tuning; acoustic/electric guitar "call and response" lines positioned to far left and center right of stereo field; keyboard to the far rear of sound stage.
03:20–03:30	Break 2	"isolation/loss"	Sudden drop in dynamic range; exposed lead vocal; sustained bass note tail; reverse reverb on cymbal tail increasing in volume; final double-tracked words "without me" dramatically foregrounded by end of the section and 6 dB louder than preceding line.
04:39–04:49	Break 3	"isolation/loss"	Vocal, synthesized organ, ride cymbal positioned to the middle of the sound stage; ride cymbal reprise from introduction.
05:13–end	Chorus / outro	"connection/unity"	Entire mix at its loudest; mix is tight and cohesive; broad stereo mix lacking in earlier extremities; tight, short room reverbs on instruments; vocal ad-lib positioned to foreground of sound stage; two harmonizing vocals positioned center left and center right; backing vocals brought level with lead vocal; "unraveling" of mix in fade out.

fore/middle ground of the sound stage. Aside from the harmonizing backing vocals in the "church" section (where they are positioned further back), this is the first time in the track that backing vocals are prominent and brought level with other instruments. That there are now two distinct lead vocals in unison, positioned just underneath the ad-lib denotes Papa Emeritus III and his protagonist as finally unified. By 05:27, the backing vocals have moved further back, and the ad-libbed vocal has merged back into the lead vocal and continues with the chorus lyric hook; the eternal "merge" described earlier in the track is complete. The very end of the track features a lengthy fade-out. Here, the mix unravels: the piano, initially buried to the middle/background and center right of the sound stage flourishes prominently into the fade at 05:40, leaving a sustained organ and keyboard chord exposed as the main instrument at 05:46. Both the piano and organ were positioned as accompanying significant vocal sections earlier in the track; by the end of "Cirice," both emerge from the mix as the "last instruments standing." The organ is, however, suppressed by two separate, sustained distorted guitar hums; one to the center and middle ground of the sound stage and another featuring more upper mid-range, stuttering as it fades to an abrupt end to the right at 05:48. All tech-processual events are summarized here in Table 16.3.

Summary

"Cirice" is a good example of a track that utilizes every area of the sound stage. The introduction sets out an extreme and ambitious, cinematic stage in its positioning of instruments to the very foreground and background, to the extreme left and right, as well as filling out the low-end with a heavily boosted kick-drum and the top end, with a boost at around 8 KHz on the ride cymbal. "Cirice" features multiple applications of dynamics and time-based signal processors to successfully construct host environments, as well as denoting environmental and emotional situ. In its tech-processual construct of the inside of a church, "Cirice" demonstrates how this is not as simple as applying a church or cathedral reverb to a single instrument. Rather, specific areas of multiple time-based signal processors, to include frequency content, diffusion, and decay times, are manipulated in order to accurately recreate such a situ. Timpani are routinely present in Western art music and their presence in "Cirice," along with piano, organ, and strings, alludes to a strong classical music influence. The strategic positioning of timpani throughout "Cirice" not only ties the music to the "heart" and "thunder" lyrical hooks but creates unique host environments within the sound stage as well as affirms other neighboring environments. In "Cirice," the timpani's tech-processual attributes result in a tri-fold meaning; they are at times a heartbeat, representative of thunder and, later, a deity. Hard rock music is often clustered in the mid-range, but in its 21st-century pop production aesthetics, as well as deliberately constructing specific host environments, "Cirice" succeeds in presenting doom metal in a highly stylized and commercially successful form.

Notes

1 The National Academy of Recording Arts and Sciences (NARAS) 58th Grammy Awards. Best Metal Performance: "Cirice"—Ghost.
2 Ghost. 2015. *Meliora*. [CD] United States: Loma Vista Recordings.
3 Klas Åhlund (b. 1972) is a Swedish songwriter and record producer. For a full discography, see Discogs: https://www.discogs.com/artist/272590-Klas-%C3%85hlund
4 Andy Wallace (b. 1947) is a US record producer and audio engineer. He is primarily known for his mix engineering, notably Run DMC's *Raising Hell* (1986) and Nirvana's *Nevermind* (1991). For a full discography, see Discogs: https://www.discogs.com/artist/59472-Andy-Wallace
5 In this instance, a professionally acoustically treated mix suite with DynAudio monitoring in Canberra, ACT.
6 For example, the average dynamic range of albums in 2010 was around 7-8dB (Vickers 2011); the dynamic range of "Cirice" is around 24dB.

References

Álvarez-Morales, Lidia, Teófilo Zamarreño, Sara Girón, and Miguel Galindo. 2014. "A Methodology for the Study of the Acoustic Environment of Catholic Cathedrals: Application to the Cathedral of Malaga." *Building and Environment* 72: 102–115.

Aplin, Karen L., and Paul. D. Williams. 2011. "Meteorological Phenomena in Western Classical Orchestral Music." *Weather* 66, no. 11 (November): 300–306.

Bennett, Samantha. 2015. "Never Mind the Bollocks: A Tech-Processual Analysis." *Popular Music and Society* 38, no. 4: 466–86.

Bennett, Samantha. 2016. "Time-based Signal Processing and Shape in Alternative Rock Recordings." *IASPM@journal* 6, no. 2: 3–21.

Bennett, Samantha. 2018. *Modern Records, Maverick Methods: Technology and Process in Popular Music Record Production 1978–2000*. New York: Bloomsbury Academic.

Brøvig, Ragnhild and Anne Danielsen. 2013. "The Naturalised and the Surreal: Changes in the Perception of Popular Music Sound." *Organised Sound* 18, no. 1: 71–80.

Cottrell, Stephen. 2010. "The Rise and Rise of Phonomusicology." In *Recorded Music: Performance, Culture and Technology*, edited by Amanda Bayley, 15–36. Cambridge: Cambridge University Press.

Dockwray, Ruth and Allan F. Moore. 2010. "Configuring the Sound-Box: 1965-1972." *Popular Music* 29, no. 2: 181–97.

Doyle, Peter. 2005. *Echo & Reverb: Fabricating Space in Popular Music Recording 1900–1960*. Middletown, CT: Wesleyan University Press.

Fernandez, Marcos, Manuel Recuero and David Cruz. 2002. "Church Acoustics." Forum Acusticum, Seville (RBA-05-015). Accessed July 12, 2021 http://www.conforg.fr/acoustics2008/cdrom/data/fa2002-sevilla/forumacusticum/archivos/rba05015.pdf

Gibson, David. 2008. *The Art of Mixing – A Visual Guide to Recording Engineering and Production*. 2nd ed. Boston, MA: Course Technology.

Hall, Edward T. 1990. *The Hidden Dimension*. 4th ed. United States: Anchor Books.

Hartmann, Graham. 2015. "A Nameless Ghoul from Ghost on 'Meliora', Capturing Religiosity + Possible Classical Album." *Loudwire*. Accessed July 12, 2021. https://loudwire.com/nameless-ghoul-ghost-meliora-religiosity-possible-classical-album/

Lucas, Olivia. 2016. Loudness, Rhythm and Environment: Analytical Issues in Extreme Metal Music. Doctoral dissertation, Harvard University, Graduate School of Arts & Sciences.

Moore, Allan F. 2012. *Song Means: Analyzing and Interpreting Recorded Popular Song.* Farnham: Ashgate.

Moylan, William. 2020. *Recording Analysis—How the Record Shapes the Song.* New York: Routledge.

Vickers, Earl. 2011. "The Loudness War: Do Louder, Hypercompressed Recordings Sell Better?" *Journal of the Audio Engineering Society* 59, no. 5: 346–51.

Wieland, Gernot R. 2017. "Language Contact: Latin" In *The History of Old English*, edited by Laurel J. Brinton and Alexander Bergs, 187–201. Berlin: De Gruyter.

Yavuz, M. Selim. 2017. "'Delightfully Depressing': Death/doom Metal Music World and the Emotional Responses of the Fan." *Metal Music Studies* 3, no. 2: 201–18.

17

THREE-DIMENSIONAL DOOM

My Dying Bride's "Your Broken Shore" (2020)

Mark Mynett

This chapter considers how the recording and mixing process of "Your Broken Shore" from My Dying Bride's (2020) *The Ghost of Orion* album informed the track's sonic signature, substance, expression, and dimensionality.

Formed in June 1990, Yorkshire's My Dying Bride (MDB) are regularly viewed as the founding fathers of the doom metal movement. Along with other associated/amalgamated subgenres (symphonic-doom; gothic-doom; death-doom; epic-doom, etc.), doom metal emphasizes slow performance tempi and rhythmic subdivisions, as well as a focus on atmosphere and ambience. To provide a degree of context here, this combination of "slow-subdivisions and atmosphere" is perfectly epitomized by Black Sabbath's (1970) self-titled track on their self-titled debut album; widely agreed as being the very first doom metal song. Despite that metal music enthusiasts and musicians often hold technical complexity as a virtue in and of itself, often claiming a level of prestige for the music based on its technical difficulty (Phillipov 2012, 64; Purcell 2003, 12–14), fans of doom metal typically pay greater attention to the manner in which the music portrays "darkness" and "melancholy."

Featuring epic/gothic/symphonic textures accompanied by violin and keyboard performance parts, MDB's slow, brooding, and often highly experimental creations have been chronicled across 12 studio albums. This has gained them widespread critical acclaim, and an impassioned international following with consistently impressive album sales. As a result, in 2017, MDB signed a new recording contract with Nuclear Blast, who have the world's most prestigious roster of hard rock/metal artists, including the likes of Slayer, Nightwish, Machine Head, and Opeth. Released on 6th March 2020, *The Ghost of Orion* was MDB's first release on Nuclear Blast, and was produced, engineered, mixed, and mastered by the author of this chapter.

DOI: 10.4324/9781003089926-24

Three-Dimensional Doom

Using "Your Broken Shore" (YBS) by MDB as a case study, this chapter's analytic goals are to examine and provide an understanding of how the "space between the notes" of slower performances can be manipulated and optimized via harmonic distortion and ambience/time-based effects processing. This will be achieved via reflective commentary on the relevant recording sessions, combined with direct analysis of the subsequent multi-track recording, and mix session processing.

Fundamental to this case study is the slower performance aesthetic involved in doom metal music, which provides greater space between the notes or musical events. The space between musical events is sometimes referred to as inter-onset intervals (IOIs), defined by Zagorski-Thomas as "the elapsed time, measured in milliseconds, between the onset of one sound and the onset of the next" (2007, 334). IOIs have a profound impact on the way we perceive the frequency content, dynamics, ambience, and effects of recorded music. Reflecting slower tempi and rhythmic subdivisions, music that has relatively large IOIs (which, in addition to doom metal, includes most hip-hop and R&B) has greater space for reverb and ambience, as well as longer and slower low-frequency wavelengths to expire before the next performance event. Concurrently, greater space is afforded for the sustained attributes of the drums, cymbals, bass, and guitar sounds.

Of relevance here is that numerous studies support the hypothesis that larger metric groupings—as with doom metal—expand the possibilities for dynamic variation, improvisation, and "swing" found in drum performance patterns. Scholars such as Collier and Collier (1997), Friberg and Sundström (2002), and Honing and Bas De Haas (2008) conclude that expressive timing does not scale accordingly with increased tempo. Nevertheless, although the drum performance on "Your Broken Shore" includes quieter (clean guitar) verse sections involving dynamics with press rolls, there aren't any areas, drums or otherwise, that could be considered as involving swing.

My Dying Bride "Your Broken Shore"

In popular music, the term "heavy" can be used to describe a variety of sound and performance characteristics in a wide range of styles. But for many, this adjective is solely reserved for the metal genre, and used to describe the music's perceived weight, size, and density (Berger & Fales 2005; Herbst 2017; Herbst & Mynett 2021a, 2021b; Mynett 2012, 2017, 2019, 2020; Williams 2014). From Black Sabbath's (1970) self-titled debut album—generally considered heavy metals first—right through to Amorphis' (2022) *Halo* and everything in between, "heaviness" is metal music's defining feature.

Although textures, dynamics, and performance styles vary, distorted guitar tones embody the fundamental identity of heavy music, consequently providing coherence to its numerous subgenres. If a group's sound does not contain this vital quality, it is unlikely their music would be classified as metal. At this stage of the genre's development, there are no other sound or performance attributes that are a prerequisite for the "metal band" moniker to be applicable and valid. It therefore makes sense to first consider the application and influence of harmonic distortion during the recording and mixing stage of YBS, not only with the guitars but also bass, and vocals, with drums, discussed subsequently.

Harmonic Distortion

Rhythm Guitars

As discussed by Logan Mader, a fundamental recording principle for the genres' down-tuned distorted guitars is reduced gain (aka harmonic distortion).

> Less gain is really important for low tuned guitars, but you need to have a player that knows how to hit the strings. Some guitarists get their tone out of the gain from their amp, but then pick really lightly – but that does not sound good. You really have to have the velocity and dig in, and that's where you get the aggression and note definition, whilst not needing a lot of gain. On Gojira's *The Way of All Flesh*, there's not a lot of gain on the guitar tracks. It's a 5150 III (the later Fender-produced revision of the earlier Peavey-produced Mk.I 5150 guitar amplifier modelled around and endorsed by Eddie Van Halen) on the blue channel.
>
> *Logan Mader in Mynett 2017, 60*

When recording rhythm guitar performances based on large IOIs that regularly feature notes/chords that sustain for several seconds, far greater gain (distortion) levels are required than with, for example, the typically far-shorter guitar performance IOIs of Gojira. Distortion provides electric guitars with an effective form of compression-like dynamic range reduction. The wide-level variations of a clean sound are modified to one with a far more even dynamic envelope, resulting in greatly extended decay times. In effect, rather than being a lot quieter in level, the further that a note/chord's sustain is from the pick strike, much greater levels of sustain are enabled at equivalent duration. Given the requirement for the rhythm guitar notes/chords of YBS to retain effective sustain characteristics within the comparatively greater IOIs, relatively greater gain levels were required. This gain level—involving a Peavey 5150 Mk.I on its red (lead) channel, with a pre-amp setting of 7.5 (i.e., creating a high level of distortion)—likewise proved effective and appropriate when tracking single-note/melodic lead-guitar style performance parts.

Concerning the impact of harmonic distortion more generally, in addition to its significant impact on dynamics, the intensity of the relevant signal's harmonics and thereby the relative balance of its harmonic series is altered. To varying degrees, the existing harmonics of the guitar become strengthened, and the accompanying natural compression affects the ascending harmonic series so that it becomes far greater in amplitude. As Walser notes, "Distortion also results in a timbral change towards brightness…since distorting a signal increases the energy of its higher harmonics" (Walser 1993, 42). Importantly, when additional spectral information, in the form of high-frequency energy, is introduced to guitar timbres, they are perceived as heavier (Berger and Fales 2005, 193–94).

To facilitate the production's tonal density and greater sense of dimension, in turn informing perceptual heaviness, the YBS production features quad-tracked guitars. Regardless of the number of guitarists in the band, or the number of microphones used to record each performance, "quad-tracked" guitars comprise four entirely separate rhythm guitar performances (broadly speaking, two in each speaker). Quad tracking is comparable to the impact of certain orchestral arrangements in the classical world. For instance, several violins being scored to play the exact same musical part provide a thicker sound and richer texture. Although metal music production typically involves guitars performed consecutively rather than simultaneously, the resulting harmonic differences similarly thicken the resulting sound, sometimes with a very slight chorused/shimmering effect. In this instance, although the same guitar was used (thereby retaining a consistent, accurate intonation), different amplifiers, cabinets, and microphone combinations were employed, not only to provide divergent frequency coverage but also to enable greater difference between the left and right guitar tones, in turn facilitating a wider sound stage. Here it can be noted that, with greater differentiation between signals panned left and signals panned right, a wider stereo image is generally afforded.

At the mix stage, the effective form of compression-like dynamic range reduction afforded by the levels of gain applied to the guitars meant that compression was unnecessary. However, the subsequent cumulative impact of high levels of harmonic distortion when combining four distinct, separate rhythm guitar recordings resulted in excessively abrasive qualities within the 2.5–4 kHz region (a regular feature of distorted guitars). This was addressed via narrow, but very deep (often referred to as "surgical") EQ cuts to 3.4k on both the left and right rhythm guitars. The first 20 seconds of the song most easily evidences the particularly dense, but non-abrasive, nature of the song's rhythm tones. Despite being accompanied by a lead guitar melody in the center of the stereo panorama, and kick/bass/cymbal accents, the hard left and hard right rhythm guitars are relatively isolated.

Worth noting more broadly about the rhythm guitar performance on YBS, is that, despite the harmonic complexity of the track—largely provided via distinct left/right rhythm guitar performances, accompanied by violin, and solo guitar—the rhythm guitar chord voicings are not complex, typically featuring root/fifth/octave voicings. The relevance here is that harmonic distortion on electric guitar becomes increasingly more apparent with an increase in notes/strings/chord voicings. Not only does this mean that the non-complex chord voicings of YBS are less revealing of the high levels of harmonic distortion applied, but that the isolated single note guitar/violin breakdown section from 2:46 to 3:06 has a smooth non-abrasive texture and fails to reveal the gain levels applied.

Bass

Other production styles sometimes introduce distortion to the bass to help separate the sound from the cleaner texture of the other instruments. Within the context of distorted rhythm guitars though, the influence of bass distortion is very different. Rather than *separating* the bass from the other instruments, distortion can be used to provide a more aggressive quality that allows the bass to *blend* more effectively with the guitars. This is an important concept. A great deal of the impact of heavy guitar tones is owed to their contrasted texture to the other instruments. However, when very clean bass sounds are involved, the contrast between the bass and rhythm guitars can be too great, with these instruments appearing detached from each other. Regardless of significant compression processing, this results in the bass being difficult to "sit" in the mix—with a restricted level that sounds protrusive when pushed any higher. Introducing distortion to the sound can provide a bass texture that is far more unified with the rhythm guitars. This unity means the bass can be provided with an increased mix level without sounding inappropriate, delivering enhanced sonic weight and a heavier production as a result. Consequently, as far back as Geezer Butler's intro on Black Sabbath's (1970) *N.I.B.*, bass sounds in metal music have regularly involved distortion, and this quality has become almost conventional in modern metal.

At the same time as distortion being an important consideration for the YBS bass sound, the relatively large performance IOIs (affording greater space for longer/slower low-frequency wavelengths) meant that significant attention to the low end of the bass was essential. To enable a more-controllable low-end, multiple bass amplifiers and cabinet set-ups (referred to as multi amp/cab recording), allowed each rig to be optimized for a given frequency role. The faster response of a solid-state Galien Krueger amp proved best suited to the role of delivering clean, "piano-like" punchy lows, with an Ampeg Classic valve amp providing the grittier, lower-mid-to-higher frequencies. These two signals were supplemented with a dedicated channel of distortion provided by a Dark Glass Alpha-Omega pedal.

During the mix stage, frequency-bracketing (heavy attenuation to either side of a desirable spectral region) enabled the clean low-end focused solid-state signal to be restricted to the lows, which was subsequently heavily compressed with VCA compression (all things broadly equal between the four compression topologies—VCA, FET, Vari-mu/Valve, Optical—VCA is generally the cleanest i.e., yielding the least harmonic distortion). The mids upwards (valve-based) track was high-pass filtered at 220 Hz, with a comparatively lower lever of FET compression (introducing comparatively more-significant levels of harmonic distortion than VCA). The distortion-focused Darkglass channel was restricted to 750 Hz–2 kHz (thereby heavily attenuating the muddy lows and abrasive highs that often results from a bass guitar being heavily distorted). As a result of the influence of extensive harmonic distortion applied at source to the Darkglass pedal, this signal had sufficiently stable dynamics as to not require compression. With a view to contributing wiry mids-focused note definition to the composite bass sound, the clean DI channel with moderate VCA compression, but no frequency bracketing, just high pass filter at 90 Hz, was also employed. All four subsequent channels were routed to a sum/group, where further VCA (SSL G-Buss), as well as (LA-2A) optical compression was applied.

Vocals

Especially with more aggressive vocal performance parts, distortion-based processing can be highly effective for thickening the texture and enhancing the character of a vocal signal. Although distortion is usually associated with signals being entirely transformed—as with guitar DIs—for vocal purposes, a moderate, relatively subtle application is required. This is important: while distortion can benefit a vocal sound to a certain extent, it can never fully compensate for a performance that lacks *natural* vocal aggression, emotion, or power. With heavy treatment, a vocal performance that lacks these features won't suddenly sound emotive, powerful, and aggressive—it will simply sound like a heavily distorted, substandard vocal performance. The processing should therefore be used to introduce musically pleasing spectral content, with coloration qualities that thicken and brighten the signal, but without the "grain" of the distortion becoming overly apparent. When applied tastefully, this affords a vocal texture that is perceived as more coherent with the rhythm guitars, allowing the vocals to "sit" in the mix in a more natural manner.

Parallel distortion processing—courtesy of Soundtoys' "decapitator"—was therefore applied to the main guttural death metal vocals, but, with a similar concept to the frequency bracketing of the bass distortion signal (attenuation of muddy lows and abrasive highs), the parallel distortion return channel was restricted to 350 Hz–2 kHz. The introduction of parallel vocal distortion provided the additional benefit of further differentiating and providing greater dynamic variation from the melodic verse vocals (which, being accompanied by clean rather than distorted guitars, were not treated with distortion).

In summarizing the application, and subsequent impact of harmonic distortion on YBS, it can be concluded that, despite comparatively radical levels, this not only facilitated effective sustain attributes but also enabled a bass and vocal texture more unified with the rhythm guitars. Furthermore, the relevant performance approach of the guitars means that the harmonic distortion is less evident and does not display abrasive qualities that would likely be far more evident if involving increased notes/strings/chord voicings, and/or faster guitar performances.

Depth and Dimension

In addition to the greater IOIs of YBS affording a greater application of harmonic distortion, it similarly afforded enhanced emphasis on ambient room mics for the drums—at both the recording and mix stage—as well as an enhanced emphasis on the mix stage application of spatial processing for the guitars, violins, and vocals.

With drums, an ambient room mic generally refers to a placement a minimum of three feet away from the drums, positioned around, rather than above the kit. This captures a relatively balanced ensemble image, with the room's ambient character "coloring" the collective textures. These qualities can afford a controllable form of audio "glue" that can deliver a valuable sense of "cohesion" to a drum sound. In this respect, an emphasis of spot mics, especially when featuring microphone sources that have had their spill (drum components other than the one that the microphone is focused on) removed or attenuated, can be perceived much like the isolated signals from a drum machine. Room mics can therefore be considered a radical alter ego to the isolated, heavily sculpted spot mics and drum samples typically required to enable the level of control necessary for this production style.

With faster drum subdivisions, there is less "space" available for ambient qualities to decay—so the mix contribution from room mic signals is likely more restricted—and importantly, this typically provides an exaggerated sense of aural intimacy (Zagorski-Thomas 2008, 204). This likely explains why, when asked about his use of room/ambient microphones, specialist metal producer Colin Richardson replied, "We don't use them. The further away you go, the less in your face it sounds. I've tried room mics, but I just get into awful trouble with them" (Richardson and Taylor 2011, n.p.). Here it can be noted that Colin Richardson has mainly produced metal acts from subgenres that focus on faster subdivisions, including Slipknot, Trivium, Machine Head, and Fear Factory.

When recording the drums for YBS, the approach with room mics was therefore to capture as profound an emphasis of ambience as possible. This involved a pair of wide-diaphragm condenser microphones set to a cardioid polar pattern placed approximately 18 feet from the drums, with a sound absorptive baffle/gobo placed halfway between the drum kit and microphone and faced away from the drums in order to focus on capturing the reflections from the wall facing the microphone. Although a microphone's polar pattern has a significant bearing, the greater the distance from mic-to-source, the greater the level of ambience captured along with the source. At sufficient distance, as with the YBS drum recording, a cardioid polar pattern with its null point aimed at the drum kit doesn't capture any direct drum sound at all, providing a "washed-out" capture of room ambience.

Room mic signals tend to be particularly well suited to radical applications of compression, especially when the drums have been tracked in an environment with pleasing acoustics, as was the case here. UAD 1176 E-revision (FET) compression was applied to both ambient room mic signals and dialed-in with a fast attack and fast release. The fast attack and release setting meant the resulting gain reduction reduced transient energy, thereby emphasizing the drum's decay portions, along with the greater levels of ambience captured in this section of the sound. To favorably accentuate the compressor's distortion artifacts, the 1176's "all-buttons-in" mode was employed with the input gain adjusted to achieve approximately 15 to 20 dB during signal peaks. The resulting distortion-based coloration provided an aggressive texture, delivering increased body to the drums, with the illusion of them being performed in a larger perceived space than otherwise.

Having applied relatively heavy-handed low pass filters at 3.5 kHz to attenuate the harsh, abrasive cymbal sounds that aggressive gain reduction to room mics often results in, the ambient room mic channels were employed at a greater level than would be the case for a metal production featuring faster performances, effectively filling the space between each drum onset. In this respect, similar to the principal classical recording technique of capturing and reproducing the sound of the concert hall and its coloration of the collective sound, the approach with the drums could be considered as what Doyle (2006) refers to as a "romanticist" recording aesthetic. In contrast, productions that tend to remove any clues as to the recording space where the performances occurred (as with the aforementioned Colin Richardson approach reference) is referred to by Doyle as a "realist" recording aesthetic.

Despite the heavy influence of radically compressed ambient room mics at the mix stage, a relatively liberal application of synthetic reverb (a bright EMT plate reverb with 850 ms decay time, as well as ReLab's LX480 "Large Wood Room" with a decay time of 2.1 seconds) was still deemed necessary for the drum sound. The snare and toms featured the greatest of reverb application, whereas the kick and metalwork featured a far lower level.

Important to note though is that the metal music production style generally derives much of its sonic impact from the listener having a sense of "in-your-face" proximity to many of the sounds involved...which is counteracted by reverb. Overly long, or excessive, reverb use typically swamps and confuses the articulation of a production, especially those that feature faster subdivisions, clogging up the sound in a way that damages the impact and clarity of the music. It is for this very reason that certain metal albums, such as Slayer's (1986) "Reign In Blood," became renowned for their spatially "dry" (minimal ambience captured during the recording stage and minimal reverb used at the mix stage) production quality. However, within the context of the comparatively far greater IOIs of the YBS mix, the application of reverb perceptually enlarged, and afforded a greater sense of panoramic width and improved texture to the drums, contributing significantly to the perceived three-dimensionality of the production.

Guitars

As a general rule, reverb treatment seldom widens, nor increases the perceived size of heavy bass and rhythm guitar tones in a pleasing manner. Instead, it tends to introduce boomy characteristics, softening note definition in a way that counteracts the "in-your-face" quality required, thereby confusing the relevant rhythm structures. Although this was certainly the case with the bass guitar, having achieved an effective level of ambience for the YBS drum sound, the guitars seemed somewhat "disconnected" from the drums when placed within the context of the mix: they lacked a vital sense of "cohesion."

The greater decay time of the "Large Wood Room" used for the drums appeared excessive when applied to the rhythm guitars, and, despite the plate reverb being far more appropriate, the fact that this had been brightened up (pre-reverb) with a high-shelf boost proved too obtrusive. A dedicated EMT plate reverb (EMT stands for Electromesstechnik, the German company who, in 1957, first developed the artificial "plate" reverb chamber) for the guitars was therefore introduced, which also proved effective at introducing a sense of cohesion to the violins. Even at a moderate level, this helped bring the drums, guitars, and violins into a more cohesive perceived performance environment. Nevertheless, after applying ambient processing, including delay, to the vocals, these delay effects also proved effective for the guitars and violins, with the longer delay lines sometimes being further accentuated, via automation, for relevant end-of-line performance sections. Later in the mix stage, broadly reflecting the separation from the performance event afforded by delay, a reduced emphasis on reverb, and greater emphasis on delay proved preferable for the guitars.

Vocals

Although the initial approach to vocal processing involved experimenting with the contribution return path levels of a short plate, a large wood room, and a church hall reverb algorithm, only the first two proved effective, and similar to the rhythm guitars, with a relatively restricted level. In contrast, the combined impact of a variety of delay times proved effective, as did a stereo widening delay and pitch processing approach. The first delay was a mono/centered slap-back echo with a 150 ms delay time, the second, a stereo quarter and eighth note delay, and lastly, a multi-tap delay line based around a half note, and (slightly longer) dotted half-note. To help gel these delay signals into the mix, thereby preventing them from standing out and being overly obvious, low-pass filters were either engaged via the relevant plug-in or via a separate EQ plug-in. Likewise, tape emulation processing was introduced, providing a warmer, somewhat less bright delay quality.

Aimed at simultaneously thickening and widening the vocals in order to contextualize them within the harmonically dense, wide-panned, quad-tracked guitars, the last component of the time-based vocal effects processing involved very short dual delays with diverse time parameters and pitch shift. Referred to as pitch thickening width enhancement, a mono aux send was taken from the vocal to two separate mono aux tracks panned fully wide. A pitch shifter was inserted on each return path, with one sharpening and one flattening their respective signals by 8 cents, so they were 16 cents apart in pitch. A mono delay was subsequently inserted after the pitch shift plug-ins, with one at 16 ms and the other at 24 ms (but with no feedback parameter value for either) thereby with the midway point between these values being 20 ms, the common value associated with automatic double tracking (ADT) echo. In effect, the pitch processing subtly blurred the vocal pitch center, which, in combination with the ADT processing, enabled a thicker vocal sound, that then, as a result of the hard-panning of the respective signals, was also more three-dimensional.

Worth highlighting is that the increased focus and application of time-based effects processing for YBS's vocals would not have been appropriate, if it were not for the heightened ambient qualities of the drums, guitar, and violin sounds: themselves directly or indirectly, reflecting the large IOIs of the performance. Likewise, the extent of effects processing would have been far more evident, and likely in an inappropriate manner, if the extent of harmonic distortion introduced to the bass and quad-tracked guitar performances was far less than otherwise. The density of the resulting sonic wall of sound had the impact of "absorbing" the extensive application of vocal signal effects processing.

References

Berger, Harris, and Fales, Cornelia. 2005. "'Heaviness' in the Perception of Heavy Metal Guitar Timbres: The Match of Perceptual and Acoustic Features over Time." In *Wired for Sound: Engineering and Technologies in Sonic Cultures*. Edited by Paul Green and Thomas Porcello, 181–97. Middletown, CT: Wesleyan University Press.

Collier, G., and Collier, L. 1997. "The Swing Rhythm in Jazz." In *Proceedings of the 4th International Conference on Music Perception and Cognition*, 477–80. Montreal, QC: Faculty of Music, McGill University.

Doyle, Peter. 2006. *Echo and Reverb: Fabricating Space in Popular Music Recording, 1900–1960*. Middletown, CT: Wesleyan University Press.

Friberg, Anders, and Sundström, Andreas. 2002. "Swing Ratios and Ensemble Timing in Jazz Performance: Evidence for a Common Rhythmic Pattern." *Music Perception 19*/3: 333–49.

Herbst, Jan-Peter. 2017. "Historical Development, Sound Aesthetics and Production Techniques of the Distorted Electric Guitar in Metal Music." *Metal Music Studies 3*/1: 24–46.

———. 2018. "Heaviness and the Electric Guitar: Considering the Interaction between Distortion and Harmonic Structures." *Metal Music Studies 4*/1: 95–113.

Herbst, Jan-Peter, and Mynett, Mark. 2021a. "Adventures in Recording Land: Engineering Conventions in Metal Music." *Rock Music Studies 8*: 1–20. Taylor & Francis Online. https://doi.org/10.1080/19401159.2021.1936410

Herbst, Jan-Peter, and Mynett, Mark. 2021b. "Nail the Mix: Standardization in Mixing Metal Music?" *Popular Music and Society 44*/5: 628–49. Taylor & Francis Online. https://doi.org/10.1080/03007766.2021.1957544

Honing, Henkjan, and Bas de Haas, W. 2008. "Swing Once More: Relating Timing and Tempo in Expert Jazz Drumming." *Music Perception 25*/5: 471–76.

Mynett, Mark. 2012. "Achieving Intelligibility Whilst Maintaining Heaviness When Producing Contemporary Heavy Metal Music." *Journal on the Art of Record Production 6*.

Available online, https://www.arpjournal.com/asarpwp/achieving-intelligibility-whilst-maintaining-heaviness-when-producing-contemporary-metal-music/
———. 2017. *Metal Music Manual.* London: Routledge.
———. 2019. "Heaviness in Three Dimensions: The Use of Sonic Space in Contemporary Metal." *The Art of Record Production: Creative Practice in the Studio.* Edited by Simon Zagorski-Thomas, Katia Isakoff, Sophie Stévance, and Serge Lacasse, 66–79. London: Routledge.
———. 2020. "Maximum Sonic Impact: (Authenticity/Commerciality) Fidelity-Dualism in Contemporary Metal Music Production." *The Bloomsbury Handbook of Music Production.* Edited by Simon Zagorski-Thomas and Andrew Bourbon, 293–302. New York: Bloomsbury.
Phillipov, Michelle. 2012. *Death Metal and Music Criticism: Analysis at the Limits.* Lanham, MD: Lexington.
Purcell, Natalie. 2003. *Death Metal Music: The Passion and Politics of a Subculture.* Jefferson, NC: McFarland.
Richardson, Colin, and Mick Taylor. 2011. "Interview: Metal Production Guru Colin Richardson." *Music Radar,* March 16. Available online, https://www.musicradar.com/news/guitars/interview-metal-production-guru-colin-richardson-400260. Accessed May 1, 2022.
Walser, Robert. 1993. *Running with the Devil: Power, Gender and Madness in Heavy Metal Music.* Hanover, NH: Wesleyan University Press.
Williams, Duncan. 2014. "Tracking Timbral Changes in Metal Production from 1990 to 2013." *Metal Music Studies 1/1*: 39–68.
Zagorski-Thomas, Simon. 2007. The Study of Groove. *Ethnomusicology Forum 16/2*: 327–35.
Zagorski-Thomas, Simon. 2008. "The Musicology of Record Production." *Twentieth-Century Music 4/2*: 189–207. Cambridge University Press.

Discography

Amorphis (2022) *Halo.* Atomic Fire. Album 13913717.
Black Sabbath (1970) *Black Sabbath.* Vertigo Records. CASS Vertigo 7138-001.
My Dying Bride (2020) *The Ghost of Orion.* Nuclear Blast—27361 51612, NB 5161-2.
Slayer (1986) *Reign in Blood.* Def Jam Recordings. 9 24131-2.

18

TRANSFORMING A POP SONG

The Journey of the Extended Club Remix (Taz Vegas' "Am I Dreaming," 2019)

Phil Harding

This chapter is supported by a webpage.

Introduction

The analytic goals of this study are to discover the aesthetic results achieved with the remix of the Taz Vegas record "Am I Dreaming" (2019). I will set out a reflective, auto-ethnographic examination of the musical, rhythmic, vocal timbre, and production elements contained within the remixed track. The analysis will also investigate the pop form compositional sections of the song. The added recorded and programmed elements of the remix were produced by myself and the other two PJS[1] Productions team members. The Moylan (2020) three-stage recording analysis framework of observation, evaluation, and conclusion will structure the study throughout and a secondary analysis will run in parallel to determine if an extended club mix can transform a pop song. Moylan states in the sub-heading for his *Recording Analysis* book "How the Record Shapes the Song" (Moylan 2020) and I hope to demonstrate exactly that in this chapter with the analysis of the extended club remix.[2] Remix culture has supported the promotion of pop and dance records since the 1980s and I was at the forefront of that movement during my time as chief engineer and remix specialist at PWL Studios in London (Harding 2010, 2020).

The client that commissioned this remix specifically requested the PWL "sonic signature" (Zagorski-Thomas 2014, 66) and style. That request will steer the direction of this aesthetic analysis and my reflective commentary throughout the three Moylan analytic framework stages. Furthermore, this essay will be supported by interviews with the agents involved in the project, including the client, Bobby Pileggi at RSM Records.

DOI: 10.4324/9781003089926-25

The subjects and topics to be included throughout this study will center around the discussion of how a pop/dance extended club remix differs from a radio version and in this case, from the originally recorded and mixed version. The traditional American pop form will be examined and compared throughout. As stated by Nate Sloan and Charlie Harding, "Form describes the large-scale musical structure of a composition" (Sloan and Harding 2020, 47).

The scope of subject coverage will range from the analysis of musical performances, influences, and rhythmic style to the effect of production and mixing decisions on the stereo spatial landscape. My discussion will focus mainly on the extended remix version.

Methodology

The first listening session to collect data for this analysis was conducted on my lounge hi-fi system, positioned to the specifications and room measurements of Moylan's "direct-field" in *Recording Analysis: How the Record Shapes the Song* (Moylan 2020, 227). The speakers were already 6 feet apart and 4 feet from the side walls. I sat in a central position 6 feet from the speakers. The Moylan recommendation to have the speakers angled at 60 degrees toward you might be contested by many audiophiles but provided an excellent stereo soundscape for this first test. My Linn Sara speakers are normally set on the linear plane and remained so for my first playback but it was enlightening for me to apply the positional changes and commence the second playback. I would recommend these specifications to all readers. I experienced a 50% increase in stereo definition and low-frequency clarity and accuracy. The whole stereo soundscape and imaging sounded better to me; this was a revelation. Home studio room specifications are often different to Moylan's recommendations due to compromised space and an abundance of equipment. However, I was interested to recognize that my home studio speakers are always angled at approximately 60 degrees but a lack of space creates a compromise of my being 3 feet from the speakers and those speakers only being 3 feet apart. There is a list of equipment in both of my listening environments for this study on the weblink to this chapter.

Using the Moylan (2020) record analysis framework I have interviewed three of the main agents from the domain and field of this record[3]:

1. **Bobby Pileggi**—the commissioning client to PJS Productions for the project via his RSM record label and Robert Sterling Publishing. Bobby is also the co-author of the song (lyricist) and original producer.

   Interviewed & recorded May 19, 2021 via Skype.

2. **Julian Wiggins**—Keyboard arrangements, programming, and sax solo/brass parts for the PJS remix.

   Interviewed & recorded June 22, 2021 via Facetime.

3. **Simon Dalton**—Rhythm arrangements for the PJS remix.

   Interviewed & recorded July 7, 2021 at Phil Harding's studio.

The Research Questions Proposed to Each Respondent Were Delivered in Advance

1. What are your thoughts on the original version before the PJS Remix?
2. Did you think the transformation to the PWL style was possible?
3. What are your comments and views on the style of the drums and bass?
4. What are your observations of the pop vocal processing and the extra backing vocals added by PJS?
5. Do you have any comments on the extended arrangement—the intro/middle break and outro, for instance?
6. What is your evaluation of the final extended mix compared to the original version and the added keyboards and orchestration especially?
7. How has the final remixed song shaped the record for you?

Observation

Main Observational Listening Session: Hi-Fi System in Lounge

This is a running commentary with the 60-degree angled speakers and my observational focus here is on the pop form and the expected "build and drop" (Sloan and Harding 2020, 47) of an extended dance remix. The many recorded elements are presented to the listener gradually during a typical formulaic Phil Harding/PWL introduction that builds every four bars up to (1:20) before introducing verse 1. These types of dynamic building blocks every four or eight bars are a common practice in the extended dance mix genre. The desired effect is to retain interest for the DJ and the dancefloor. The complete musical structure and bar lengths are presented in Table 18.1. I will describe my observational discoveries chronologically.

TABLE 18.1 Pop form musical structure including timeline & bar count: "Am I Dreaming" by Taz Vegas (PJS Extended Dance Mix)

Section	*Intro*	*V1*	*PC1*	*C1*	*Link 1*	*V2*	*PC2*	*C2*	*Link 2*	*Solo*
Bars	40 ext	4	9	8	4	4	9	8	4	8
Time	0	1:21	1:30	1:47	2:03	2:11	2:19	2:37	2:53	3:00
Section	*Bridge / M8*	*Break*	*Link 3*	*V2*	*PC3*	*C3*	*C4*	*Outro*	*End*	
Bars	17	16 break	4	4 break	9	8	8 break	29 ext		
Time	3:17	3:50	4:22	4:30	4:38	4:56	5:12	5:27	6:27	

From the start of the extended introduction the vocals with their quarter beat delay effects (with a long feedback tail) leading into the rhythmic elements are the first things to draw the listeners' attention. The drum loops and percussion establish themselves for four bars with tremendous drive and intensity. This is punctuated at the end of bar 4 with the "Let's Go" vocal sample and then a 1980s disco-style 4s kick drum enters with a heavy bottom-end bass boost (around 60–80 Hz) plus agogo bells percussion with a Latin-style pattern. Guitar is introduced and something new is presented to the listener every four bars. A typical PWL-style tom-tom fill takes us into a beat 1 brass sample stab that punctuates yet another change with more vocal sample locs from (one expects) the chorus of the song. Many musical elements are added (often in pairs) as pad keyboards come in plus keyboard bells that support the main vocal hook "Am I Dreaming." Strings and brass riff entries follow. Vocal sample locs of "Am I Dreaming" and "You're the one for me" also feature to take us into a radio-style intro with a full brass melody line and solo sax riffs. The bass part is also driving the track along with an emblematic PWL-style bass line.

Finally, at (1:21) of the extended mix we arrive at verse 1. The lead vocal immediately commands prominence above the music backing track. Taz's voice is high but with a relaxed timbre as we enter the pre-chorus after just a four-bar verse; a verse this short is unusual in the verse–chorus pop form. The backing vocals enter and lay down a bed of "oohs" during the pre-chorus. An extra bar at the end of this section introduces us to Taz's high falsetto voice, an excellent example of his variable timbre, also the rhythm elements drop out for bar 9 of the pre-chorus, creating an exciting drop-dynamic for the song.

The main "Am I Dreaming" vocal hook of the song comes on bar 4 of the chorus (often in pop music it is on bar 1) and there is subtle movement of the strings and pads in the three bars that lead up to this. At bar 4 of the chorus the hook line is supported by the strings and synth bells plus harmony backing vocals, enhancing the lead vocal beautifully. The same happens at bar 8 of the chorus.

The brass riff that features in the last eight bars of the extended introduction comes back after the chorus. This brass is presenting a different hook line to any of the song vocal lines—this is a commercial bonus for the song's catchiness to the listener and we hear high falsetto ad-libs from Taz for the first time, again displaying his excellent vocal timbral variation.

During verse 2 the chords are noticeably different to verse 1 and this is discussed in the following evaluation section. This is a musical example of how Sloan and Harding describe pop-form chorus lyrics: "choruses almost always use the same lyrics each time, except when they don't" (Sloan and Harding 2020, 48).

Chorus 2 is a repeat of the first chorus and in the second link or post-chorus we hear high falsetto ad-libs from Taz in long reverb-plate echo with quarter beat delay repeats (40% feedback) again. This leads us into a very dominant, rhythmic, and melodic sax solo, presented over the verse chords. The rhythm and music carry on behind the alto sax solo. A brass arrangement is introduced in a supportive role behind the sax solo.

For the next section a new musical and vocal section is introduced, different from anything in the song so far. Backing vocals sing "oohs" again throughout this section, similar to the pre-choruses. In classic American verse–chorus form this is called a bridge; in the UK this is generally called the middle eight. An extra bar at the end contains the same vocal hook that appears at the end of the previous two pre-choruses.

Middle Breakdown Extended

Here we witness the build-and-drop form used in many 21st-century pop records such as "We Found love" by Rihanna featuring Calvin Harries (2011). As Sloan and Harding declare, "The build generates pent up energy and the drop releases it" (Sloan and Harding 2020, 47). This starts with just drums and percussion. The feature here is a new conga sample and pattern which sounds very Latin-influenced to work alongside the agogo bells. The breakdown continues to build with the introduction of guitars, chorus keyboard chords, brass, and string parts, all very similar to the extended intro. This is followed by a third link section and the intro brass hook comes back again, leading us into the second verse repeated but this time delivered as a breakdown with just vocals and the musical elements—no rhythm. The bass, drums, and keyboards come back in for the pre-chorus. The vocals come across as very prominent and the low frequencies of the bass and kick sound very solid and knit together tightly as a unit on this sound system.

The third chorus is an exact repeat of chorus 1 and we finally reach a breakdown chorus, which is a typical feature at this point for an extended dance mix. Another set of examples of the pop form build-and-drop feature throughout this outro. The initial round of choruses feature all of the vocals and drums but no musical elements except the guitar from bars 5 to 9, a clap-along/sing-along section for the dance floor. This leads into a classic Phil Harding/PWL extended outro with bass, musical, and rhythm elements coming in and out. Here we can really notice the beat 4 hand clap in echo feature that exists throughout the whole song but is not often heard clearly. Table 18.1 shows a table to support the observations noted in the commentary of the song segments.

Further listening sessions took place in my recording studio, both with and without the remix rhythm programmer, Simon Dalton.[4]

Lyrics

This chapter is not offering a major focus on the record's lyrical content or analysis thereof. The book chapter website contains the lyrics of "Am I Dreaming" and a synopsis of their meaning and background from the songwriter/lyricist Bobby Pileggi.

Evaluations

As a practicing music engineer and producer with over 40 years' experience,

I very rarely listen back to my own records and certainly not in my lounge environment described in the methodology section. If I were to listen to music I have worked on it would generally be for a "professional listen" in my studio, hence the second phase of studio listening sessions to both the stereo mix and the multitrack Pro-Tools and Logic Pro sessions to examine some of the individual parts.

The recorded and re-recorded elements that initially sounded prominent to me during the first playback in my lounge were the vocals, kick drum, and bass. As a sweeping generalization one could say that vocals and drums are the priority musical parts of a modern hit pop song. The vocals clearly consist of a commanding lead performance by Taz and the backing vocals "wrap around him like a duvet" (Dalton 2021, personal interview). The drums and bass sound as though they are referenced to the retro PWL "sonic signature" sound and style of the 1980s and there are comments on these from my interview respondents at the end of this section.

The extra bar (1:45) at the end of the first pre-chorus section introduces the listener to the wonderful timbre of Taz's high falsetto voice. Although we have heard some of his vocal samples in this register during the extended introduction of the remix, these are supported by the female backing vocals. At (1:45) we become aware of his falsetto range in solo for the first time. Interestingly I discovered that this vocal line is supported by a monophonic synth playing the same lead phrase but this is really only detectable on the instrumental version.

Production Dynamics

This chapter and its focus on an extended dance remix affords me the opportunity to highlight the use of "dynamic contours [...] as great changes in the overall dynamic level of a recording" (Moylan 2015, 45). The *dynamic contours* created by remixers and producers for an extended dance mix are crucial to its success because the audience wants to be taken on a dynamic musical and rhythmical rollercoaster ride. The "build-and-drop" method I have previously described here really comes into play on an extended remix. It is part of the goal of the remixer to take the listeners on that journey. The space created by breakdowns (drops) where either the musical elements drop out at (3:50) on this track or the rhythm drops out (4:30) and the 1-bar music and rhythm breaks into every chorus are all good examples of programming and mixing for dynamic effect. Guy Fletcher, former Chairman of the UK performance collection society PRS, suggested to me that in his view, the most important aspect of writing a successful pop hit is for it to contain *dynamics*. Here we see an example where many of the dynamics of the song have been transformed by the record and particularly on this remix version.

Further production and compositional dynamics investigation at (2:11) will raise a question for the listener; are the chords of verse 2 different to the chords in verse 1? I had to ask myself this question during the pre-production preparation for the remix and confirmation via the interview with the songwriter and client, Bobby Pileggi, will tell us yes, the chords in the verses are deliberately different. This is very unusual for a pop song where for the last ten years or more, the majority of worldwide best-selling pop songs are based around four-bar musical loops of the same chords in the verses. This would be confusing for listeners trying to understand what section of the song they are in but Bobby assures us that this chord difference is set up as a deliberate surprise to maintain the interest of the listener. It was no accident.

This leads us onto further compositional form discussion for this song. The traditional verse–chorus form is basically thrown out of the window on this record as it adopts very unusual metric durations within the musical structure. The four-bar verses and nine-bar pre-choruses are very unusual in pop, as previously discussed. Further evidence of moving away from the verse–chorus form can be identified in the bridge section on the original version[5] of Taz's "Am I Dreaming" but for this extended remix the beginning if the bridge is simplified by the introduction of an 80s-style Sax solo (eight bars), which is followed by the original 17-bar bridge, then the middle breakdown. Whichever version of this record is reviewed, it defies the standard pop verse–chorus form and the result is very effective and original.

Production and Spatial Analysis of the Recorded Novel Elements

The sound stage dimensions achieved on this mix are created by the sound source positions and relationships within the stereo soundscape presented (see Figure 18.1). Examples of how this is achieved within the rhythmical elements are not easy to define until the multitrack session was scrutinized by the rhythm programmer, Simon Dalton, and myself. Here we discovered the percussive elements of the record were carefully set in their stereo image locations: We isolated the agogo bells and confirmed they are a sample from the original Taz version, with a pattern re-programmed by Simon. They are panned half left. We also discovered 4× hi-hats–2× offbeat open hats panned hard L&R, a 16s in the middle and a PWL/Rick Astley-style pattern panned half left. To match that last hi-hat on the left we identified the PWL/Rick Astley-style Roland 727 conga pattern panned half right. Finally, we discovered in the middle eight (bridge) breakdown a new conga that was sampled from the Taz original and re-programmed by Simon, positioned centrally but with a large amount of stereo room ambience generated by the AMS RMX16 Reverb plugin. There was also an 8s tambourine sample from Simon which is auto-panning on the mix. Finally, a cabassa panned to the right to match the main PWL hi-hat pattern on the left.

Spatial fulfillment and power in the stereo soundscape are created on this record by the drum and percussion ambiences, the mono delays on the lead vocals throughout, and the stereo panning delays of the vocal samples featured

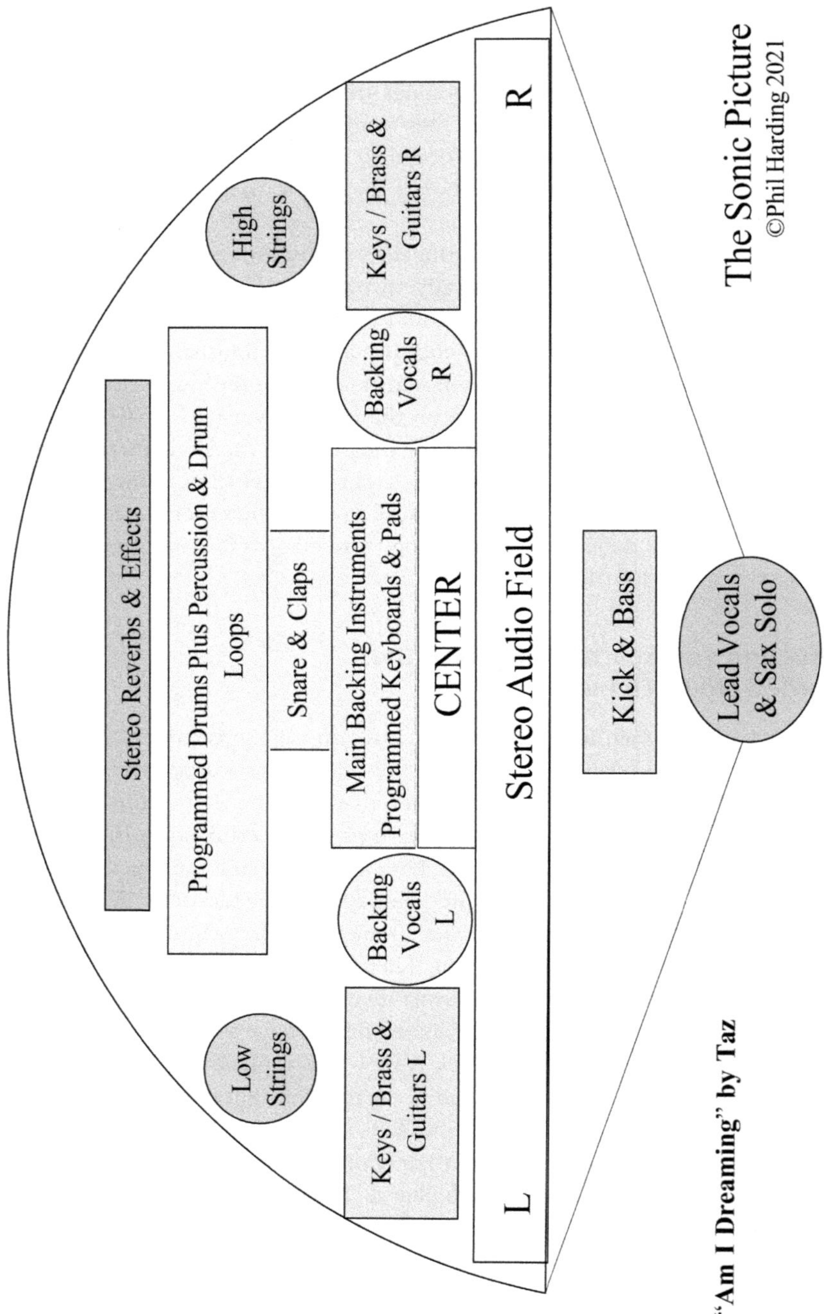

FIGURE 18.1 The sonic picture (Phil Harding 2022).

on the extended intro, middle breakdown, and the extended outro. These sections in particular, which feature on most extended pop dance mixes, offer the listener the opportunity to discover individual musical, rhythmical, and vocal parts that may be hard to define throughout the busy and exciting choruses. The loudness peaks of the drums, musical elements, and the vocals are all presented to us during the choruses.

Throughout my production career spanning from the 1980s to present I have utilized a systematic approach to planning my productions and remixes, usually within a team environment and therefore, in my experience, extended dance remixes are the result of an intensive period of group creativity (Sawyer 2003). To assist the PJS Production team on this record I would have utilized my "sonic picture" (Harding 2020) based on the theory of "sonic cartoons" (Zagorski-Thomas 2014, 49). Figure 18.1 offers a visual representation of the recording elements and their multi-dimensional positions within the stereo soundscape of this record.

Pitch and Frequency Highlights of the Recorded Novel Elements

The high pitch register of the strings throughout the choruses delivers an aesthetic lift and movement to the whole song as they emphasize the rising chords of the first three bars that culminate with their direct support of the vocal melody in bar 4 and is reminiscent of the Trevor Horn ABC (1982) productions of the 1980s (see "The Look of Love" at 0:45). It is also noticeable that the strings on bar 4 of the chorus are further enhanced by a typical 1980s synthesizer bell sound in a high pitch range and enriched by a long hall reverb.

I asked the musical arranger of the remix, Julian Wiggins, for his observations about the effect of his orchestral string lines in the chorus. These elements were added for the remix to enhance the chorus vocals:

> The chorus strings are in "contrary motion" when Taz is singing. They're static and filling the gap at the end of each phrase so they're weaving in and out but they come together on the title track line "Am I Dreaming". Maybe "contrapuntal" is the correct term—where he [Taz] moves the strings are static and they respond in a similar rhythm when he's not moving.
>
> *Wiggins 2021*

I think these types of comments summarize the emotional context of the recorded elements and their functions.

Loudness Levels and Relationships

The loudness levels on this record are provided by a number of elements that all stem around the heavy kick drum(s) and the brass stab on beat 1 of every fourth bar of the extended intro, middle breakdown, and extended outro. This brass stab is providing a dynamic contour to the listener and highlights that something new is arriving each time at these points; it also emphasizes the arrival of the choruses throughout the song. The ambience, reverb, and delays added to the brass stab provide a timbral relationship to other elements around it such as the drums and vocals. A vocal element that shapes the dynamic contour of the extended sections is the "Let's Go" and "Hit It" samples. These re-recorded samples may be familiar to 80s dance floor fans and appeared on records such as "Wap Bam Boogie" by Matt Bianco (1988). They warn the listener (and DJ) that the elemental characteristics of the next section are going to change.

In terms of example frequency relationships and levels during my first listening session, the bass is sitting just on top of the kick drum throughout, which is just as I like it and I was pleased to be able to identify that relationship on my hi-fi system. Full details of my "top-down" mixing techniques and processing can be found in the mixing chapter of my recent book *Pop Music Production* (Harding 2020).

This record is 121BPM and at the height of the disco music boom in the 1970s, and its transformation into the many genres of dance music in the 1980s, 120BPM was considered an ideal tempo for people to dance to. This positions the remix into the very popular 1980s retro dance genre and is clearly specifically targeted at DJs and fans of the PWL-style sonic signature from that era. The rhythm elements of the remix are unmistakably aimed in that direction, from the choice of sounds and samples through to the programmed drum and percussion patterns such as the 4s kick drum and the energetically stuttering percussion, hi-hats, and hand-claps.

Further study and analysis of the record's musical and rhythmical elements are highlighted by some of the responses from my interviewees:

> [This remix] was typical PWL out of Phil Harding and PJS, pure energy, every time I hear a Phil Harding induced drum and bass line, I get goose bumps and chicken skin at the same time on my birthday, that's the only way I can describe it. You gave me what I wanted and I don't get that [from other people usually] and I know I seem overwhelmingly appreciative but I am.
>
> *Pileggi 2021*

> Not just the drums and bass but it's busy from the start in a good way, fantastic energy and that is going to have someone listening from the get go. I thought it was fantastic having a busy/driving bass but not too busy to detract from other elements on the track. I also loved how early the backing vocals come in, I'm a big fan of that. I always liked that on *Waterloo Sunset* by the Kinks (1967), it always amazes me how well it works to start with backing vocals before anything else.
>
> *Dalton 2021*

I asked Bobby Pileggi for his reaction to hearing the backing vocals on the pre-chorus and chorus sections for the first time (they were PJS additions for the remix):

> I was shocked, unbelievably shocked, it's not every day that a remixer goes the distance to put his heart, his time and talent into producing a project as you did with "Am I Dreaming". Sarah Greene's background vocals were stellar, Julian Wiggin's sax riff, ripped and gave the track another heartbeat and Simon's sequencing was outstanding and then top it off with a foundation of genius in the magic touch of Phil Harding; we now have a smash hit in process/progress it was a blessing from the heavens of music for me. I tell you—this is the way it should be done and all the stars connected, the dots were all in line and I was just in awe!
>
> *Pileggi 2021*

I think these types of comments summarize the emotional content and contextualize the recorded elements and their functions. Importantly the reaction of the client and the parts that he chose to identify and praise are noteworthy. In response to the question, has this final remixed record re-shaped the song for you?

> I agree 100%, the song had legs but you gave it wings! with the remix—I'm serious. With wings you can take off and for me it was one of the highlights of my music career. I've done a lot, I've been around for many years, I've been doing this for 40 years from when I first started spinning [as a DJ] in '75 until now and to me this was one of the highlights and thanks to you.
>
> *Pileggi 2021*

The final comments from my PJS collaborators are worth noting:

> If I had to sum up the record in one word it would be excitement! The final remixed record was full of happy energy right from the start, lots of elements for the listener to enjoy, taking the original song from good to great. One of my favourite projects we were involved with.
>
> *Dalton 2021*

> We've really helped all those emotional moments and the sax helps that and you [Phil Harding] were right to ask for that. And coming back to the foundation of it [the record] the PWL bass style with the 4 on the floor [kick pattern] all of those elements are all channeling people towards the song. Including adding Sarah on backing vocals. It's a beautiful, energetic, emotional epic boy/girl song and to have that female vocal on [...] she [Sarah Green] did a great job.
>
> *Wiggins 2021*

The aesthetics of a great remix are difficult to replicate on paper, so I shall leave the last word on this mix to the PJS client, songwriter, publisher, and executive producer:

> This mix was the final destination in the project! I kept searching for 5 years from 2016 -21. I kept looking for the path and it wasn't happening I had these guys before you that went in and remixed it and it sounded alright but it was nothing that I wanted, nothing that it was potentially capable of being and the final project [the PJS mix] gave me the vision that I had for it and now as a publisher, I have the tools to sell [the song].
>
> *Pileggi 2021*

Conclusions

It may be difficult for the reader to understand why the client did not convey to the original producers and mix engineers working on the record prior to the PJS Remix discussed in this essay that he wanted the PWL/Rick Astley-style of musicality, production, and mix. One can only surmise that sometimes commissioning clients need to hear what they do not want to realize what they do want—maybe a "light-bulb" moment in their minds?

This chapter has described and evaluated the journey of a record going through a transformation process by way of an extended dance remix. Dance and club remixes have been part of pop culture since the disco boom in the 1970s and 12″ vinyl records were introduced for DJs. An alternative chapter title could have been "The Loneliness of the Long-Distance Remixer." I have often felt that way sitting at mixing consoles in recording studios, constructing the final mix with its many layers and arrangements, generally alone. While I have stated that this mix was a PJS collaboration and team effort, it started with myself as the team leader and ended with me as the mix engineer/producer. This workflow process is the foundation of my "Service Model for (Pop Music) Creativity and Commerce" (Harding 2020; Harding and Thompson 2019, 155; Thompson and Harding 2019, 300).[6]

The music mixing process has never been underestimated by the recording audio industry. The reward of achieving a personal creative *flow* has largely come about during my time spent in the mixing stage of the field process. I would generally be alone and isolated for the majority of a typical mixing day, and as Csikszentmihalyi states, "when we live creatively, boredom is banished" (Csikszentmihalyi 1996, 344).

One of the unexpected discoveries for me was to evaluate and reflect upon the unusual musical structure of this composition. Pop form is always under intensive scrutiny and discussion with producers and remixers when embarking on a project and this record is a fine example of moving away from the standard verse–chorus form. The formulaic musical structure of most PWL-style records was: Intro–Verse–Pre-Chorus–Chorus–Verse–Pre-Chorus–Chorus–Bridge–Chorus. The intro would generally reflect the chorus hooks and the bridge was called a middle eight in the UK. We also used to call the pre-chorus a bridge but 21st-century pop terminology has fallen into line with the American system.

The concept of analyzing a record brings into question the legitimacy of someone being able to do this without being part of that record's process. I have used this opportunity to present an overall stereo analysis following the Moylan framework plus the addition of analyzing the multitrack session due to my involvement in the record. This opens a much broader question about practice-based research, which is quite a new concept in the academic research of the arts and creativity. I would suggest that to successfully research and analyze a record requires the researcher to interview those involved in the recording and mixing process. This can include three categories of contributors to be certain the agency in influencing the final record is fully described and captured:

1. The artists and musicians.
2. The producer(s), engineers, and studio personnel.
3. The record company A&R (Artist & Repertoire), artist managers, and music publishers.

Therefore, this chapter includes extracts of interviews with the record's main contributors other than myself, as discussed in the methodology section.

There is little academic discussion about how the credits of songwriters and producers can crossover and often become blurred:

> Sometimes producers and singer-songwriters share responsibilities to the extent that they credit themselves as co-producers, and sometimes as co-writers.
>
> *Burgess and Toulson 2017, 99*

There can be occasions where the producers' creative contributions and suggestions can be construed as co-songwriting but unless an "open" policy of the possibility of this happening and a pre-agreed songwriting split is agreed in advance of the sessions, this will very rarely be agreed upon post-session.

This study also opens the interesting question about the blurred boundaries of remix and production credits in this genre and whether or not remixers deserve a production credit for their recorded parts. It has been a traditional music industry standard that whoever records the master vocals is given the production credit but this is open to on-going debate. In this case the client and original (executive) producer, Bobby Pileggi, assigned equal production credits to himself, Harding/Wiggins, and Dalton. That would be considered generous but fair to most industry practitioners.

Glossary of Terms

Auto-panning: This is where an L-R panning effect is applied, usually in time with the song BPM.

Feedback: This occurs when digital delay effects are "fed-back" into themselves, creating more than one delay. This is generally set by percentages such as 40% to produce the "long tail."

Locs: These are generally vocal samples "locked" into a sample program and re-triggered via a midi keyboard.

Plugin: A digital effect app that can be inserted during a DAW mix, such as reverb or compression.

Notes

1 PJS Productions consists of Team-leader and mix engineer—Phil Harding / Keyboard arrangements & music programming—Julian Wiggins / Rhythm arranger & programmer—Simon Dalton.

2 Remixes are essentially a re-recording process in the dance music genre and the roots of the remixing process are in Dub Reggae, which emerged in Jamaica in the 1960s: "Dub" refers to a process of deconstruction by which the engineer strips music down to its basic rhythm components, introduces novel elements, and thereby provides a new interpretation of the material. While such processes are very common within contemporary electronic music, their origins are not well known and often misunderstood (Partridge 2007, 309). The remix team (in this case, PJS MusicProductions, which specializes in "80s retro pop-dance") manipulates and re-arranges the originally recorded material (often retaining just the vocals). New instrument and vocal recordings plus computer music programming can transform the track into an artifact often suitable for an entirely different music genre and marketplace. 1980s remix culture also rides on the back of the 1970s boom in disco music and the extended mixes that appeared on 12″ vinyl to feed the DJ culture in dance clubs.

3 Links to "Am I Dreaming" (PJS Extended Mix) and the audio files discussed throughout this study can be accessed on a number of web open-source platforms: Spotify: [https://open.spotify.com/track/1XFhjhYPsWvkAfCULZpk8V]; Apple: [https://music.apple.com/jp/album/am-i-dreaming-phil-harding-pjs-extended-dance-mix-single/1475605076]; Amazon: [https://www.amazon.com/Dreaming-Phil-Harding-Extended-Dance/dp/B07W7HL9DH]; YouTube: [https://www.youtube.com/watch?v=IJpX6vst1Pc].

4 Notes from those sessions can be found on the webpage for this chapter.

5 The original version of "Am I Dreaming," discussed for comparison purposes, can be found on the webpage for this chapter.

6 Further details about the workflow and the journey of a record going through a transformation via the process of an extended dance remix can be found on the book webpage under the heading "Work-Flow Analysis" for this chapter.

References

Burgess, Richard and Toulson, Rob. 2017. "Singer-Songwriter Meets Music Production." In *The Singer-Songwriter Handbook*, edited by Justin Williams and Katherine Williams, 91–111. New York: Bloomsbury.

Csikszentmihalyi, Mihaly. 1996. *Creativity: The Psychology of Discovery and Invention*. New York: Harper Collins.

Dalton, Simon. 2021. *Personal Interview.* Recorded July 7, 2021 via i-phone. UK.
Harding, Phil. 2010. *PWL from the Factory Floor.* London: Cherry Red Books.
Harding, Phil. 2020. *Pop Music Production: Manufactured Pop and BoyBands of the 1990s.* New York: Routledge.
Harding, Phil and Thompson, Paul. 2019. "Collective Creativity: A 'Service' Model of Commercial Pop Music Production at PWL in the 1980s." In *Innovation in Music: Performance, Production, Technology, and Business*, edited by Jay Hodgson, Russ Hepworth-Sawyer, Justin Paterson, and Rob Toulson, 143–59. New York: Routledge.
Moylan, William. 2015. *Understanding and Crafting the Mix: The Art of Recording.* 3rd ed. Burlington, MA: Focal Press.
Moylan, William. 2020. *Recording Analysis: How the Record Shapes the Song.* New York and London: Routledge, Focal Press.
Partridge, Christopher. 2007. King Tubby meets the Upsetter at the grass roots of dub: Some thoughts on the early history and influence of dub reggae. *Popular Music History,* 2(3).
Pileggi, Bobby. 2021. *Personal Interview.* Recorded May 19, 2021 via Skype. UK.
Sawyer, Keith. 2003. *Group Creativity: Music, Theatre, Collaboration.* New York: Taylor & Francis.
Sloan, Nate and Harding, Charlie. 2020. *Switched on Pop: How Popular Music Works, and Why It Matters.* New York: Oxford University Press.
Thompson, Paul and Harding, Phil. 2019. "Collective Creativity: A 'Service' Model of Creativity in Commercial Pop Music at P&E Studios in the 1990s." In *Journal on the Art of Record Production—Proceedings of the 12th Art of Record Production Conference, 2017,* edited by Jan-Olof Gullo, 287–302, Stockholm, Sweden: KMH/JARP.
Wiggins, Julian. 2021. *Personal Interview.* Recorded June 22, 2021 via Facetime. UK.
Zagorski-Thomas, Simon. 2014. *The Musicology of Record Production.* Cambridge: Cambridge University Press.

Discography

ABC, 1982. *The Look of Love.* [vinyl 7″ single]. UK: Mercury Records.
Matt Bianco, 1988. *Wap Bam Boogie.* [vinyl 7″ single]. UK: Warner Records.
Rihanna Featuring Calvin Harries, 2011. *We Found Love.* [single]. USA: Def Jam Records.
Taz Vegas, 2019. *Am I Dreaming.* [single]. USA: RSM Records.
The Kinks, 1967. *Waterloo Sunset.* [single]. UK: Pye Records.

SECTION VII

Voicing Identity through Genre

19

FRAMING THE FEMALE VOICE IN DOOM METAL

Compositional and Sonic Elements in The Gathering's "Strange Machines" (*Mandylion*, 1995)

Lori Burns

Influenced by the death-doom stylings of bands such as Paradise Lost (formed in 1988), My Dying Bride (1990), Anathema (1990), and Amorphis (1990), The Gathering (1989) featured male vocalists on their first two albums, with death metal vocals and gothic death/doom stylings on *Always* (1992) and clean vocals on *Almost a Dance* (1993). The introduction of female vocalist Anneke van Giersbergen on their third album, *Mandylion* (1995), is marked as an innovative moment in the history of doom metal. Distinct in style from the gothic death/doom bands that were emerging in the early 1990s, such as Theatre of Tragedy (1993), Draconian (1994), and Tristania (1995), The Gathering did not feature Van Giersbergen in the "Beauty and the Beast" vocal style, with an ethereal soprano set against a deep growling male vocal. On the contrary, Van Giersbergen's role in The Gathering was remarkable for her primary vocal position: not pitted against a male voice, she carried the full musical authority of lead vocalist within the heavy stylings of doom metal.[1]

James Bickers of *AllMusic* includes these comments in the artist biography entry for The Gathering:

> It would appear that the seed had been planted in the bandmembers' heads to pair their uniquely compelling songwriting style with a passionate female vocalist; the result was the addition of the incomparable Anneke van Giersbergen, an incredible singer capable of stretching miles of emotion out of each and every syllable. The next album, 1995's *Mandylion*, remains one of the very best Dutch heavy metal albums ever recorded. Van Giersbergen's poetic and haunting lyrics, together with the band's orchestral ruminations, resulted in darkly important songs that seamlessly blended from one to the next.
>
> *Bickers n.d.*

DOI: 10.4324/9781003089926-27

Beginning with *Mandylion* (1995), Van Giersbergen was then featured on *Nighttime Birds* (1997), *How to Measure a Planet?* (1998), *if_then_else* (2000), *Souvenirs* (2003), and *Home* (2006). By the end of her time with the band, they had shifted from the doom/progressive stylings of *Mandylion* to progressive metal and then progressive rock, ultimately toward an art rock or alternative rock classification. Van Giersbergen was replaced by Silje Wergeland in 2009, releasing *The West Pole* (2009), *Disclosure* (2012), and *Afterwords* (2013).

Metal critics and enthusiasts receive Anneke van Giersbergen's work in The Gathering as having broken down some of the barriers for female metal vocalists. In a fascinating blend of genre references, *Mandylion* draws upon doom, goth, and progressive metal.[2] The album is received as a hallmark of doom but also as a highly influential moment in the development of gothic and symphonic metal, genres that were significant for the advancement of female-fronted bands such as Lacuna Coil (1994), After Forever (1995), Nightwish (1996), Within Temptation (1996), Epica (2002), Leaves' Eyes (2003), and later bands. And yet, rather than staying within a doom/gothic style, they carved out a unique and singular path, developing a body of work that gradually shifted toward progressive rock and alternative rock. Van Giersbergen's distinctive vocal style was catalytic in that innovative development.

Music critics celebrate the album and Van Giersbergen's contributions in particular. For instance, one critic writes,

> The opener, Strange Machines, must surely provide some of metal's most distinctive and instantly recognisable six minutes. An oversized, lumbering riff crawls through the speakers, but is transmogrified by a crystal clear, radiant vocal line that swerves from high to low effortlessly and magnificently. The combination is immense; a dramatic, swooping melody coupled with real doomy weight, and Mandylion's most instantly recognisable track.
>
> *Charles n.d.*

This interpretation of Van Giersbergen's voice in relation to the doom stylings of the band emerges frequently in the critical reception of The Gathering and points to the genre ambiguity of their work. *Mandylion* is often heralded as a gothic album, while critics simultaneously (and one could say, contrarily) acclaim Van Giersbergen's potent and uplifting sound.[3]

In the context of the overwhelming reception of this album as a hallmark of metal and specifically for its striking vocalist, it is an interesting analytic exercise to consider Van Giersbergen's contributions to one of The Gathering's well-known songs. This chapter explores how the opening track of *Mandylion*, "Strange Machines," is formally structured and sonically shaped. I will illustrate how the vocal and instrumental gestures are presented to occupy specific registral positions in the full sonic spectrum and how they are framed within the stereo field. Using sonic visualization tools (spectrograms) from iZotope RX, I will illustrate my analyses in a dynamic engagement with both formal and sonic attributes. My aim is to demonstrate how sonic elements—timbre and space—intersect with the formal features of the song—form, temporal elements, and pitch materials. My analysis explores the emergent dialogue between sonic attributes and musical structure, bringing forward moments where sonic space and timbre are paramount to the crystallization of musical structure and expression.

Overview of Sonic Elements

Mandylion was recorded at Woodhouse Studios in Hagen, Germany, engineered by influential metal producers Siggi Bemm and Waldemar Sorychta who are known for their work with other female-fronted bands such as Theatre of Tragedy and Lacuna Coil.[4] "Strange Machines" is produced in a style that allows for sonic differentiation: the main idea is always given space. As Martin Popoff writes, "Century Media regular Waldemar Sorychta produces the album with a lucid mind firmly bent on clarity, and the result is a well-crafted piece of oxymoronically refreshing, replenishing doom …" (2004).[5] At the same time that clarity is achieved, the track is dark, atmospheric, and theatrical.[6] There is an austerity and simplicity to the materials, with a unity of gesture that allows for a dramatic delivery of ideas and a drawn-out vocal presentation.

The lyrics, written by Van Giersbergen, invite imaginative reflection as a first-person subject expresses desire for time travel. That desire is expressed as a wish to confront major cultural figures and societal movements and crises (Cleopatra, Beethoven, Gershwin, Chaka Khan, Jesus, the Renaissance, WWII, Russian Revolution) as well as a wish to experience this sensorially (to feel and to see).[7] Making an explicit connection to science fiction, the track embeds a sample from the 1960 film adaptation by George Pal of H. G. Wells' *The Time Machine*, conjuring the mid-range sonic world of the classic film style.

To communicate this story, in the duration of 6 minutes, Van Giersbergen's voice is centered in the stereophonic mix, occupying a strong presence in the mid-range, with a dark-hall reverb yielding warmth. Although her voice does not carry a lot of detail at the top, it clearly emerges in the mid-upper register of the mix. A rounded bass occupies a very low octave, sustained guitars (hard-panned left and right and down-tuned to Drop D tuning) cut through clearly in the low-mid range with a short-hall reverb, atmospheric keyboards shimmer in the high register, and the kick and snare are tight and articulate in the low and mid registers.[8] As my analysis unfolds, I will illustrate how these registral and spatial positions emerge in a dynamic expression of the song.

Overview of Compositional Elements

Table 19.1 tabulates the form of "Strange Machines," with details about meter/tempo, riff/phrase, and mode/harmonic progression. It is a productive analytic exercise to describe and reflect on these formal (compositional) attributes without mentioning any of the sonic (timbral and spatial) qualities. I am deliberately

TABLE 19.1 Compositional analysis of "Strange Machines"

Song form	*Meter/tempo (BPM)*	*Riff/phrase structure*	*Mode/harmonic progression*	*Modal center*
Intro (0:01)	4/4 63	2-bar riff X 5 (10 bars), 8th notes	G dorian G–F–E power chords closes C–B♭–F	G
Verse 1 (0:39)	4/4 63	based on Intro 4-bar vocal phrase X 2 (8 bars), 16th subdivisions	G dorian G–F–E power chords	G
Chorus (1:10)	4/4 63	5-bar vocal phrase (5 bars)	G phrygian/C natural minor C–B♭–F–E♭–B♭–A♭–F	G/C
Bridge 1 (1:43)	4/4 72	1-bar riff X 6 (6 bars) 16th notes	D phrygian dominant (prolonged dominant of G minor) static D harmony	D/G
Verse 2 (1:28)	4/4 63	4-bar vocal phrase, plus 2-bar link to Br2	G dorian G–F–E power chords	G
Bridge 2 (2:04)	12/8 132 [metric modulation]	4 bars X 3 [X 2] (24 bars) Syncopated triplet patterns	F natural minor F–A♭–E♭–G–A♭–G	F
Verse 2 (2:48)	4/4 63	2-bar riff 4-bar vocal phrase X 2 (8 bars)	G dorian	G
Chorus (3:19)	4/4 63	5-bar vocal phrase (5 bars)	G phrygian/C natural minor C–B♭–F–E♭–B♭–Ab–F	G/C
Interlude (3:38)	4/4 63	4 bars X 4 (16 bars plus pause)	D minor	D
Chorus (4:11)	4/4 63	5-bar vocal phrase plus 3 bars (8 bars)	G phrygian/C natural minor C–B♭–F–E♭–B♭–Ab–F	G/C
Bridge 1 (4:40)	4/4 72	1-bar riff X 6 (6 bars)	D phrygian dominant (prolonged dominant of G minor) static D harmony	D/G
Bridge 2 (5:01)	12/8 132 [metric modulation]	4 bars X 3 [not repeated] (12 bars)	F natural minor F–A♭–E♭–G–A♭–G	F
Outro (5:23)	4/4 132	4 bars X 4 (5th repetition as fade out sustain) (20 bars)	D dorian	D

withholding such comments for the moment but will soon put the sonic into dialogue with the compositional elements.

The song is framed by an intro and outro marked by contrasting materials: the intro sets the doom-inspired pace of repeated eighth notes at 63 BPM, with a 2-bar guitar riff (heard five times) in G dorian, characterized by a descending pattern of power chords on G, F, and E (the E declaring the dorian quality of the mode). The outro, by contrast, moves at a rapid pace of 132 BPM, as guitars, kit, and keyboard deliver a repeated 4-bar pattern (20 bars) in D dorian, with a striking melodic B in the keyboards (5:42) to mark the dorian quality, now a fifth higher than the originating G dorian.

The main body of the song comprises a verse–chorus–bridge design that offers gestural contrast and harmonic development. The verse continues with the G dorian power chords of the intro, maintaining the slow tempo of 63 BPM; the voice mounts two 4-bar vocal phrases over the repeated 2-bar guitar riff. The chorus also maintains the tempo but introduces an irregular five-bar vocal phrase that articulates—in its range and patterning—the G phrygian mode, although it is supported by a harmonic progression that avoids stating G and that might be heard more easily in C natural minor. The first bridge section (labeled Bridge 1 in Figure 19.1) increases the pace slightly to 72 BPM, with unison guitars introducing a 1-bar riff in the D phrygian dominant mode. In the first two bars of the section, a vocal solo that is also grounded in the D phrygian dominant mode is supported by the guitar riff with its harmonic stasis on D. The D phrygian dominant mode connects to the G harmonic minor scale, with D creating a dominant effect. In this regard, it could be understood to represent a return to the G modal tonic of the opening, however, with greater tension created by the dominant function. The second bridge section (Br2) is entirely instrumental. The passage immediately prompts a metric modulation from the 4/4 of the previous sections to a compound 12/8. Two bars of the previous 4/4 become one bar of 12/8, effecting a tempo increase (132 BPM). The harmonic orientation is now F natural minor, in a progression that emphasizes a shift from F minor to E♭.

The overarching progression of the verse–chorus–bridge structure falls to the flat side in the circle of fifths: the descending fifth shift from G to C is felt as a move in the flat direction and the harmonic progression signals that shift with harmonies on E♭ and A♭. The D phrygian dominant mode,[9] with its pitch collection in common with G minor, represents an interesting moment of potential return to the tonic G of the opening, however, while still contributing to the increase of flats (with E♭ staying prominent in the story). Next, the move to F minor represents another falling fifth in the chain from the C minor of the Chorus, taking the harmony further into the flat sphere.

This formal design articulates directional movement that is then elaborated. The second part of the song repeats the verse–chorus–bridge structure, however, with an interlude section that is nested within two chorus statements, expanding the length of the structure. The harmonic design and metric content are repeated from the first iteration, with the addition of an Interlude, which adds a D minor

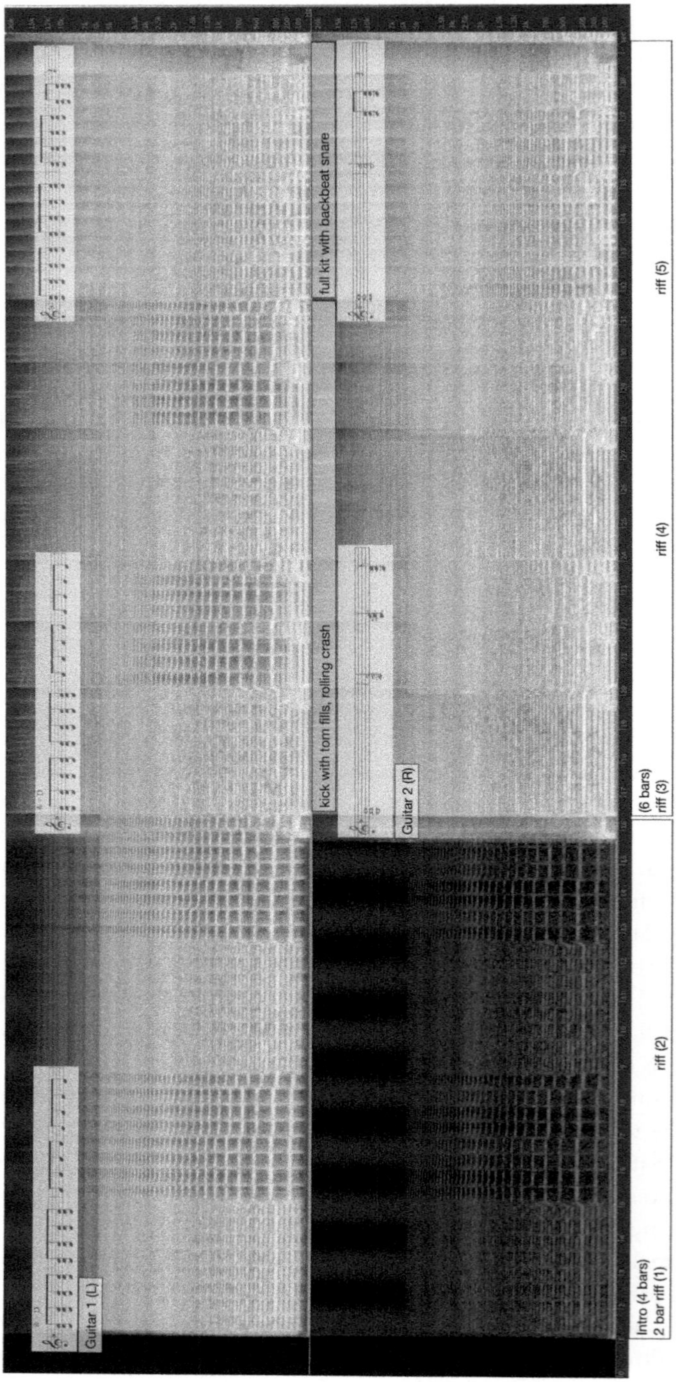

FIGURE 19.1 "Strange Machines," annotated spectrogram, Intro.

interruption to the falling fifth design. D counters that falling-fifth design by representing a rising fifth from the original G minor and foreshadowing the ultimate move to D dorian to close the song.

With this design comes opportunity: the selection of modes allows for an overall progressive approach to pitch and harmonic emphasis; the chromatic story has potential for elaboration as the harmony falls to the flat side; the rhythmic and metric design creates directional change that allows for intensification; the riff structure and instrumental effects open up the space for timbral shifts and production effects; and ultimately this complex song design offers immense potential for the vocalist to interact with these elements.

Sonic and Compositional Elements in Dialogue

With a formal analysis in place, and in order to explore William Moylan's notion of "How the record shapes the song" (2020), let us proceed to a more detailed illumination of the formal materials in dialogue with sonic content. As I conduct this analysis and focus my listening energies, my central analytic questions and organizing principles are as follows:

- How are formal elements framed and illuminated by the sonic elements?
- How is clarity achieved and space created for the voice in relation to the band?
- Compositional elements to be considered: formal (song sections, design); temporal (meter, rhythm, tempo); and matters of pitch (melody, harmony, mode).
- Sonic elements to be considered: timbral and spatial elements that frame the height, breadth, and depth of the sonic field.[10]

Still referring to the formal content in Table 19.1, I will introduce figures that reproduce spectrogram data, aligned with music transcriptions and annotations of selected passages. In order to tell the sonic story of this song, I focus primarily on the first statement of the formal materials, that is, the intro–verse 1–chorus 1–bridge 1 and 2.

Intro (0:01–0:39). Figure 19.1 is an annotated spectrogram derived from iZotope RX 7, with left channel at the top and right channel at the bottom of the image.[11] I use such images to illustrate and map the structural and sonic content of the song.

The two-bar riff of the intro, with its simple G dorian gesture (G–F–E–D), is introduced by the first guitar, hard-panned to the left, delivered on the middle pickup, using a Marshall amp with a distortion pedal to create a warm, but raw distortion.[12] The second guitar enters, hard-panned to the right, delivering power chords on the bridge pickup to create a heavier and more direct tone. Both guitars are saturated, forward in the mix, and relatively dry (with a low level of reverb). The extremely slow tempo, Drop D tuning, and wide stereo field open up both temporal and sonic space to create an austere doom-metal effect. The kit, in contrast to the guitars, is centered in the mix, building a column of sound in the middle of the stereo field. Anticipated by a tom fill at the end of riff statement 2, the next two riff statements 3 and 4 include ominous downbeat bell strikes and rolling crash cymbals, while riff pattern 5 is enhanced by punchy kick, hard snare, and distant China cymbal keeping steady eighths. The final backbeat of riff 5 is an extremely hard snare shot, enhanced by the crash, effectively announcing the close of the intro. The sonic intensification created by the kit activity in riff 5 coincides with a change in harmony presented by the guitars as they modify the progression to power chords C–B♭–F: the first two chords take the previous pattern (G–F) up a fourth, while the last chord functions as a natural seventh scale degree of G dorian, facilitating a return to G for the start of the verse. The harmonic shift to F happens precisely at the hard snare shot and crash cymbal that articulate the close of the intro, an instance of compositional elements, and sonic design merging to mark a significant moment of the form.

Verse 1 (0:40–1:09). Figure 19.2 maps the first verse. As the voice enters, the guitars continue with the original hard-panned riff of the intro, with the left again distinguished by its raw, warm distortion while the right remains aggressive. The kit maintains its central placement, now delivering a prominent syncopated kick pattern with backbeat snare on beats 2 and 4 in the first bar of the riff, but on beat 4.5 in the second bar, contributing to the syncopation. The kick downbeats (|) and snare backbeats (★) are annotated in the centre of the image, beneath the vocal line transcription for the first two riff statements. (The syncopated rhythms of the kick drum are evident in the lowest frequency register of the spectrogram.) The vocal pitches are aligned (to the extent that it is possible) with the sonic data. In order to maintain the temporal alignment, there are no barlines in the transcription; however, I have marked the downbeat of each 4/4 bar with a bar number (1 through 7).

Against this riff structure, the voice sits in the center of the mix, with a long reverb creating a cavernous distance. Her voice is dark, with a prominence of mid and mid-upper-range frequencies. The vocal melody occupies a range from G_3 (196 Hz) to C_5 (523 Hz). However, Van Giersbergen's voice is equalized to bring forward the mid-upper frequencies, especially around 1000 Hz. The lower brace in the right channel (lower section of Figure 19.2) marks the mid-upper frequencies where her vocal line emerges with intensity. The upper brace in the image of the right channel highlights the band of energy in the high register (2000–5500 Hz level) where the guitar cuts through, above the voice. In the left channel (upper section of Figure 19.2), the spatialization of guitar and voice now places the intensity of the guitar's middle pickup, cutting through in the mid register (as marked by the oval), apparently beneath the voice (as marked by the brace). In addition to this complex pitch treatment, the stereophonic mix presents a rich listening experience: the left and right channels offer different ways of hearing the guitar riff in relation to the centered voice. The depth of texture also presents a vivid spatial arrangement as the guitars are relatively dry and forward

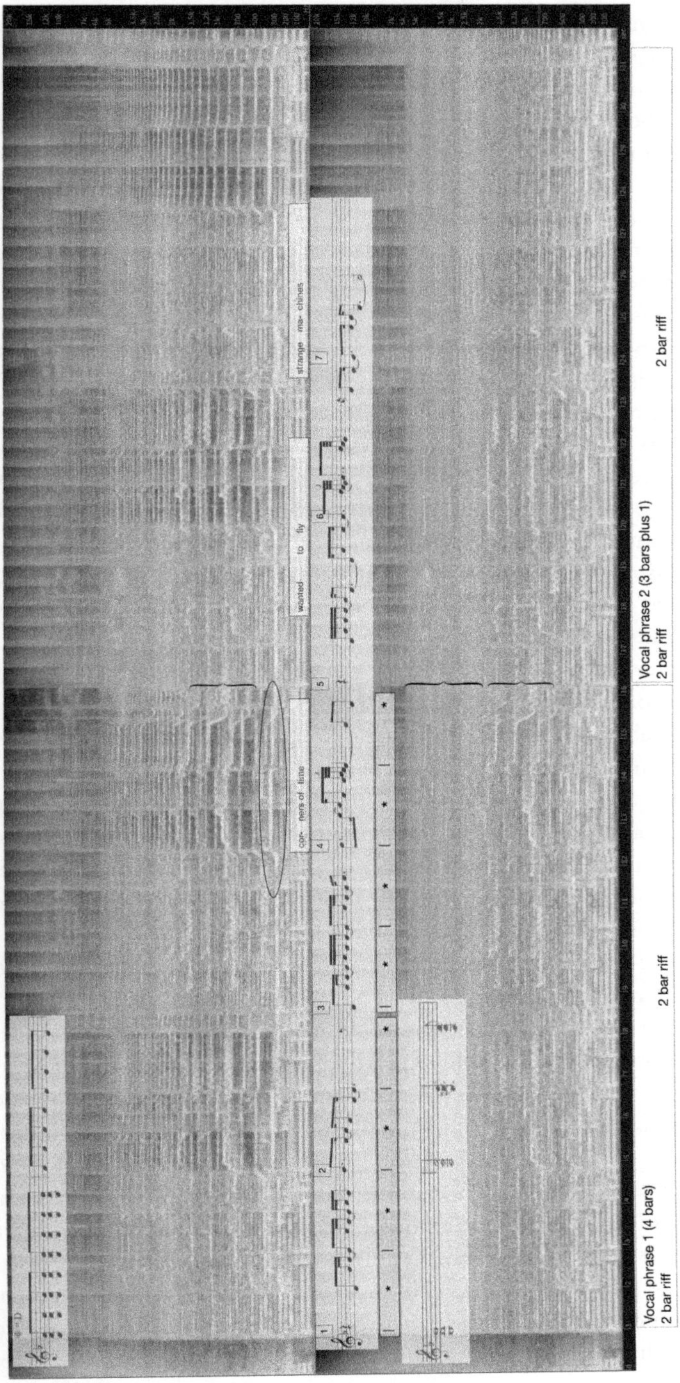

FIGURE 19.2 "Strange Machines," annotated spectrogram, Verse 1.

while the voice carries a great sense of expanse and distance. Another effect worthy of mention is the lack of heavy compression on her voice. There is no sizzle in the high register; the level of compression is relatively mild, not breathy, and not overly enunciated.

Van Giersbergen delivers her vocal with a powerful chest belt, transitioning smoothly to a nasal mask (tilting her palate to the nasal bones) as she enters the upper frequencies of the vocal line (starting around A_4).[13] She never occupies a head voice (with a raised palette) and does not sing pitches that are high enough to necessitate that vocal shift. The impact of her presentation is the illusion of chest resonance from the lowest to the highest note of the passage. Her thicker vocal quality is very well produced and controlled so that the listener would not perceive her to be struggling to travel from low to high. In this regard, her vocal timbre matches perfectly with the lyrical sentiment of freedom and ease of flight. And yet it is not an ethereal or delicate illusion of flight that she presents; rather, it is a weighty and commanding impression of temporal travel. One moment where a degree of effort might be perceived is the final low G_3 of the Verse, on the last syllable of "machines." Here, she broadens her vowel in order to adjust to the lower point of the melodic range.[14]

The kit also develops a distinctive dialogue with the vocal line, as evident in the annotated passage beneath the transcription of Figure 19.2. (The annotations, as mentioned earlier, are only provided for the first two statements of the riff.) The voice enters on beat 2 of the vocal phrase 1, aligning with the backbeat for the first word of the phrase. Thereafter, Van Giersbergen begins bars 2, 3, and 4 on the downbeat. Although that first entry aligns with the backbeat, thereafter, she typically anticipates the backbeats (beats 2 and 4) with a sixteenth note tied over the beat. Moving ahead to vocal phrase 2, although it also begins squarely on the backbeat, the tendency to anticipate beats a sixteenth note early seems to pervade the phrase. Her delivery of "fly" is ornamented to place accents on beats 2 and 3 of that bar, a moment of clear beat emphasis that is then immediately replaced by syncopation for "strange machines."

Chorus 1 (1:10–1:42). Figure 19.3 maps the irregular five-bar phrase that comprises the chorus. This passage is marked by a significant change in the guitar texture that is apparent in the spectrogram: the guitars, still panned left and right, now both play on the bridge pickup, creating a heavier and more aggressive tone on both sides of the stereo spectrum. In unison, they play power chords in the progression C | B♭–F | E♭* | B♭–Ab | F* (the lines (|) in Figure 19.3 indicate barlines). The E♭ and F chords (marked *) in bars 3 and 5 of the chorus are treated to palm-muted (staccato), pulsing eighth notes. The palm muting marks a significant moment of harmonic change as the E♭ harmony is the first indication that we are leaving the G dorian mode. The palm muting also opens the texture to allow the keyboard to make its presence known with an atmospheric, shimmering sound in the upper register, offering a moment in the texture when the gothic reference comes through. The kit complements these textural changes: against the power chords, the drummer plays on the bell of the ride

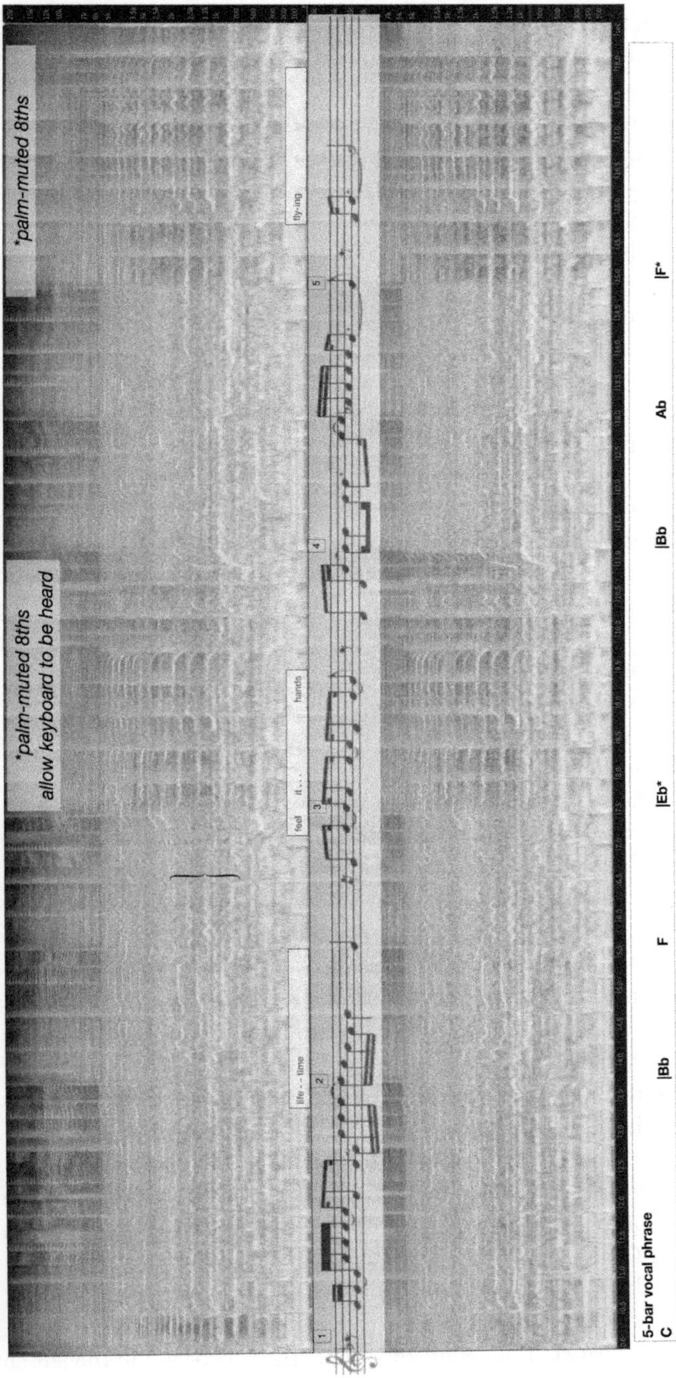

FIGURE 19.3 "Strange Machines," annotated spectrogram, Chorus 1.

cymbal and on open hi-hats during the palm-muted chords, with a cymbal crash to start each of the palm-muted bars.

At the center of this heavy and panned texture, the vocal line shifts to a higher tessitura, with the range of D_4 to D_5. Despite this registral shift, Van Giersbergen maintains the previously established thick vocal quality, holding presence within the heavy texture. The brace in the left channel highlights the second, third, and fourth partials that cut through with intensity. She places her pitches in her nasal mask register for much of the line, shifting to chest register for the lower pitches. Her tone quality and dynamics remain consistent throughout this passage, seeming not to back away even when the palm muting thins the guitar texture. In terms of her melodic pitch content, the vocal line places emphasis on the pitches G–B♭–D, creating a triadic orientation to the earlier modal tonic (G). Her melodic line offers both harmonic consonance and dissonance with the guitar harmonies. The descent from B♭ to G in bar 2 introduces a dissonance against the harmony F but creates a consonance with the E♭ to which the harmony is heading. As the melody moves toward a close, she leans into A♭, in keeping with the guitar harmony below, clearly leading A♭ down to G at the end of bar 4 and then reinforcing G with her cadential pattern F–G in bar 5, once again creating the dissonance of G against the harmony F. The passage does not unequivocally present a new modal tonic. The orientation is clearly no longer G dorian as soon as E♭ emerges, and the A♭ confirms a change of orientation. The guitar harmonies might point us in the direction of C natural minor; however, the voice seems to suggest a G phrygian scale. Her role here is significant in the pitch materials that organize the song's modality. It is an agential role in relation to the band and the song structure.

Bridge 1 (1:43–2:03). The first bridge passage, represented in Figure 19.4, offers a number of remarkable formal and sonic changes.[15] A temporal intensification happens with the increase in tempo (to 72 BPM) as a new 1-bar guitar riff delivering a pattern subdivided at the 16th-note level. During a brief passage with no kit activity, the first guitar (left channel) delivers two statements, using the bridge pickup for a heavier sound. The riff pattern emphasizes D, inflected by E♭ and C; the modal profile of that pattern takes on greater clarity through the vocal line, outlining the pitches of the D phrygian dominant mode (scale form: D–E♭–F♯–G–A–B♭–C–D). Once again, the voice claims agency within the formal structure through the articulation of the mode and within the sonic structure through the clear placement within the texture and riff structure.

As the voice closes on D_4 with a rich vibrato in the chest register, the third statement of the riff welcomes the second guitar (right channel) and the two guitars then continue in unison. The drummer keeps time on the crash cymbal, using the China cymbal for beats 1 and 3. During the third through sixth statements of the riff, a whip sound moves sharply across the stereo field, hard-panned, heavily filtered, and processed to create a hollow, ominous effect.[16] The swoosh (drawing back) of the whip anticipates the backbeat 2 in the left channel, while the resulting crack of the whip occurs on the second half of that beat in

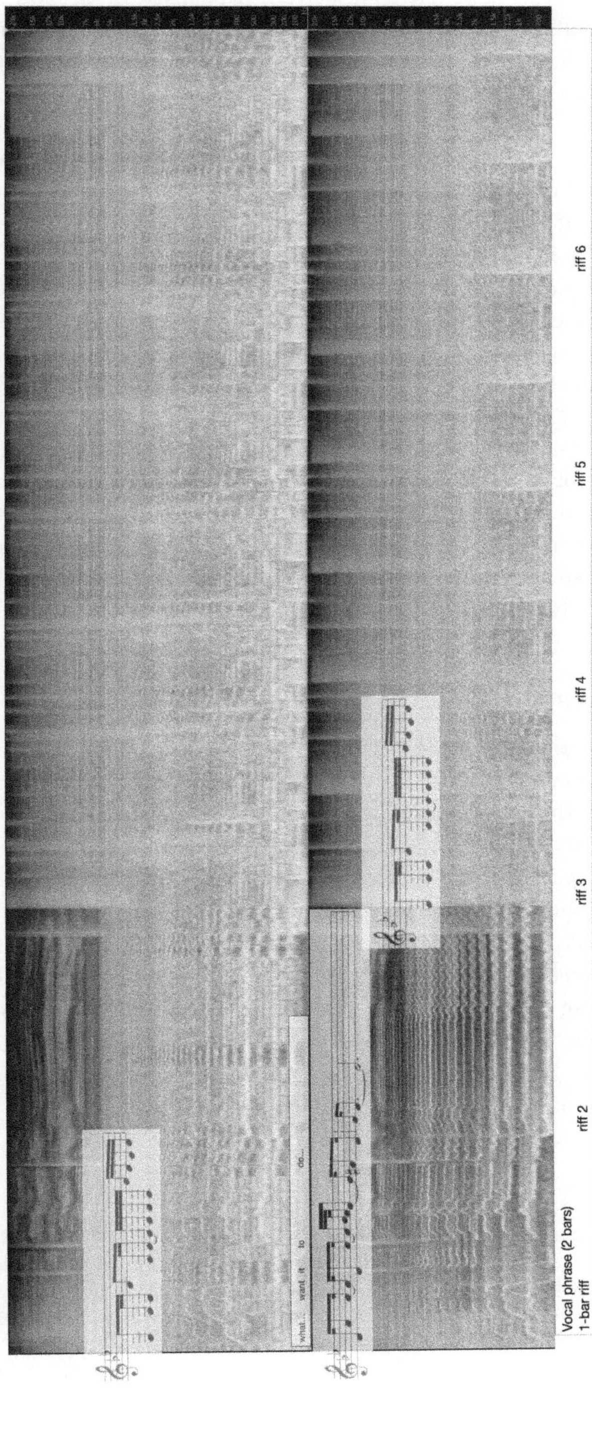

FIGURE 19.4 "Strange Machines," annotated spectrogram, Bridge 1.

the right channel; this same left-right pattern happens again around backbeat 4 and its second half. The pace of the whip delivery is therefore quite intense and insistent, conjuring a mechanical (industrial) effect that connects to the theme of a "strange machine."

Bridge 2 (2:04–2:47). The second bridge section is characterized by a heavy guitar, kit, and keyboard texture in a chugging or galloping riff structure that effects a metric shift to 12/8 time at 132 BPM. The listener enters a completely different temporal and harmonic space here as a 4-bar pattern is delivered three times to create a 12-bar section that is then repeated. The harmony functions in F natural minor, representing another descent in the falling fifth cycle. The passage offers the band the opportunity to drive the riff-based structure, without accommodating to the voice. The speed, intensity, and heaviness might connote the theme of travel, however, without a sense of lightness to this strange ship.

Returning to Table 19.1, it is easy to follow the remainder of the song form, as the verse–chorus–bridge structure is, in effect, repeated, although with an interlude that draws upon the passage "Testing the Time Machine" from *The Time Machine* film. With chugging D minor chords holding the foundation, the sample from the film features Rod Taylor's voice and orchestral effects that mimic the machine, including a rising pitch sequence in the theremin to close the passage that leads—with climactic effect—into the final chorus.

Concluding Remarks

My intention with this analysis of "Strange Machines" has been to interpret the structural or compositional features in dynamic conversation with the sonic developments of those features. This style of analysis promotes engaged listening and interaction with the immediate gestures and effects of the recording. An acute awareness of space—breadth (stereo), depth (reverb), and height (frequency)—can illuminate aspects of the gender dynamics of this remarkable track that initiated and influenced a long genealogy of female-fronted metal bands. Anneke van Giersbergen's voice is presented and framed as a resilient and compelling force, capable of holding space within the aggressive and heavy stylings of doom metal. At the same time, a certain level of genre fluidity—evident in the references to gothic, industrial, and progressive genres—further breaks down barriers and disallows any sense of monolithic genre and/or gender identity.

Notes

1 For more on gender and voice in metal, please refer to Burns (2020a, 2020b, 2021), Heesch (2019), and Ritlop (2020). Weinstein (2016) discusses gender roles in metal subgenres more broadly.

2 Hillier (2020) and Smialek (2016) both offer thorough taxonomies of metal subgenres. For an overview of extreme metal, see also Kahn-Harris (2007).

3 Fans often receive the work as gothic metal, as indicated by reviews; see, for example, Sargon the Terrible, (n.d.) The album is classified as gothic, progressive, doom, and

ethereal wave on *Rate Your Music* (n.d.) and as experimental/post metal on *ProgArchives* (n.d.).

4 For more on Woodhouse Studios, see https://www.woodhouse.de and https://de.wikipedia.org/wiki/Woodhouse_Studio.

5 *Mandylion* appears at position 315 in Popoff's list of the top 500 metal albums of all time, based on a polling of metal fans (2004).

6 Isabella Van Elferen (2017) discusses the importance of "dark" timbres to gothic music genres.

7 The lyrics to "Strange Machines" can be accessed at https://genius.com/The-gathering-strange-machines-lyrics.

8 For a clear chart of these frequency registers, please refer to Moylan (2020, 246).

9 Moore (2011) and Scotto (2019, 259–26) both discuss the prevalence of the phrygian dominant mode in metal.

10 See Chapter 3, "Domain and Elements of Music," in Moylan (2020, 69–104).

11 The spectrogram type is a short-time Fourier transform (STFT), using a Hann window setting and a Bark frequency scale, both of which allow for greater clarity in the black and white color scheme.

12 For more on the role of distortion in hard rock and metal, please see Scotto (2017).

13 For a thorough discussion of vocal timbre in popular music, please see Heidemann (2016).

14 Van Giersbergen does sing lower than this on many tracks, for instance, E_3 in "The Earth is My Witness" (*Nighttime Birds*) by The Gathering. She also sings with head voice on many tracks, for instance, $E\flat_5$ in "Leaves" (*Mandylion*) by The Gathering.

15 This image of the spectrogram data has been slightly rebalanced in iZotope to lighten the instrumental texture for the purpose of visual clarity. The graph would be much more blurry due to the heaviness of the passage.

16 This whip sound was an industrial loop common in the 1990s.

References

Bickers, James. n.d. "The Gathering Biography." *AllMusic*. Accessed August 8, 2021. https://www.allmusic.com/artist/the-gathering-mn0000061949/biography.

Burns, Lori. 2020a. "Female-Fronted Death Metal: Jinjer, Gender, and Genre Norms in Sound & Image." Paper presented at the Society for Music Theory Annual Conference, November 2020, virtual. https://www.academia.edu/video/lg9qAj.

———. 2020b. "Femininity and Vocal Subjectivity in Metal: Floor Jansens's Negotiations of Expressive Space." Keynote address at the IASPM-UK & Ireland Conference, June 20, 2020, virtual. https://www.youtube.com/watch?v=1Vo7FRCl3j0.

———. 2021. "Sharing Sonic Space in Metal Duets: Floor Jansen's Vocal Collaborations." Keynote address at the South Central Society for Music Theory Conference, February 20, 2021, virtual. https://www.academia.edu/video/lmKqej.

Charles. n.d. "The Gathering—Mandylion." *Metal Review*. Accessed August 8, 2021. https://www.metalreviews.com/reviews/album/5561.

Heesch, Florian. 2019. "'Voice of Anarchy': Gender Aspects of Aggressive Metal Vocals. The Example of Angela Gossow (Arch Enemy)." *Criminocorpus*. https://journals.openedition.org/criminocorpus/5726.

Heidemann, Kate. 2016. "A System for Describing Vocal Timbre in Popular Song." *Music Theory Online* 22, no. 1. https://www.mtosmt.org/issues/mto.16.22.1/mto.16.22.1.heidemann.html.

Hillier, Benjamin. 2020. "Considering Genre in Metal Music." *Metal Music Studies* 6, no. 1: 5–26. https://doi.org/10.1386/mms_00002_1.

Kahn-Harris, Keith. 2007. *Extreme Metal: Music and Culture on the Edge*. Oxford: Berg.

Moore, Sarha. 2011. "Metal, Machismo and Musical Mode: How the 'Feminine' Phrygian Second Has Been Appropriated and Transformed." *Networking Knowledge: Journal of the MeCCSA Postgraduate Network* 4, no. 1. https://doi.org/10.31165/nk.2011.41.67.

Moylan, William. 2020. *Recording Analysis: How the Record Shapes the Song*. Waltham, MA: Focal Press.

Popoff, Martin. 2004. *The Top 500 Heavy Metal Albums of All Time*. Toronto: ECW Press.

ProgArchives. n.d. "Mandylion." Accessed August 8, 2021. http://www.progarchives.com/album.asp?id=7582.

Rate Your Music. n.d. "Mandylion." Accessed August 8, 2021. https://rateyourmusic.com/release/album/the-gathering/mandylion-4.

Ritlop, Tina. 2020. "'You Scream Like a Girl': Growling and Screaming Female Voices in Metal Music." In *Words, Music and Gender*, edited by Michelle Gadpaille and Victor Kennedy, 149–57. Cambridge: Cambridge Scholars Publishing.

Sargon the Terrible. n.d. "Review: The Gathering—Mandylion." *The Metal Crypt*. Accessed August 8, 2021. https://www.metalcrypt.com/pages/review.php?revid=1895.

Scotto, Ciro. 2017. "The Structural Role of Distortion in Hard Rock and Heavy Metal." *Music Theory Spectrum* 38, no. 2: 178–99. https://doi.org/10.1093/mts/mtw013.

———. 2019. "System 7." In *The Routledge Companion to Popular Music Analysis: Expanding Approaches*, edited by Ciro Scotto, Kenneth Smith, and John Brackett, 249–64. New York: Routledge.

Smialek, Eric. 2016. "Genre and Expression in Extreme Metal Music, ca. 1990–2015." PhD diss., McGill University.

Van Elferen, Isabella. 2017. "Dark Timbre: The Aesthetics of Tone Colour in Goth Music." *Popular Music* 38, no. 1: 22–39. https://doi.org/10.1017/S0261143017000551.

Weinstein, Deena. 2016. "Playing with Gender in the Key of Metal." In *Heavy Metal, Gender, and Sexuality: Interdisciplinary Approaches*, edited by Florian Heesch and Niall Scott, 11–25. London: Routledge.

Discography

The Gathering. 1995. *Mandylion*. Century Media, 77098-2, compact disc.

20

MASKING

Queer Aesthetics and Production Tricks in Orville Peck's "Hope to Die" (2019)

Stan Hawkins and Zack Bresler

This chapter is supported by a webpage.

Introduction

The enigmatic country-pop artist, Orville Peck, has produced some of the most awe-inspiring sound recordings during the second decade of the 21st century. Born somewhere in the Southern Hemisphere, around 1987/1988,[1] he has been fastidiously elusive about his origins and personal life, covering his face up with elaborate masks.[2] From the little he has revealed, he is son of a sound engineer and spent a good deal of his childhood doing voice-overs for cartoons. He also trained as a ballet dancer for twelve years during his youth, studied acting at The Royal Academy of Music and Dramatic Art in London, and took part in West End musicals.[3] Residing in Canada at the time of conducting this research, Peck has established an international career as a queer country singer, always confirming his gay sexuality. So large is his following that on the 50th anniversary of the LGBTQ Pride events in 2020, *Queerty Magazine* would rank him among the top fifty heroes who have fought for liberty, dignity, and acceptance for all people.[4]

During the course of this chapter, we concentrate on the correlation between sound production, imagery, and subjectivity in the track "Hope to Die," from Peck's debut album *Pony*, released in 2019. Notably, Peck played all the instruments himself and had a major say in the production. While the predominant stylistic trait is standard country music, it is mashed up with numerous other references. Our prime purpose is to examine the sonic details of production alongside audiovisual representation, with a focus on the official track and video of "Hope to Die." With the analytic methods we have chosen to employ, we seek to foreground the aesthetic effects of production, audio engineering, and compositional design. We also intend to uncover the congruences between technologies

DOI: 10.4324/9781003089926-28

of music production and performativity; sonic devices such as reverb, delay, stereo panning, balance, vocal compression, and instrumentation immerse and encourage the listener/viewer to see country music in a new light. Our argument is that this constitutes a resignification of country music by challenging the sociocultural stereotypes of the genre.

By harnessing an Americana aesthetic, a style defined by its influence from the sounds of early American genres like folk, country, and blues, Peck's musical signature comprises a range of stylistic and technical codes, with evident influences of artists such as Chris Isaac, Elvis Presley, Dolly Parton, Steven Morrissey, Whitney Houston, Prince, and others. He employs a range of innovative technologies as part of stylistic resignification. Moreover, the homoerotic codes he draws on visually in the video of the song have a resonance in the legendary art of Tom of Finland, whose trademark was built upon exaggerated traits of tight and partially removed clothing, sexual allure, emphases on muscles and genitalia, and tough and tender depictions of S&M.[5] In our analyses, we highlight a plurality of listening strategies and critical analytical methods, maintaining that the queer narrative of the song has its corollary in the video. Confronting the heteronormative stereotypes of country music, Peck simultaneously engages with the politics of his own queerness, escorting us on a journey that starts with a discussion of the compositional design, elements of recording, and production and culminates with a consideration of his visual performance.

Compositional Design in the Pop Score—Stylistic and Technical Coding

Musical features in the track "Hope to Die" disclose what we define as the *pop score*, namely the totality of the sound recording and all the complexities that are invested in engineering and production.[6] In this case, the score includes a conglomeration of musical codes that entice the listener into a process of cognizing, enjoying, and relating to the artist or band in question. It also comprises the audiovisual design, whereby its *syntagmatic primary codes* can be broadly categorized as stylistic and technical.[7] These shape the design of a track structurally, mediating the experience of artistic expression. Engaging with the composition's stylistic and technical elements—structure, space, rhythm, timbre, and production—we set out to explore how they contribute to the recording. Our aim is to draw attention to sonic material that aspires to what William Moylan identifies as the "qualities and subtleties of recorded sound" that "pull the listener into understanding and perceiving recorded sounds for their unique individual characteristics, relationships, and the sound qualities they form when combined" (Moylan 2020, 191). Illustrations and transcriptions are provided in this study to highlight the unique attributes in "Hope to Die," where the focus on country as a genre is articulated through the syntax of recorded elements that function according to listener perception and competence. An underlying critical task is to consider how recorded sounds function in a way that proffers "insights into the ways in which musical codes are manipulated to create expression through invocations of resistance, compliance, and pleasure" (Hawkins 2002, 12).

"Hope to Die" also raises important questions of vocal presence and authority that draw the listener into a specific space.[8] For instance, during the verse, the vocal part establishes the mood (and, in our reading, the sentiments of heartbreak) through a sense of intimate spatiality. By this we are referring to the minute characteristics of the voice that communicate to the listener a sense of mournfulness. Heavily processed in the mix, Peck's low bass register is ornamented with devices, such as portamento, vibrato, and a subtle use of compression that heightens the physiology of the vocal folds; the shaping of mouth and throat in recording and microphone techniques magnify a wealth of details. On closer inspection, it is also his vibrato that tailors his timbral coloration; the sobbing, trembling quality at the end of most of the phrases on long sustained pitches exemplifies this well. The spectrograph in Figure 20.1 illustrates the slow and wide vocal vibrato on the words "way" and "were" at the end of the first line. Evident in the spectrograph is the intensity and the strength of the note, with the lowest visible line indicating the fundamental note while the parallel lines above represent the overtones that constitute the timbre for the voice. There are occasional subtle shifts away from the pitch toward a more guttural enunciation at the end of each line; the vocal tone here assumes a sobbing quality, notably in the fading-out intensity at the ends of words like "were" and "burn."

Spatiality

Spatiality in the recording and the placement of voices in the mix is shaped by a heavy use of reverb. Unlike the guitar and drums which are awash in a spacious reverb, the minimal use of reverb on the voice creates a close and intimate spatiality. The reverb itself is lower in the mix on the voice than it is in the guitars and drums, and the subtle delay at the start of the reverb creates even more space at

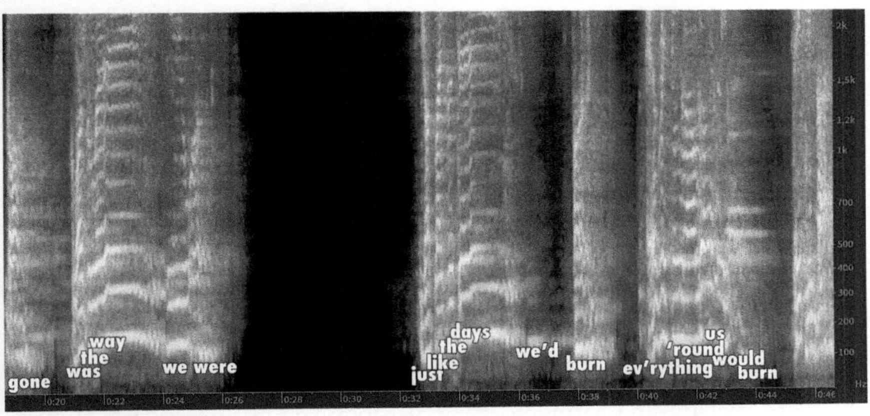

FIGURE 20.1 Spectrograph of "Hope to Die," first verse, voice isolated.

the front of the image for the dry vocal sound. The quality of the reverb warrants examination, since its general lack of early acoustical reflections and long decay time give the impression of a large space, while its low volume compared to the dry voice implies intimate proximity. At the same time, the reverb of the guitars reminds one of the Hawaiian pedal steel in the *hape haole* style of the 1940s, which Doyle described as "lament[ing] the departure of the visitor, call[ing] vainly for his or her return" (2005, 132). This is one of the numerous instances where reverb takes over as a dominant compositional device, leading to the impression that the "acoustical environment, in essence, becomes the sound source" (Moylan 2002, 266–67). Vocal presence is felt through the long pauses that not only create a vast sense of acoustic space in the song's mix but also strengthen the relationship between the song's musical and lyrical structure. In the pre-chorus, the vocal track gradually increases in volume in anticipation of the chorus. During the second phrase, Peck's voice rises in pitch for the first time on the phrase that begins with "take me back." At this moment, the pause following the word "back" is elongated for almost a full measure. This arrival point follows approximately one minute of melodic material on the first five notes of the A-major scale and involves an abrupt octave leap. In our reading, the effect of this leap is to increase the dramatic quality.

Vocal techniques such as these and their audio engineering also accentuate the sense of "vulnerability-on-display"[9] drawing attention to the role and arguably fragility of the singer. The vocal styling reflects the lyrics: "Take me back, the words I'd say. I had to whisper, because you liked it that way;" the earnestness in his vocal expression is exaggerative. This raises the matter of vocal strategies and the employment of stylistic codes within the track's overall compositional design. Indeed, Peck's melodic lines are delivered in a quasi-operatic style through the use of deep vocal vibrato and a low bass register; this style of vocal enunciation is reminiscent of *Sprechgesang*, as well as *recitativo secco*, where the singer liberates themself rhythmically. In "Hope to Die" this singing technique harks back to traditional styles of performance, finding an intertext, when it comes to tone, tempo, and timbre, in the vocal style of Johnny Cash, whose ability to delve into the low register while controlling long held notes with wide vibrato (a very difficult task) accomplished a profound sense of heartache, sorrow, and regret.[10] In Peck's vocal style we also perceive other intertextual referencing to, for example, Depeche Mode's Martin Gore. His song, "To Have and To Hold," involves a vocal delivery that comprises similar strained vocal folds. In the verses of "Hope to Die," Peck turns to vocal folds that emphasize exhaustion and heartbreak, where the voice almost cracks at the end of each phrase. Virtually evaporating, it leaves us, the listeners, hung and dry with little more than the remnants of a faint sound of a scratchy throat.

Structure, Melody, and Harmony

Structure, melody, and harmonic flavor are dominant features in the compositional design of "Hope to Die." The main melody (see Figure 20.2), first played by the electric guitar in the introduction, is repeated by Peck in each of the

FIGURE 20.2 Primary melody in "Hope to Die" with harmonic labels.

verses. Commencing with a simple structure, it is elongated and languid: the first notes of each measure are sustained, with an emphasis on the first, fifth, second, then fourth degrees of the tonic key, A major. Upon Peck's entry, the harmonic accompaniment is taken up by the electric guitar. In the second measure the mixolydian flavor is created by the minor dominant (Em), further reinforced by a resolution from v to ii (Em to Bm) in the third measure, before the chord progression, IV–V–I (D major–E major–A major) firmly establishes an A-major tonal center in the fourth measure.

We would suggest that the interplay between mixolydian and major-mode progressions for much of the track provides the song with its specific ambience.[11] To explain: while the song is in a major key, most of the chords are minor, creating a pull toward the mixolydian where the emphasis falls on the dominant minor and supertonic minor (v and ii). Driven by a lethargic tempo, the sequence of chords, A–G♯m(C♯m)–F♯m–E suggest an Andalusian cadence,[12] commonly borrowed by blues and blues rock music, albeit in the major key. Such harmonic elements are common in traditional blues, R&B, and country, and in the track under analysis, experienced as overtly dramatic due to the slow tempo, major/minor tonal ambiguity, and long pauses, which create a sense of striking a harmonic pose.

In terms of the song's overall style, structural and harmonic expectations are regularly broken; this is discernible in the pre-chorus (0:43–1:16 and 2:05–2:39) that begins with the standard chord progression, I–iii–vi–V. On repetition, however, it is cut short by one measure with a long IV chord that extends over two bars, followed by the dominant leading into the chorus. The effect of this break is to jolt the sense of motion, further heightened by the slowness in tempo, with the duration of each section fading out before the end. Interestingly, these unexpected breaks in the song's formal structure are eased by predictable chord progressions, another tactic of interest or surprise. The four measures following this break involve exaggerated pauses, filled in only by the voice and a single guitar strum that offers slight harmonic support. In Figure 20.3, the track's harmonies are placed within the song structure to illustrate the subtleties of compositional design.

Toward the end of the song (from measures 62–71), yet another departure from traditional formal devices occurs now in the form of an instrumental bridge lasting approximately 28 seconds, preceded by an unexpected break as the chorus is cut short by one measure and put on pause. In the music video, such gestures are used to dramatic effect, as we will turn to later. While the harmonic rhythm in the bridge (mm. 62–71) is consistent with the previous sections, the guitar reiterates the chords from the verse (mm. 34–42), but this time omitting any modal

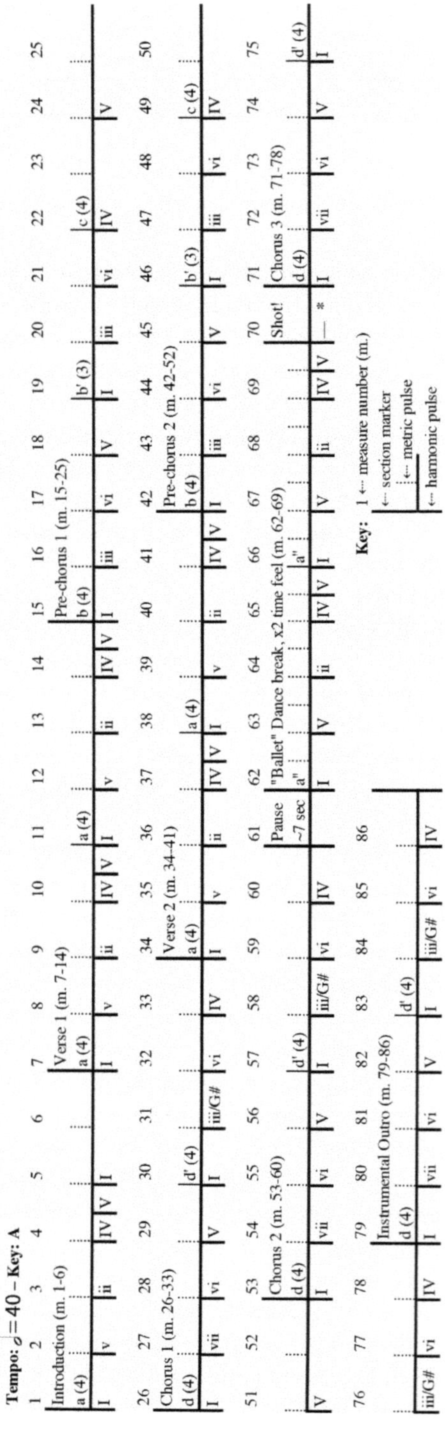

FIGURE 20.3 Formal structure of "Hope to Die."

flavor by use of a major dominant (see Figure 20.3). Rhythmically, though, both the guitar and percussion convert into double-time, with a quasi-Latin strummed dance groove alternating between the guitar and rim shots.

Modulation and Tempo

As the section draws to a close, the dominant chord rejects resolution, cut short by a pause that is filled by loud drum parts leading into the final chorus, modulating from A major to C major. Melodramatic and likely tongue-in-cheek, this modulation is a stylistic code commonplace in the songs of female pop singers, such as Whitney Houston, Madonna, and Celine Dion, for whom such theatrical harmonic devices are standard practice.[13] While the opening part of the chorus melody throughout the song bears a striking resemblance to the chorus of the song "I Will Always Love You" by Dolly Parton, the elevated modulation, pause, and drum lead-in is in no uncertain terms a more striking reference to Whitney Houston's version of the same song, with a similar modulation announcing the final chorus.[14]

The slow tempo of "Hope to Die" (with a half-note pulse of 40 bpm) influences the timbral quality of the recording, the pace of which poses inevitable vocal challenges. Not unlike Steven Morrissey, who is known for slow, mournful songs, Peck's approach to melodic tempo raises a host of issues that relate to control and regulation.[15] His pacing is regulated by lengthy pauses, erotic gaps, that result in astute dramatic intensity. Accordingly, the sound production is greatly sculpted by tempo and effects, such as reverb; in a sense, this has a manipulative effect by queering a range of norms. Through the elements of the mix, we gain an insight into how Peck masters the technologies of music production, which create the narrative effect. Comprising relatively sparse instrumental material, the arrangement consists of two electric guitars, kick and snare drums, bass guitar, and voice. In the verse sections there are two main guitars: a rhythm guitar outlines the harmonic rhythm, while the other guitar extracts the melody at the end of each phrase. The guitars are awash with reverb, giving the sensation of a large and open space, a sonic character which Doyle has described as endemic in the American country and Western canon, denoting the sense of the wide-open frontiers of the West (Doyle 2004, 32–33). The rhythm guitar line is double-tracked in stereo, while the melodic guitar is mono, centered within the mix alongside the lead vocal, processed through a stereo reverb that increases its apparent size within the mix.

So far, we have referred to the control of the voice, its tempo and timbre, as a central stylistic device in "Hope to Die." Peck's implementation of tempo and timbre stylistically imitates operatic style. Similarly, the ways in which the electric guitar's sound is perceived are contingent on the tempo. The rhythm guitar, emphasizing each chord change with slow strumming, is an electric guitar with reverb and chorus effects. While the sound is evocative of the classic pedal steel guitar tone found in much country music, in this case it is not performed by a pedal steel. In sounding out the chords, the guitar, as with the voice, leaves a large amount of space filled by the sounds not only by reverb but in the case of the guitar a subtle chorus effect which would be otherwise almost imperceptible but for the sparse nature of the arrangement. In staking out the melodic material in tandem with the vocal part, the lead melodic guitar is prominent in the mix, panned in mono but with a large stereo reverb, occasionally competing with the voice for space in the mix. We should add that the guitar timbre at this point contains very little harmonic distortion and is equalized with an emphasis on the upper midrange frequencies, a tone that might be described as *clean* with a heavy *twang*—characteristic of guitar in country music to be sure. The heaviness of the pick by the performer, combined with the often-audible portamento between pitches in the melody, and the subtle boost in the lower mid-range combine to give a sound that is reminiscent of classic Americana, such as that in Johnny Cash's "I Walk The Line" or The Highwaymen's self-titled single "Highwayman."

Drum Parts and Rhythm

We also want to make mention of the drum parts and their recording: the rhythm established at the beginning persists throughout the entire track, with little variation except in the short bridge toward the end. The effect of reverb on the drum sound produces a long decay, made all the more audible by the sparse nature of the beat. The beat itself is compatible with the overall aesthetic of the track, and the slow, repeating dotted eight-note pattern of the kick is reminiscent of a heartbeat, which can be interpreted as somber and relaxed. Its dramatic function is important, the slow tempo enhanced by the high reverb on the snare drum that fills the audible empty space. Throughout the song, this reverb has a decay that lasts almost the entire duration of the measure between each snare attack, which is particularly audible in the pre-chorus when the rhythm guitar drops out of the arrangement. The timbre of the snare drum also changes subtly in the chorus sections, increasing in volume and thus driving the compressor to a greater extent which has the effect that the buzzing of the snares sounds to be of a slightly longer duration, and thus allowing more of its higher frequency characteristics to be amplified by the reverb. These intricate changes in the sound of the drums greatly increase the level of timbral interest, and hence the energy and dramatic qualities between the verse and chorus.

Sonic Imagery into Visual Dramaturgy

Let us now consider the relationships between sound and visual staging in the video of "Hope to Die" (Orville Peck 2019), where our focus falls on equating elements of sound imagery with visual dramaturgy.[16] As such, we delve into the storyline, recording elements, and compositional elements to bring to the surface the aesthetics of Peck's performance. To begin, we suggest that the visualization of human performance in recorded format is charged with impressions of the audio track. Lori Burns and Stan Hawkins have stated: "The sight of the performing body invites intensified reflexivity on the part of the viewer, characterized by embodiment, simulation, cognition, and agency" (Burns and Hawkins 2019, 3). As we have hitherto pointed out, Peck's vocal performance in "Hope to Die" is awash in timbral detail, resonating in a manner that impacts the visual effects through a catalogue of movements. Indeed, every section of the song projects its own sonic characteristics, which are visually enhanced by the intensity of every shot in the video; the elements of audiovisuality thus have a powerful bearing on one's perceptions of sound. With reference to features, such as color, lighting, choreography, fashion, props, and scenery, we have extracted four heartfelt moments in the video that support our analytic observations: the introduction, the first chorus, the instrumental bridge, and the final chorus and outro.

Introduction (0:00–0:55): The video commences with two men silhouetted against a white backdrop. We first gain sight of Peck having his mask adjusted in preparation for the performance. In terms of subject-positioning, there is a reference to Jim French's iconic portrait of two cowboys in the same stance, naked from the waist down, from 1969, which would be later popularized by the Sex Pistols' bassist Sid Vicious (Gorman 2011). The opening shot is homoerotic, reinforced just 30 seconds later by a low shot angle from the floor in between the bare muscular legs of another cowboy. The second cowboy, credited as Adrian Nallo, has a candid expression and is dressed in loose-fitting yet suggestive clothes that obfuscate to some degree his physique. The masculinity on show, inspired by Tom of Finland, queers the hetero norm, a common trope in Peck's other videos, such as "Queen of the Rodeo" and "Dead of Night." The first frame comprises twenty seconds with the instrumental guitar melody, heavily reverbed, creating a high sense of expectation. Peck's costume is tight fitting, resembling that of a Spanish matador (apart from the cowboy hat and mask). His first scenes function as a short prelude to the commencement of his dance, whose choreography draws on the traditional Paso Doble, with an emphasis primarily on hand and arm gestures (see Figure 20.4); they are circular, extending above the head and

FIGURE 20.4 The art of masking as vogue.

forming a rounded contour as he starts singing. With little warning the camera angle suddenly changes (0:40) panning to floor level as we view Peck through the legs of what first seems to be another male about to move into combat mode. The homoeroticism of this medium zoom shot of Peck is short-lived though, as the lens angle is changed to close-up. At this point the lighting alters as Peck is silhouetted against a black backdrop with one spotlight pointed over him. Shots of him singing with a stand microphone are then juxtaposed with ponies (in close-up), as he completes the phrase, "take me back to the time you were mine" (0:45–0:55). On the words "you are mine" there is a costume change with new head-dress (a brim shaped hat with bull horns protruding) as he is filmed holding two ponies by their reigns. In our reading of the material so far, it is important to stress that the visual text is constantly polysemic; this stems from the synergy between image and sound.

Chorus 1 (1:18–1.42): Peck faces the camera through an open leg arch, with his shirt unbuttoned, exposing a muscular torso. The blatant reference to gay cowboy pornography is indisputable, and as we read it this constitutes a confrontation of the male, cisgender gaze. In this shot (1:18–1:27), now in color, Peck sports a long, blue fringe mask that matches the denim shirt and trousers that have a double clasp belt, with a silver sheriff badge. Notably, the rest of the chorus is filmed in monochrome, its visual aesthetics defined starkly by silhouette shots against a white screen, including the protagonist on his own, then with the two brown ponies, and, finally, him standing back-to-back with the other cowboy. The visual hue alternates between shades of color, matching the subtleties of sound production. Overall, an increase in rhythmic visual movement characterizes the choreography, particularly in the hand, arm, and torso gestures, with the contrasting textures and timbres in the mix. On close inspection, Peck's repertoire of movements suggests a stance of defiance through an arching of his back and head backward and a raising of his right arm slowly upward. His choreography involves agile yet slow motions of stretching the upper torso that are reminiscent of flamenco; the dramatization of this spectacle is intensified by the drawn-out melodic lines (vocally and instrumentally) and heavily reverbed production. In many ways the musical material complements the feline quality of Peck's movements. Poignantly, this chorus dissolves with Peck dropping his head, as he belts out the mournful phrase, "cross my heart, now I hope to die" (1:38–1:42), a hard downbeat accent placed on the word "die."

Instrumental Bridge (3:09–3:36): The vocals drop out as the guitar and percussion take over. Peck's visual agency is at its most powerful here as a solo dance ensues in a dimly lit barn with the light streaming in from a skylight in the ceiling. In this scene the details of his face are obscured. Something chilling in his aura hints at the film music from *Brokeback Mountain* (2005), evident in the instrumentation and epic cinematic feel. Another costume change occurs with Peck in a white short-sleeve t-shirt, an extra-long fringe mask, and tight-fitting trousers with *chaps*—coverings for protecting the legs made in leather-like material named after the Spanish *chaparral* for brush, thorny, and thick, designed to

FIGURE 20.5 Mid-pirouettes during the barn ballet dance.

protect the legs when riding horseback and bull-riding in rodeo man culture (see Figure 20.5). Significantly, this accessory, while playing a major part in the assimilation of cowboy culture into the American West, falls outside the conventions of Western classical ballet, hence its disruption of a set of conventions. Importantly, Peck's use of this costume is an example of queering, where the functions of the chaps and their frills have become repurposed for their aesthetic qualities and flowing movements. For the first time, the camera focuses on Peck's lower torso and legs, as he executes his own version of a barn dance. The sequence starts with him stomping his feet like hooves in the straw and dust, revving up for a genre-bending repertoire of moves that stylistically challenge anything associated with cowboy traditions of moving the body to music: we witness a crossover of ballet, line dance, disco to MJ pop.

Starting with the sound of his stomp on a wooden floor, the rhythmic part gives way to a rapid fire of *palillos* or castanets, which belong to the clapper group of percussion idiophone. This figure, which accompanies the main guitar melody (Figure 20.6), is the most frenetic in the entire track matching the dance movements that showcase the virtuosic skills of Peck. While some of his foot movements might well belong to dance routines associated with cowboys, such as the hands on the hips with step movements, the pirouettes derived from classical ballet certainly do not, where the dancer rotates on one leg with the other off the ground. In order to maneuver this exceptionally difficult move, dancers need to be graceful, flexible, and sturdy, as it involves a complete turn of the body on one foot (*en pointe*). Peck delivers three pirouettes (3:17, 3:29, and 3:36), the final one culminating with his arms reaching a wide-open gesture through circular motion. The effect of such visual spectacle is to prepare us for the climactic point (Figure 20.5), which starts with a long pause and then an elevated modulation.

Final Chorus and Outro (3:36 to end): A shot of Peck's hand imitating a revolver shot triggers the scene change to the artist on stage, with upper torso and head on display. This signals the climactic point of the song, following the elevated modulation, with a rise in intensity as Peck uses wide open arm gestures that

FIGURE 20.6 Transcription of the bridge in "Hope to Die," mm. 62–69.

circle above his shoulders and head. The lighting creates a semi-silhouette effect, with the red mask and inner lapel contrasting with his black costume. As he sings, "cross my heart now I hope to die" (3:59–4:03), his costumes change twice from the blue denim and blue mask to a rhinestone leather jacket. The scene with blue denim refers back to the first moments in the video, when he is framed by the legs of his partner. Beating his breast in one of the shots (4:06–4:11), his hands are filled with a white milky substance that splatters over his black shirt; the scene is rife with sexual innuendo. Then, yielding to an earlier image of him with the two brown ponies, the next scene includes shots of him standing back-to-back with Adrian Nallo. In the final scene he has collapsed, lifeless on the floor with stars dangling down in front of the black curtain drop; his name PECK is positioned at center top. The audiovisual tempo of this final chorus is painfully slow—the long-sustained notes brim with vibrato, lush and open reverberation, slow and deliberate movements, dramatic and long-held poses—all of which support the visual representation of a tormented soul with a broken heart. Highly dramatic, this scene (4:23–4:38) culminates with the main guitar melody and then a few seconds of silence, giving the viewer pause for thought. The imagery is rendered all the more powerful by the resonance of the highly reverbed final note (pitch C) and its slow diminuendo into nothing but silence. The final shot intensifies Peck's vulnerability and resilience, with the crushed cowboy striking back at the restrictive gender roles queer people are subjected to. A complicated assimilation of the queer cowboy into country culture is epitomized in these final seconds of sheer abandonment. Such closure is transgressive in intent.[17]

In these four moments, we are mindful of the multimodality in close readings (Burns 2018, 95–97). Our analytic observations concur with Burns' multimodal theory, where "the artistic integration of multiple semiotic modes within one media text" remain distinct from the concept of multimedia wherein multiple texts are present in one setting (Burns 2016, 96–97). Applying this to our interpretation of "Hope to Die," its semiotic heterogeneity becomes apparent—as much in the modes of sound and image as in text, dance, fashion, and vogueing.[18]

Art of Masking: Aesthetics and Production

Contemplating visual features is an all-defining element of gaining meaning from the pop score. Connecting recording elements in the track's structure to the imagery of Peck's performance in the video is relevant to the listening process in much pop music. At the core of our quest is the matter of understanding the intentions of the artist, and it is also the multidimensionality of the audiovisual production that prompts observations and then evaluations. We concur with Moylan that "a multidimensional, multi-domain texture exists at the highest dimension of the recorded song; this is the overall sound quality of the track" (2020, 197). By investigating the wealth of domains in "Hope to Die," our aim is to unravel the aesthetics of production that are attributable to an array of features: sonic invention, production techniques, compositional design, lyrical treatment, and performance strategies, which we now turn to.

Our analyses of "Hope to Die" have produced the following findings:

- Vocal presence and attributes of singing, stylistically and technically, shape the mood of the track, with Peck turning to a range of techniques to convey a high level of pathos.
- The artist's palette of stylistic references is integral to the sonic inventions in compositional design.
- The dramatic quality of delivery is harnessed by the arrangement, where varying degrees of spatiality in the production heighten the levels of interest on the part of the listener.
- The arrangement and mix impact on harmonic and melodic structuring, where subtleties of coloration are cognized alongside rhythmic details, such as the guitar and percussion parts during the final measures of the track (Figure 20.6) in the form of the quasi-Latin groove.
- The creative regulation of the sound stage situates the voice in relation to the surrounding textures and timbres as the stage undergoes continual transformation. The recording and sound production shapes the narrative.
- The vitality of Peck's vocal performance is a direct result of detailed sonic processing, which provides an illusory sense of reality, a sense of masking,

that is conditional on the content and character of what Moylan terms the "holistic environment," which "establishes an expanse that complements the expression of the track" (2020, 478).
- Integral to the sonic environment is the inventive property of spatiality that is moderated creatively to form a rich aesthetic backdrop.

All in all, then, the deceptively simple arrangement strikes an intricate balance with a sophisticated production. Consider that the reverb and delay effects not only serve as a placeholder for the long gaps between rhythmic and harmonic events in the score but also amplify the performative drama and vogue-like posturing in both the music recording and music video.[19] The theatrical use of reverb is most salient at the moment of modulation going into the last verse (3:35–3:45), when the progression is cut off by a measure, creating a silence filled only by reverb before the drums pound into the last section.

Of significance is the close-up sense of vocal staging and signal processing, which establishes a high degree of intimacy in stark contrast to the epic-like spatiality afforded by reverb. Clearly audible in the vocal track especially at the ends of phrases in the verses is the tapering of long-sustained tones that give way to scratching *fry* sounds that are emulated through vibrato. This gets us to think of masking, which in music production can refer to the effect where "a sound (or portion of a sound) is not perceived because of the qualities of another sound" (Moylan 2002, 32). Similarities in pitch range in the guitars and the voice in such moments of vocal fry create tension, intensifying the voice as it is slowly masked by the sounds which come to dominate. Such sonic effects are a result of the combination of close-microphone techniques, processing through vocal compression, and clever mixing. As such, highly produced vocal sounds evoke the emotional sense of anguish and vulnerability and reinforce visual interpretations of masking.[20]

In keeping with many pop music recordings and videos, "Hope to Die" is replete with signifiers that titillate the listener/viewer. In our analytic approach we recognize that audiovisual codes coalesce at the moment of reception to create a propensity for immersion. In particular, Peck's approach when it comes to vocal styling functions as a primary teaser: his low register, combined with deep vibrato and close-microphone techniques, often blurs the line between singing, speaking, and crying, making intertextual reference not only to the vocal stylings of mid-century country and folk singers, such as Johnny Cash, Dolly Parton, and Willie Nelson, but also, as we have shown, to other traditions, such as opera and musicals. In terms of the track's compositional design, the lethargic rhythmic and harmonic pace taunts the mixolydian major/minor uncertainty through the use of the blues Andalusian cadence. Also, we have observed how the standard verse-chorus structure is cleverly broken up at times, establishing dramatic and unexpected shifts between sections and making space for pregnant pauses. Added to this is the slow tempo that highlights timbre, spatial sound, and balance as compositional elements in the track.

As a rule, vocal style is an outgrowth of recording techniques and decision-making. In "Hope to Die" an intensity of expression feels constructed in a way one never encounters in concert experiences. One might argue that a sense of over-production makes the song knowingly artificial and non-real. Yet, as Virgil Moorefield insists, in pop recordings realism is not really the point: "What matters is the sonic experience the record offers, on its own terms, as *sound*" (Moorefield 2005, 55; author's emphasis). Thus, the recording and its very concept draw attention to not only the meaning of what is being sung but also to its aesthetic effects. We have noted how poignancy is rendered mainly by the chorus hook, where the final words "hope to die" are delivered with soft dynamics in the lowest register. Notably in the second chorus the word "die" is omitted as Peck delivers "now I hope to …" (3:02), which is followed by a long pause, with only the sound of footsteps on the barn floor (3:03–3:10), which then leads to the instrumental passage with tap-dancing sounds complementing the guitars. Following this passage, the climactic moment, "I'm still undone," leads into the final chorus where the phrase "now I hope to die" (4:02–4:14) is repeated twice before the track ends. Quivering on the word "die," the vocal part gives way to the guitar which plays the melodic hook one last time. The crooning quality of Peck's voice help defines his persona, conveyed by shades of timbral coloration that fluctuate between a sense of tranquility and urgency that impact the narrative of the song. In the end, it is Peck's highly produced voice that encapsulates what Moylan has described as the "performance intensity" (2020, 451), where the verbal space of the mix is contingent on the timbral and textural employment of distance positions. All this occurs within a holistic environment, where the technicalities of stereo positioning, effects, compression, equalization, and choice of microphone spell out the seductive attributes of the pop recording.

Conclusion

Peck's remarkable score of "Hope to Die" articulates a touching story of queer sensibility; its success is highly contingent on the multiple details of the production process. The sound production complements the video, which contains a wealth of elements that result in mediating the antics of the gay cowboy, something we interpret as a liberation from stereotypical strictures of representation. However, any notions of utopia are short-lived by the protagonist's pursuance for dying. Nestled within the spatial environment of the recording is the moving display of queer performativity with a telling tinge of ironic intent as Peck sings, "But I, I still try, cross my heart, now I hope to die." It is as if he pleads for a social and political sense of becoming; as such, he taunts difference as a facet for openness and promise. These sentiments are delivered passionately in "Hope to Die." Drenched in innuendo, Peck's performance calls into question an alternative political impasse for understanding the past by mobilizing an ideology of gender difference, all of which is achieved in large part by the spatial features of the mix, production, and stylistic codes of country music. Alas, freedom lies in the hope to die with a gravitational pull away from utopian longing. Cruising a narrative in such a sophisticatedly controlled audio space might be a response to not only repressive heteronormativity but also a bolstering of pop ephemera through the melancholy of the queer cowboy!

Notes

1 As revealed in an interview with *L'Officiel Magazine* in March 2019, on the day of the release of the record *Pony*, on which the song we analyze in this piece is found: https://www.lofficielusa.com/music/orville-peck-interview-2019.

2 While many have speculated about Peck's real name, he has been insistent in maintaining his privacy, saying in a statement in *The Guardian*, "there is a temptation to try and unmask what I do, but to do so would be to miss the point entirely." In the previously referenced interview with *L'Officiel*, he has stressed his study of "mask as an art form … the method made famous by Jacques Laqoc." While speculation surrounding his "true" identity is rife on the internet, we choose not to engage with it explicitly in this essay. https://www.theguardian.com/culture/2019/nov/19/orville-peck-i-grew-up-feeling-alienated-so-i-became-a-lone-cowboy.

3 This biographical information about his parents, education, and West End experience was revealed in an interview on the podcast "Sloppy Seconds with Big Dipper and Meatball" on May 1, 2020: https://foreverdogpodcasts.com/podcasts/sloppy-seconds/.

4 https://www.queerty.com/pride50/.

5 https://www.tomoffinland.org/about-tom-of-finland/.

6 By *pop score* we refer to Stan Hawkins' theory of the recorded format of the pop song. See Hawkins (2002).

7 See Hawkins (2002, 10).

8 Lori Burns' theory on vocal presence and vocal authority is exemplary in understanding this phenomenon (Burns 2010).

9 See Hawkins' concept of vulnerability-on-display and masculinity, where it is suggested that the exhibition of vulnerability occurs as much off-stage as on-stage, where it is a "prerequisite of stardom to appear to bare all" (2009, 186). For a critical development of this theory, through an in-depth study of "fake naivety," illusions of authenticity, and the traits of the "true self" in pop personas, see Ålvik (2014).

10 See Askerøi (2017) for a detailed interpretation of Cash's voice in the track "Hurt" (2002). Note that with "intertext," we draw on literary criticism to describe the way in which one text's meaning is derived from others as an aid to new interpretations. See, for example, the music analytic work of Burns and Woods (2018).

11 For a discussion of modal and tonal ambiguity, see Hawkins (1992, 2002).

12 The Andalusian cadence is a descending tetrachord with its origins in the Flamenco tradition, either as iv–III–II–I (phrygian) or i–VII–VI–V (minor), typically signaling the end of a long section or a piece of music in an ostinato form. For more on the cadence and Flamenco guitar playing specifically, see the thorough guide "Music Theory for Flamenco" by Chuck Keyser (1998). Also, for a discussion on the appropriation of Flamenco's sounds in popular music generally, see Folch (2013).

13 Modulations of this ilk have often been disparaged and dismissed as kitsch, trivial, and overkill. See the writings on "bad music" by writers Robert Walser, Carl Wilson, Simon Frith, Dai Griffiths, Stan Hawkins, Washburne, and Derno.

14 Orville Peck has acknowledged the profound influence of Whitney Houston on his musical style and identity.

15 See Hawkins (2011) for a detailed analysis of vocal poignancy and vulnerability in the performance style of the English artist Steven Morrissey.

16 To date much has been written on audiovisuality in pop music, where emphasis is placed on the creative enterprise of performance. See Auslander (2008, 2009, 2021); Burns and Hawkins (2019); Burns and Lafrance (2014,2017); Burns and Watson (2010); Burns and Woods (2018, 2019); Hawkins (2004, 2009, 2016); Korsgaard (2013, 2019a, 2019b); Lafrance, Burns, and Woods (2017); and Vernallis (2008, 2020).

17 For a discussion of the appropriation of the hegemonic masculine cowboy figure by queer culture as an intended transgressive act in itself, see Hawkins (2016, 203–205).

18 Similarly, Mathias Korsgaard has insisted that music video is "defined by its very heterogeneity, its wide range of audiovisual expressions" (2017, 30).

19 For a detailed discussion of vogueing as part of ball culture with direct links to queer agency, see Hawkins (2016, 205–206).

20 A theorization of masking as an entity of the art of studio production and shaping of musical style and performance practice, and integral to recorded musical expression, is discussed in Hawkins (2009, 153–82).

References

Ålvik, Jon Mikkel Broch. (2014). *Scratching the Surface: Marit Larsen and Marion Raven–Staging Authenticity, Faking Naïvety and Flaunting Banality.* PhD dissertation, University of Oslo.

Askerøi, Eirik. (2017). "Spectres of Masculinity: Markers of Vulnerability and Nostalgia in Johnny Cash." In S. Hawkins (Ed.), *The Routledge Research Companion to Popular Music and Gender* (pp. 63–76). New York: Routledge.

Auslander, Philip. (2008). *Liveness: Performance in a Mediatized Culture* (2nd ed.). New York: Routledge.

Auslander, Philip. (2009). "Musical Persona: The Physical Performance of Popular Music." In D. Scott (Ed.), *The Ashgate Research Companion to Popular Musicology* (pp. 303–15). Surrey, UK: Ashgate.

Auslander, Philip. (2021). *In Concert: Performing Musical Persona.* Ann Arbor, MI: University of Michigan Press.

Burns, Lori. (2010). "Vocal Authority and Listener Engagement: Musical and Narrative Expressive Strategies in the Songs of Female Pop-Rock Artists, 1993–95." In M. Spicer & J. Covach (Eds.), *Sounding Out Pop: Analytical Essays in Popular Music* (pp. 154–92). Ann Arbor, MI: University of Michigan Press.

Burns, Lori. (2016). The Concept Album as Visual-Sonic-Textual Spectacle: The Transmedial Storyworld of Coldplay's Mylo Xyloto. *IASPM Journal, 6*(2), 91–116.

Burns, Lori. (2018). Interpreting Transmedia and Multimodal Narratives: Steven Wilson's "The Raven That Refused to Sing." In C. Scotto, K. Smith, & J. Brackett (Eds.), *The Routledge Companion to Popular Music Analysis: Expanding Approaches* (pp. 95–113). New York: Routledge.

Burns, Lori, & Hawkins, Stan. (2019). Introduction. In L. Burns & S. Hawkins (Eds.), *The Bloomsbury Handbook of Popular Music Video Analysis* (pp. 1–9). New York: Bloomsbury.

Burns, Lori, & LaFrance, Marc. (2014). Celebrity, Spectacle, and Surveillance: Understanding Lady Gaga's "Paparazzi" and "Telephone" through Music, Image, and Movement In M. Iddon & M. L. Marshall (Eds.), *Lady Gaga and Popular Music* (pp. 117–47). New York: Routledge.

Burns, Lori, & Lafrance, Marc. (2017). Gender, Sexuality, and the Politics of Looking in Beyoncé's "Video Phone" (Featuring Lady Gaga). In S. Hawkins (Ed.), *The Routledge Research Companion to Popular Music and Gender* (pp. 102–16). New York: Routledge.

Burns, Lori, & Watson, Jada. (2010). Subjective Perspectives through Word, Image and Sound: Temporality, Narrative Agency and Embodiment in the Dixie Chicks' video "Top of the World." *Music, Sound, and the Moving Image*, *4*(1), 3–37.

Burns, Lori, & Woods, Alyssa. (2018). Rap Gods and Monsters: Words, Music, and Images in the Hip-Hop Intertexts of Eminem, Jay-Z, and Kanye West. In L. Burns, & S. Lacasse (Eds.), *The Pop Palimpsest: Intertextuality in Recorded Popular Music* (pp. 215–51). Ann Arbor, MI: University of Michigan Press.

Burns, Lori, & Woods, Ayssa. (2019). Humor in the "Booty Video": Female Artists Talk Back Through the Hip-Hop Intertext. In T. M. Kitts, & N. Baxter-Moore (Eds.), *The Routledge Companion to Popular Music and Humor* (pp. 310–320). New York: Routledge.

Doyle, Peter. (2004). From "My Blue Heaven" to "Race with the Devil": Echo, Reverb, and (Dis)ordered Space in Early Popular Music Recording. *Popular Music*, *23*(1), 31–49.

Doyle, Peter. (2005). *Echo and Reverb: Fabricating Space in Popular Music Recording, 1900–1960.* Middletown, CT: Wesleyan University Press.

"Flamenco Chuck" Keyser. (1998). Music Theory for Flamenco [web page]. https://www.flamencochuck.com/files/Music%20Theory/Theory.pdf. Accessed June 28, 2021.

Folch, Enric. (2013). At the Crossroads of Flamenco, New Flamenco and Spanish Pop: The Case of Rumba. In S. Martinez & H. Fouce (Eds.), *Made in Spain: Studies in Popular Music* (pp. 33–43). New York: Routledge.

Gorman, Paul. (2011). The Cowboys Came from Colt Studio. *Paul Gorman*. https://web.archive.org/web/20110527092942/http://www.paulgormanis.com/?p=2935.

Hawkins, Stan. (1992). Prince: Harmonic Analysis of "Anna Stesia." *Popular Music*, *11*(3), 325–35.

Hawkins, Stan. (2002). *Settling the Pop Score: Pop Texts and Identity Politics*. Burlington, VT: Ashgate.

Hawkins, Stan. (2004). On Performativity and Production in Madonna's "Music." In S. Whiteley, A. Bennett, & S. Hawkins (Eds.), *Music, Space and Place: Popular Music and Cultural Identity* (pp. 180–90). Surrey, UK: Ashgate.

Hawkins, Stan. (2009). *The British Pop Dandy: Masculinity, Popular Music and Culture*. New York: Routledge.

Hawkins, Stan. (2011). "You Have Killed Me"—Tropes of Hyperbole and Sentimentality in Morrissey's Musical Expression. In E. Devereux, A. Dillane, & M. J. Power (Eds.), *Morrissey: Fandom, Representations and Identities* (pp. 307–23). Bristol: Intellect.

Hawkins, Stan. (2016). *Queerness in Pop Music: Aesthetics, Gender Norms, and Temporality.* New York: Routledge.

Korsgaard, M. B. (2013). Music Video Transformed. In J. Richardson, C. Gorbman, & C. Vernallis (Eds.), *The Oxford Handbook of New Audiovisual Aesthetics* (pp. 501–21). Oxford: Oxford University Press.

Korsgaard, Mathias. B. (2017). *Music Video after MTV: Audiovisual Studies, New Media, and Popular Music*. London: Routledge.

Korsgaard, Mathias. B. (2019a). Changing Dynamics and Diversity in Music Video Production and Distribution. In L. Burns & S. Hawkins (Eds.), *The Bloomsbury Handbook of Popular Music Video Analysis* (pp. 13–26). New York: Bloomsbury.

Korsgaard, Mathias. B. (2019b). SOPHIE's "Faceshopping" as (Anti-)Lyric Video. *Music, Sound, and the Moving Image*, *13*(2), 209–30.

Lafrance, Marc, Burns, Lori, & Woods, Alyssa. (2017). Doing Hip-Hop Masculinity Differently: Exploring Kanye West's 808s & Heartbreak through Word, Sound, and Image. In S. Hawkins (Ed.), *The Routledge Research Companion to Popular Music and Gender* (pp. 285–99). New York: Routledge.

Moorefield, Virgil. (2005). *The Producer as Composer: Shaping the Sounds of Popular Music.* Cambridge, MA: MIT Press.

Moylan, William. (2002). *The Art of Recording: Understanding and Crafting the Mix.* New York: Focal Press.

Moylan, William. (2020). *Recording Analysis: How the Record Shapes the Song.* New York: Routledge.

Peck, Orville. (2019). "Orville Peck—Hope to Die (Official Video)." YouTube video, 4:41, posted by "Orville Peck." https://www.youtube.com/watch?v=60MHmrtEuRY

Vernallis, Carol. (2008). Music Video, Songs, Sound: Experience, Technique and Emotion in Eternal Sunshine of the Spotless Mind. *Screen, 49*, 277–97.

Vernallis, C. (2020). Writing about music video: Tracing the ephemeral. In L. Patti (Ed.), *Writing About Screen Media* (pp. 100–103). Routledge.

21

"WHAT ARE YOU GONNA TELL HER?" (2020)

Mickey Guyton's Advocacy and Protest for Equality in Country Music

Jada Watson

In February 2020, Capitol Nashville recording artist Mickey Guyton made a surprise appearance at the Ryman Auditorium for UMG's luncheon at the annual Country Radio Seminar convention to perform "What Are You Gonna Tell Her?" An indictment of gender and racial disparities that pervade society, the song spoke pointedly to her audience of radio programmers about their role in perpetuating practices that marginalize women within the industry. Written three weeks earlier with co-writers Victoria Banks, Emma-Lee, and Karen Kosowski (who also produced the demo), the song came out of a moment of personal realization for Guyton of the depths of control that white men have over the careers of women in the music industry. After an emotionally draining songwriting session, Guyton laid down the vocal track for the demo, which Capitol Records released—unchanged—two weeks after her CRS performance.

The writing and recording of "What Are You Gonna Tell Her?" offers a critical opportunity to think about breaking points for women in the music industry, of considering the ways in which women respond when pushed to extremes. My analysis is indebted to Sara Ahmed's (2017) work on the "feminist snap," which offers a framework for thinking about the paths that lead to breaking points, of the narratives that encapsulate them, and of the messages that they communicate about an artist's unwillingness to meet the conditions placed on them by patriarchal structures. A "snap," as Ahmed (2017, 190) indicates, "is one moment of a longer history of being affected by what you come up against," a moment (or an accumulation of moments) that pushes someone so far that they push back. Drawing on interviews with Guyton, her co-songwriters and producer, and close analysis of the recorded demo released in March 2020, this chapter addresses how the writing and recording of this song was part of a larger and painful process of recognition for these collaborators—of the burdens of exhaustion, pressure, harassment that women endure within society (in general) and

DOI: 10.4324/9781003089926-29

the country music industry (in particular). I consider how the lyrics function as a form of what bell hooks (1989) terms "talking back" to the industry as a critique of the genre's white patriarchal structure, capturing her moment of pushback. Influenced by Moylan's (2020) work on the use of recording strategies and effects to elaborate different zones of perspective for listeners, the analysis will focus on how Guyton's vocal performance captures the emotional intensity of the January 2020 songwriting session, as she boldly demands a response from the industry on their discriminatory practices.

Accumulating Moments

To understand the context of this breaking point for Guyton, it is imperative to outline the path that led to this moment. Snaps are not the starting point, as Ahmed (2017, 189) notes, they are not actions, they are *reactions* to various forms of pressure and stress that individuals accumulate over time. And sometimes, she argues, snaps can be inherited from those who came before them. For Guyton, the only Black female artist signed to a major Nashville label, her experience has been defined not just by contemporary issues plaguing women in the industry but by the inherited system that has rendered Black artists—women, in particular—invisible within the country music industry.

The country music industry has long been defined through a "Southern thesis" (Malone 1968; Malone & Laird 2018), a narrative trope that privileges the contributions of white, male southern US-born artists and ignores the multiracial and multiethnic roots of the genre. The recording industry developed this identity construct in the 1920s to categorize both musicians and market records, segregating music into distinct musical genres linked to race, gender, and class (Miller 2010). Black musicians were shut out of a music genre that emerged from their cultural traditions, and women—white women, to be more precise—were relegated to subordinate positions within the industry, occupying a secondary position to the genre's white male stars (McCusker 2017). Not only have women consistently represented less than 30% of the long-running *Billboard* Hot Country Songs chart (Watson 2019a), but radio has historically censured or been reluctant to play songs by women that depicted responses to abusive husbands or that were perceived as "male-bashing" (Keel 2004, 76). This racialized and gendered industrial configuration has been maintained through the emergence of Nashville as the center of the country music industry in the 1950s (Stimeling 2020), and embedded in the practices of country format radio, label, publishers, and trade organizations (Watson 2019a/b, 2020, 2021). As my own work shows, the industry's longest-running chart (Watson 2022) reveals a deep form of cultural redlining within the country music industry—a process that has continually relegated the songs by Black and of color artists to the margins of the industry. This is particularly true for Black women, in a position of "double jeopardy" (Beal 1969), who have been relegated to a position of acute social invisibility within the country music industry.

It is within this sphere of racialized and gendered structural discrimination that Mickey Guyton has been building her career. When Guyton signed with Capitol Records in 2011, female artists were struggling to get their music placed on radio playlists—a historic issue for women in country music that had worsened since 2000 (Watson 2019a/b). Capitol's President Mike Dungan made a point to acknowledge this struggle when Guyton signed her contract, telling her that with "hard work and perseverance, [they could] make this work." At the time that she signed, she was just one of forty solo female artists signed to one of the three major Nashville labels, who combined represented just 26.0% of signed artists that year (Watson 2021). But unlike 98.0% of the women in the Nashville-based industry, Guyton was the only Black female artist (and just one of four Black artists signed to a mainstream label).[1]

Guyton moved to Nashville after signing with Capitol and began participating in writer's rounds and immersing herself in the community. Often the only Black woman in predominantly white (male) spaces, she relayed that she "wanted to be respectful of these people because essentially, I am an outsider. They're not used to seeing me, but I'm used to seeing white people and being around them and being the only Black person in a room." Guyton felt the added burden of working to make sure that those around her in the industry were comfortable. Over time, this environment led her to feel as though she didn't belong in these spaces, magnified by conversations surrounding her "cultural fit" and authenticity, often initiated by her label, about how she would be perceived by those around her. She recalled being told early on that she would "have to try harder to show people that [she's] really country," because fans would feel she's "inauthentic." These conversations served to make her feel even more self-conscious about being an outsider within the industry, putting her in a position of having to prove her "country authenticity" all the while silencing her own authentic country story. These experiences only served to render Guyton invisible within the writing and recording spaces in which she worked.

Guyton released her debut single in 2015. "Better Than You Left Me" was widely praised by critics and broke a record for the most-added song to radio playlists by a new artist, but after five months of climbing the country airplay charts, Guyton's song was dropped from radio playlists in favor of Cam's "Burning House." Programmers were not going to play two ballads by two female artists, she was told, and Cam's song had just been selected as part of iHeartCountry's "On the Verge" program (Bjorke 2015), which meant that iHeart stations had to play a song. Guyton's song dropped off the charts by August 2015 and then fell out of radio programming entirely, while "Burning House" went on to reach No. 1 on December 20, 2015. By 2017, radio failed to support Guyton's singles, which had been repackaged as two different EPs, and her label stalled in releasing a debut album.[2]

While Guyton was struggling to find her lane in the industry, there was increasing attention to the declining presence of women on industry charts and radio playlists. Guyton's debut ballad was dropped from radio as (what is now known as) #TomatoGate erupted in the industry in reaction to a radio consultant's statements about programming songs by female artists at 13–15% of radio paylists. Play them sparingly and avoid playing them back-to-back, space them out like tomatoes in a salad, was the message relayed to the industry—a public declaration of a behind-the-scenes practice that had been ongoing since the 1960s (Hill, qtd. in Penuell 2015, 8). A movement emerged to address the declining presence of women in the industry. But these campaigns focused on gender only (Watson 2019a), centering the narratives of white women and campaigned on a platform of white feminism that failed to consider intersectional identities. As a result of the lack of support from radio and her label, Guyton had been rendered invisible by the system and she was largely forgotten by this movement in Nashville.

After seven years of struggling in Nashville, the singer was at a crossroads and asked her husband why country music wasn't working for her. He told her that she was running away from everything that made her different (Oliver 2021). She began writing from her experiences, reshaping her output to address a range of social issues, centered around advocacy for marginalized voices in society. While this narrative stretches back to her 2019 single "Sister"—a song of sisterhood that celebrates a diverse range of female voices, her 2020 demo for "What Are You Gonna Tell Her?" became part of an unfolding body of music and advocacy around the marginalization of women and systemic racism within the industry.

Writing and Recording for a Breakdown

Guyton began writing with Canadian singer-songwriter Victoria Banks shortly after she moved to Nashville in 2011. Banks, who had moved to Nashville in 1997 to pursue a career as a songwriter, had witnessed the declining presence of female artists on country format radio, which had by 2011 resulted in decreased opportunities for female songwriters as well. Not only was Banks herself impacted by this changing ecosystem in Nashville, but she then had a front-row seat to Guyton's career and remembers her label constantly second guessing how to record and promote the singer's music. Around 2019, they began writing with Canadian songwriter and producer Karen Kosowski, who changed the game for the duo. Before Kosowski joined them, Banks and Guyton were often unable to get their songs "across the finish line as a demo" because they didn't write with track builders (as had become the norm in Nashville). Because their publishers couldn't justify spending money on demos that might never be recouped, Banks told me, they created work tapes that would (in her words) never go anywhere. With Kosowski, their sessions resulted in the production of demo recordings.

Integral to their relationship was how they worked together in their writing sessions. Both Banks and Kosowski relayed that Guyton typically enters sessions with an idea or a title, usually "something really personal that was going on in her life" that they would turn into a song. These songs "always had a woman's spin," Banks relayed, and sessions would often result in tears, as they were "so cathartic, real, and honest." On January 27, 2020, the day "What Are You Gonna Tell Her?" was written, another Canadian songwriter joined the trio. Emma-Lee (who had been writing with Guyton and Kosowski separately) entered the writing session with the idea that they would be writing a "hit" that day. "When writing with a signed artist in Nashville and they're going for commercial country radio, we're usually thinking 'let's write an up-tempo hit!'" But when the session started that afternoon, Guyton entered the room "feeling pretty down and disillusioned about everything—about being a woman in the industry," Emma-Lee recalls.

Guyton had hit a breaking point in the industry; despite critical acclaim of her debut single, she had been floundering since 2015—responding to criticisms about her "fit" in the industry and frustrated with being left out of the growing movement to fight for gender equality in the industry. She was also living between Nashville and Los Angeles, splitting her time between a community that made her feel like an outsider and her home with her husband. The evening before the session, Guyton had been at her label's Grammy after party, hoping to speak with Dungan about her career. She relayed to her co-writers that she was scanning the room, watching young female artists move through a space dominated by men in suits, and thought, "we don't own ourselves. All we are, is the change in their pocket." Devastated by the events of the evening, she left the party early and went home to pack for her early morning flight to Nashville to write with Banks, Emma-Lee, and Kosowski. When she got to the session, she was "fired up about it" (Guyton relayed) and vented to her co-writers about what had happened the night before. Exasperated, Guyton turned to them and said:

> What are we doing? All of us? What is every one of us doing in this room? We are talented. We are excellent songwriters and singers, what are we doing? And what are we going to do for the future because it doesn't matter how good you are, if you're a woman it's like they don't care! What are you going to tell your daughters, Vic? What are you gonna tell them?
>
> *Guyton, in conversation*

Emma-Lee, listening to this conversation unfold, turned to her co-writers, and said, "why don't we write that?"

They wrote the chorus first, as they all remember it, but they struggled with how to approach the verses. Everything they came up with sounded sugar-coated, as Banks relayed. "But that's what you do when writing for radio," she said. "It's like you have to censor everything and you have to just make it a soft-enough punch that it doesn't offend anybody." Kosowski saw two ways forward with this song, they could be "really brutal," or they could "shave off the corners a little bit, round it out," and make it more palatable. When Banks said the line "her friend's older brother's gonna keep his hands to himself," they realized that this was the tone they needed—the verses, as Kosowski relayed to the group, needed to "be ugly."

The song is written in the standard verse-chorus-bridge structure alternating between two narrative stances: a third person, intimate conversation in the verses and bridge and direct public address in the chorus. Written in what Lori Burns (2010) refers to as an authorial voice, the song's narrative addresses the consequences of raising girls to believe in fairness and equal opportunity. The verses are structured as an intimate conversation about the lies fed to young girls: that life is fair, that everyone succeeds with hard work, that they can love who they want, and that skin color doesn't matter. The lyric that sets the tone of the writing session then interjects, shifting the tone of the lyric from "innocent" lies that society tells little girls to the inevitable gaslighting that occurs when we ask them to tell the truth, but then fail to believe them when they've been abused. The second verse draws out these statements, from the youthful innocence of little girls in the first verse to the hopes and dreams of teenagers, who are led to believe that the world is open and accessible to everyone regardless of their gender identity, sexual orientation, or race and ethnicity in the second verse. In this way, the narrative shows how this messaging is ingrained in every phase of development—that these lies persist beyond childhood and into teenage and early adulthood for women. The underlying message of the song suggests that there are varying levels of access based on these intersections of identity: while all women face barriers in society, these barriers are far greater for Black and of color women as well as LGBTQIA+ individuals—as is the case in the country music industry (Watson 2021).

The chorus shifts to a direct and public form of address, demanding a response to the title question: "what are you gonna tell her?" The "you" in this narrative stance has multiple identities. The addressee is, first and foremost, the systems of power that serve to oppress women in society. For the writers, "you" is the country music establishment—the white, cisgendered, heterosexual male-dominated system that has heavily controlled the careers of women in Nashville for decades. For the implied authors (i.e., the women hearing this song in allegiance with the writers), "you" are the systems that govern society—from the government to educational systems, and to the medical, legal, and other social systems that are designed to support but instead present often insurmountable barriers to women in society, magnified for Black, of color, queer, and trans women. While the title question is rhetorical, it increases in intensity through its continual repetition, becoming a demand for truth and justice.

But women are complicit in this narrative, and that complicity emerges in the bridge. The bridge comes from Banks' experiences as a mother. Just as their writing sessions typically begin with Guyton relaying her experiences, Banks, too, shares her experiences as a mother of two young girls. When it came to the bridge, then, the songwriters drew from Banks' experiences: "The bridge was me," she says, "because that's where I was like, 'do I tell my kids not to fight it? Do they just go along with it? It's a lot of work to fight it. You suffer when you fight it.'" The bridge is a pivotal transition in the narrative—that shows the dilemma of women and mothers in society who are tasked with raising girls in a world in which they themselves have struggled. The women writing the song—and the women listening to the recording—are then forced to question the role they play in perpetuating lies and weighing out the value of truth and honesty when there is no sign of change.

Musical Snap!

The musical setting and recording of the song highlight this dual narrative stance, capturing the intimacy of the shared conversation in the verses and bridge and the public address of the chorus through instrumentation, recording strategies, and Guyton's vocal performance. The morning of the recording session, Kosowski was playing around with a chord configuration before her co-writers arrived to write. She did not have plans to bring the progression into the session, but after they started writing, Kosowski played through the chord progression she had been working on that morning. Emma-Lee sang the words "what are you gonna tell her?" over top and the writing process flowed like "dominoes," Kosowski says, each songwriter interweaving their ideas into the narrative. That musical idea, notated in Figure 21.1, is a G-major block chord progression that opens the song and supports the verses. Through a combination of open voicing of the chords, the inversion of the penultimate chord with F♯ in the bass (rather than the chord root), suggests a "youthfulness" in its simplicity. In the introduction and opening verses, Kosowski plays the chord progression on a Korg minilogue, using a patch that sounds like a music box—an effect that reinforces the simplicity of harmonic progression. The sound created by the minilogue patch conjures images of a little girl playing with a music box, evoking a range of emotions about the innocence of the young girls at the center of the lyrical narrative. When the music box fades away after the first verses (perhaps a symbolic gesture of the loss of innocence), a piano enters, alternating between broken chords in the chorus and the main chord progression in the verses. In this way, the instrumentation captures the passage of time in the life of the young girl at the center of the song narrative, from the youthful music box to the more adult piano.

The recorded vocal performance of the demo serves to elaborate two narrative stances in the lyrical narrative. Moylan's (2020, 314–18) typology of understanding physiological context of sound is helpful for understanding these stances in relation to perceived distance from the recorded voice and, by extension, the song messaging. There is an immediate sense of intimacy in the recording, established through close-miking of Guyton in Kosowski's recording studio. The

FIGURE 21.1 Korg minilogue chord progression in the introduction and verses of "What Are You Gonna Tell Her?."

recording captures the breath and the timbral detail of her vocal performance creating a sense of closeness for the listeners. In so doing, the listener is drawn into the singer's personal space (or personal zone [316]), permitting them to hear and even sense the fragility of Guyton's voice as she unpacks these societal lies. These recording strategies enhance the conversational feel between Guyton and the listener, bonding them over the singer's personal experiences (perhaps experiences that are shared by the listener).

In the chorus, Guyton pulls back from the microphone singing louder and in a fuller chest voice, shifting from the personal to social zone and directing her message to a broader audience—including the industry in which she works. Despite the more social stance (or what Moylan [2020, 316] calls a social zone), there is still a fragility to the vocal performance, captured in the breathy voice with which she concludes the upward melodic gesture at the ends of each chorus question. Guyton shifts between this intimate conversational setting in the verses and the fuller voice of the social zone in the chorus in a way that conveys change in the intensity of the messaging of the lyrical narrative toward the opening questioning of the way societal structures render women's dreams impossible. Singing in a fuller chest voice with wide vibrato, her voice increases in volume and intensity and begins to break under the emotional weight of the narrative, which is heightened in the final line of the second chorus. As Guyton pushes through the anger of the title question, her voice catches on the "onn" of "gonna," creating a raucous vibration on the open syllable. Here, the use of reverb and delay emphasizes the intensity of the performance. As Moylan (2020, 353) notes, manipulation of spatial properties of the recorded track through reverb and delay can elicit powerful effects within listeners, while also positioning the singer's voice within close proximity to the listener. The sparseness of the recorded setting gives both effects greater spatial weight, extending the final word beyond its articulation, directing the listener's attention to words, forcing deep reflection on their meaning and on the emotional weight of the question being asked of the industry.

Following the intensity of the second chorus, Guyton pulls inward to a more intimate zone at the start of the bridge for a moment of self-reflection on the role women play in perpetuating these lies in their teaching and mothering. Here, she returns to the intimate breathy voice for the first three lines of the bridge (Figure 21.2), displayed in the first half of the spectrogram. This is a moment of private dialogue between women—asking if they can continue to look girls in the face and promise them fairness and justness in society. On the final line of the bridge, a moment of utter heartache, Guyton's voice becomes at once rough and brittle, audible on the words "promise *her*" (marked in Figure 21.2) Guyton's voice slides downward on the word "her," registering the emotional pain line and of the moment, sounding as though she is choking back a sob.

The climactic vocal snap then arrives in the final chorus on the second time she sings forcefully "what *the hell* do you tell her?" The line has the feel of a guttural yell (market in Figure 21.3), of a person that has been pushed to the absolute edge of what she can endure as a woman in society, as the only Black woman signed to a label in an industry founded on white supremacist heteropatriarchal values.

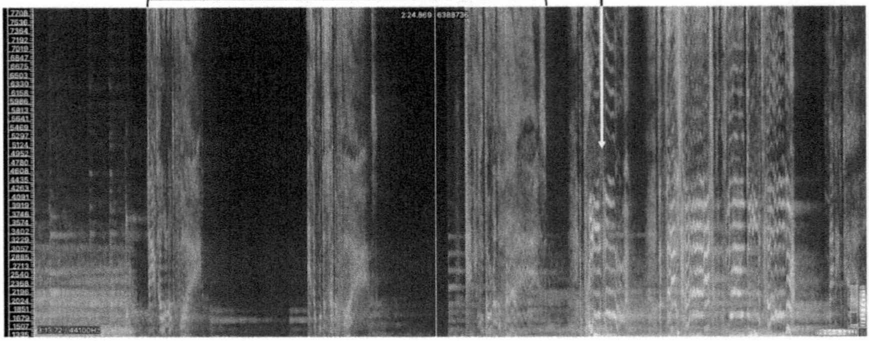

FIGURE 21.2 Spectrogram of Guyton's vocal performance in the bridge.

But Guyton's snap isn't a breakdown: this is a release of the pain, frustration, and trauma of her experience and she finds her ground—and *builds new ground* in the final two lines, and she asks the title question two times in full chest voice. Despite the power of the vocal demand, the ending is unresolved. As Guyton stands her ground and pushes back at the system that has for a decade pushed her down and to the side, she demands an answer over an inverted D major (V chord), leaving the listener in suspense—or in anticipation—of a response the song's title question.

Talking Back, Pushing Back

Country music was an integral part of Guyton's upbringing. She has relayed in numerous interviews that her love of the music came from her grandmother, with whom she watched Dolly Parton and Kenny Rogers VHS tapes (Petrusich 2021). But the reality of being a Black female country artist had become unbearable for

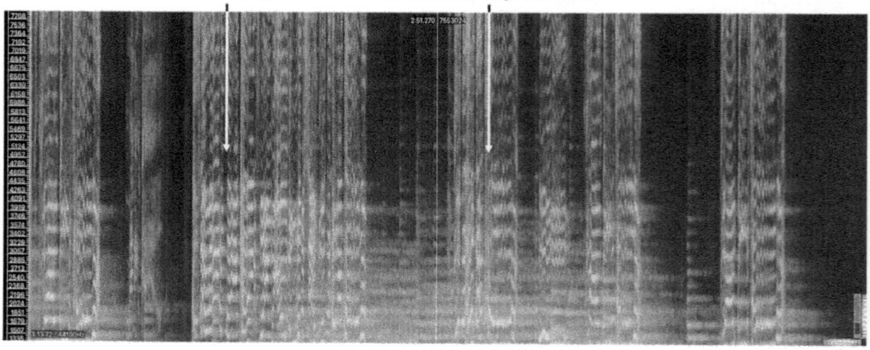

FIGURE 21.3 Spectrogram of Guyton's vocal performance in the final chorus.

the singer. Sitting in Kosowski's studio that January 2020 afternoon, Guyton and her co-writers spoke about the "moments" that women accumulate in their lives. These stories, interwoven with each other, become what Ahmed (2017, 202) terms a "feminist tapestry," a collection of experiences that become part of their bodies, that chip away at their self-worth, that break them down. While for Banks, the song came from thinking about her daughters, this song was also deeply personal:

> [T]here was this vitriolic anger that I was putting in this song that had built up from all these years [of being in the industry] and all the times that I have not had an opportunity because I was a woman, or all the times I've had to sit in the songwriting room with guys and hear them make sexual jokes about women and laugh along like it's cool. And all that stuff just kind of built up and it all came out that day in the song. (in conversation)

An expression of anger and solidarity, the song weaves these stories together to challenge established societal structures and demand for change in an industry that has for a century limited opportunity for women, and pushed Black, of color, and LGBTQIA+ artists to the margins.

What happened in the vocal booth that day, that Kosowski then emphasized in production, was an embodied feminist snap, of the breaking and rebuilding of a Black woman pushed to her very extreme. Guyton didn't want to sing the song that day, but her co-writers encouraged her to do it, knowing that the heightened emotion of the session would be integral to the recording. As Kosowski recounted, "there's nothing like that magic of the same-day vocal. It's the most intimate you will ever feel with a song—that moment just after writing it. We'd all been singing it all day long so it was alive in us in a way that it never would be again." Her co-writers prepared the space for Guyton to lay down her vocals—getting her tea and turning down the lights. While the session allowed Guyton and her co-writers to push back through writing, the vocal booth afforded Guyton her own moment to feel the weight of the lyrics, to navigate the emotions of the moment (and the decade of accumulated moments that preceded it), and to use her voice to push back against the industry. "I remember I was crying throughout," Guyton told me. "I had to keep stopping. And I didn't even know what I sounded like. I had my hands in my pockets of my sweatshirt and I had my eyes closed thinking how I sounded like shit." The demo recording captures the emotional release of the burden accumulated moments and micro-aggressions that Guyton has endured for a decade in Nashville. Ahmed (2017, 190) writes of the intonation of the feminist snap: "Speaking sharply, speaking with irritation. Maybe we can hear her irritation; a voice that rises, a voice that sharpens. A voice can lose its smoothness; becoming rougher, more brittle." The emotional weight of this release is made audible in both the fragility of the intimate conversation of the verses and the frustration and anger of the vocal breaks, the choked-back sobs, and the guttural yell in the final chorus.

But this is not a breakdown or a collapse, this is a musical "revolt against what women are asked to put up with" (Ahmed 2017, 210). This moment of snap became a new start for Guyton. In *Talking Back,* bell hooks (1989, 8) writes, for Black women, "speaking is not solely an expression of creative power; it is an act of resistance, a political gesture that challenges politics of domination that would render [them] nameless and voiceless." This demo recording captured Guyton's pushback against the system, a message that she was no longer willing to work within the boundaries placed on her, no longer willing to be silenced or invisible to the industry's structures. This pushback emerges most prominently in the recorded demo, which centered on Guyton's voice and the rawness of her emotions after that January 2020 writing session. Close-miking techniques in the verses and the bridge capture the detail and breathiness of Guyton's voice, creating a sense of intimacy between the singer and her listeners. But it also highlights the singer's fragility, her vulnerability, picking up on the subtle nuances of the breaks in her voice that capture the emotional weight of this message on the singer—suggesting to the listener that this message is indeed personal to Guyton. But this message is also inherently social and directed publicly toward the industry that has pushed her—and the women around her—to the margins. Through the growing roughness and raucous power of her full vibrating voice in the second and final chorus of the song, Guyton revolts against the industry's white supremacist heteropatriarchal structures and values, speaking out—speaking and pushing back—and demanding to be heard.

Acknowledgements

Thank you also to James Law for spending time with me to discuss the recorded track and miking techniques.

Notes

1 Guyton was the third Black woman to sign to a major label in Nashville: Lenora Ross was signed to RCA in 1975 and Nisha Jackson was signed to Capitol Records in 1987 (Williams 2021).

2 Guyton released her debut album, *Remember Her Name,* on September 24, 2021.

References

Ahmed, Sara. 2017. *Living a Feminist Life*. Durham, NC: Duke University Press.

Beal, Frances. 1969. "Double Jeopardy: To Be Black and Female." From the Third World Women's Alliance; New York. Available online, http://www.hartford-hwp.com/archives/45a/196.html.

Bjorke, Matt. 2015. "Cam Is 'On The Verge' at Radio with 'Burning House'." *Roughstock*, July 6. Available online, http://roughstock.com/news/2015/07/36125-cam-verge-radio-burning-house.

Burns, Lori. 2010. "Vocal Authority and Listener Engagement: Musical and Narrative Expressive Strategies in the Songs of Female Pop-Rock Artists, 1993–95." In *Sounding Out Pop: Analytical Essays in Popular Music*, edited by Mark Spicer and John Covach, 154–92. Ann Arbor, MI: University of Michigan Press.

Hill, Keith. 2015. Interview with Russ Penuell. *Country Aircheck*, vol. 449, pp. 1, 8. Available online, https://www.countryaircheck.com/pdfs/current052615.pdf.

hooks, bell. 1989. *Talking Back: Thinking Feminism, Thinking Black*. Boston, MA: South End.

Keel, Beverly. 2004. "Between Riot Grrrl and Girl: The New Women's Movement in Country Music." In *A Boy Named Sue: Gender and Country Music*, edited by Kristine M. McCusker and Diane Pecknold, 155–77. Jackson, MI: University of Mississippi Press.

Malone, Bill C. 1968. Country Music, USA. Austin, TX: University of Texas Press.

Malone, Bill C, and Tracey E.W. Laird. 2018. *Country Music, USA*. 50th Anniversary Edition. Austin, TX: University of Texas Press.

McCusker, Kristine M. 2017. "Gendered Stages: Country Music, Authenticity, and the Performance of Gender." In *The Oxford Handbook to Country Music*, edited by Travis D. Stimeling, 355–74. New York, NY: Oxford University Press.

Miller, Karl Hagstrom. 2010. *Segregating Sound: Inventing Folk and Pop Music in the Age of Jim Crow*. Durham, NC: Duke University Press.

Moylan, William. 2020. *Recording Analysis: How the Record Shapes the Song*. New York, NY: Routledge.

Oliver, Codie Elaine. 2021. "Mickey Guyton Is My New BFF." My New BFF with Codie Elaine Oliver, May 17. Black Love Podcast Network. Available online, https://www.iheart.com/podcast/269-my-new-bff-with-codie-elai-78324948/episode/mickey-guyton-is-my-new-bff-82634662/.

Petrusich, Amanda. 2021. "Mickey Guyton Takes on the Overwhelming Whiteness of Country Music." *The New Yorker*, June 7. Available online, https://www.newyorker.com/magazine/2021/06/14/mickey-guyton-takes-on-the-overwhelming-whiteness-of-country-music.

Stimeling, Travis. 2020. *Nashville Cats: Record Production in Music City*. New York, NY: Oxford University Press.

Watson, Jada. 2019a. "Gender on the *Billboard* Hot Country Songs Chart, 1996–2016." *Popular Music & Society* 42 (5): 538–60.

Watson, Jada. 2019b. "Gender Representation on Country Format Radio: A Study of Published Reports from 2000–2018." *SongData Reports;* Ottawa, Ontario, Canada. Prepared in consultation with Woman of Music Action Network. Available online, https://songdata.ca/wp-content/uploads/2019/04/SongData-Watson-Country-Airplay-Study-FullReport-April2019.pdf.

Watson, Jada. 2020. "*Billboard's* Hot Country Songs Chart and the Curation of Country Music Culture." *Journal of Popular Music History*, Special Issue on Popular Music Curation 13 (1–2): 168–90.

Watson, Jada. 2021. "Redlining in Country Music: Representation in the Country Music Industry (2000–2020)." *SongData Reports*. Ottawa, Ontario, Canada. Available online, https://songdata.ca/wp-content/uploads/2021/03/SongData-Watson-Redlining-Country-Music-032021.pdf

Watson, Jada. 2022. "A Double-Edged Sword: Industry Data and the Construction of Country Music Narratives." In *Whose Country Music? Genre, Identity, and Belonging in Twenty-First Century Country Music Culture*, edited by Paula J. Bishop and Jada Watson, 55–70. Cambridge, UK: Cambridge University Press.

Williams, Andrea. 2021. "What Reckoning? Country Music Race Reckoning, One Year Later." *Vulture*, June 28. Available online, https://www.vulture.com/article/essay-country-music-race-reckoning-one-year-later.html.

22

FORM, GENRE, AND VOCAL PERFORMANCE IN NICKI MINAJ'S "STUPID HOE" (2011)

Olivia R. Lucas

This chapter is supported by a webpage.

Introduction

"Stupid Hoe" is one of Nicki Minaj's most controversial singles.[1] The music video, posted to YouTube in January 2012, maintains a nearly 1:1 like/dislike ratio,[2] offering a snapshot of the song's polarizing quality. Even with the mixed reception, the song charted, peaking at 59 on the Billboard Hot 100. Its music video set an early Vevo record of 4.8 million views within 24 hours of release (Corner 2012a).[3] At face value, the song is understood as a diss track about Lil' Kim and/or Nicki Minaj's other female "haters" (detractors),[4] but a closer listen to the song's self-consciously obnoxious, aggressive, and unpredictable sounds reveals unexpected commentaries on genre and mainstream palatability. Minaj's exploratory vocal performance, with its sobs, nasalized growls, vocal fry, falsetto, and deliberately off-key pitches, combined with the beat's iconic whoops, squeaks, sirens, and whistles, merge into a song that uses pop and rap conventions in surprising ways that render the familiar alienating and offer sardonic commentary on the nature of the diss track and commercial success.[5] In this chapter, I analyze how the song's genre references, form, instrumentation, and vocal performance work together to create the track's iconic aesthetic of frantic overwhelm, as understood through the song's polarizing and evolving reception history. A full vocal transcription of the track is available on the chapter's accompanying web page.

Genre: Diss Track Meets Bounce

Nicki Minaj is known for embracing both hip-hop and pop; Sören Schoppmeier analyzes her genre and image fluidity in terms of racial code-switching, as she has

DOI: 10.4324/9781003089926-30

achieved commercial success in both white-coded and Black-coded genres (2015). This pre-existing fluidity is important to understanding how "Stupid Hoe" plays with various genre conventions. In the track, her self-consciously unstable singing mocks Western pop conventions of stable intonation while resonating with similar moments in hip-hop such as Biz Markie's off-key singing in "Just a Friend" (1989). The song's unusual form, as it builds toward this sung conclusion, further subverts the conventions of both pop and hip-hop. However, it is the way "Stupid Hoe" merges two seemingly incompatible strands of hip-hop—the diss track and New Orleans Bounce—that yields the song's strongest shape-shifting qualities.

"Stupid Hoe" is generally understood as a diss track attacking rapper Lil Kim and/or Nicki Minaj's female haters. Anthropology and African American studies scholar Christian Free defines a "diss" as "a form of insult and rejection commonly employed by Hiphop artists … to highlight some issue one claims to have with another person or institution, typically another artist" (Free 2016, 1). He writes that disses tend to beget further disses,

> opening up a discourse space … known as "beef." Beefs are characterized by sustained dissing … between feuding interactants in a fluid, scalable participatory frame. The most iconic form of beef is a back-and-forth exchange of "diss tracks," or full rap songs purposed as a cotextual assemblage of various types of disses.
>
> *Free 2016, 1–2*

Tia Tyree and Melvin Williams further analyze "how women rappers [including Nicki Minaj] have used diss songs to achieve commercial success, respect, and voice in rap" (2021, 65). Tyree and Williams' article includes coverage of the lengthy feud and exchange of diss tracks between Nicki Minaj and Lil' Kim, of which "Stupid Hoe" is one example; online fan communities have also tracked the history of Minaj's musical and extramusical beef with Lil' Kim.[6] In Mike Barthel's review of the "Stupid Hoe" music video, he describes hip-hop's diss tracks as "a lawyerly form," in which the artist must present evidence against the person they are attacking, in a way that displays their own verbal dexterity (Barthel 2012). In "Stupid Hoe," however, Minaj eschews this format, opting instead for unvarnished insults and relentless repetition of the song's title. As Barthel points out, Minaj declines to build "a specific case for why [Lil'] Kim sucks … because Nicki is not trying to convince us rationally that she's right." Barthel goes on to argue that the track's music video carries the diss-related information that the lyrics don't. I agree that Minaj is deliberately subverting diss track conventions, but I believe more of this comes through in the track itself than Barthel acknowledges, specifically through the sonic details of her vocal performance. As Minaj spends the song exploring her vocal-expressive range, she subverts and transforms the diss track's conventions of verbal dexterity into a performance of sonic/vocal dexterity, playfully displaying her superiority through sonic brashness instead of wordplay.

This rejection of diss track norms overlaps with the song's embrace of Bounce music aesthetics. Bounce emerged in the early 1990s in New Orleans as a form of "dance-oriented hip hop based on up-tempo beats, repetitive hooks, and call-and-response sequences with the audience" (Casey 2018, 5). Matt Miller further characterizes Bounce's lyrics as frequently having an "emphasis on dancing, pleasure, and commodified sexual relations" (2012, 157). While eschewing mainstream hip-hop's tradition of first-person storytelling, and the diss track's conventions of lyrical dexterity, "Stupid Hoe" simultaneously adopts Bounce's use of a simple, raunchy hook over a hard and fast dance beat; the extended repetition of "you a stupid hoe" presents a variation on Bounce's characteristic repeated exhortations to "shake that ass like a saltshaker."[7] In live performances of the song, Minaj also explicitly draws on Bounce's call-and-response tradition, trading off lines of the hook with the audience through the classic gesture of turning the microphone toward the crowd. The song's form, especially after the second hook (sections E–Outro in Table 22.1), can also be viewed through the lens of Bounce's characteristic "cellular structures" (Miller 2012, 75), in which performers chain together sections that support "continuities as well as disjunctures in musical style and lyrical content" (Miller 2012, 76).[8] It is also worth noting that "Stupid Hoe" was released by Cash Money Records, a New Orleans-based label that Miller associates with the Bounce scene.[9]

The extreme repetitiveness of "Stupid Hoe" manifests Bounce's dance-directed energy, while also pointedly upending expectations of lyrical virtuosity. This fosters the song's prioritization of vocal timbre and style over lexical variety, which works to both mock diss track norms and adopt Bounce's approach to rhythmic and lyrical repetition.

Form

Analyzing the song's form offers a bird's-eye view of the collision of typical hip-hop segmentation with elements of the "cellular" organization associated with Bounce, as well as a terminally climactic approach more commonly associated with rock and metal (Osborn 2013). Table 22.1 shows a form chart of the song's sections, with brief descriptive commentary that feeds into each section of the analysis.

The song's segmentation is aurally clear throughout, with strict 8-bar units that sometimes repeat or group into larger 16-bar units. The clarity of this segmentation is driven by changes in the beat. For example, the drums drop out for the last bar of Verse 1 (0:17), demarcating the boundary with the pre-chorus-type section that follows. Despite this clear demarcation, however, the function of the segments can be unclear, and these ambiguities increase over the course of the song.

The first ambiguity emerges in section B, which I have labeled as a "detached pre-chorus." In this section, the vocal shifts into a sing-song delivery with relatively clear fundamental pitches and a repetitive rhythmic pattern. This

TABLE 22.1 Form chart of "Stupid Hoe"

Time	*Section*	*Name*	*Vocal characteristics*	*Lyric*	*Alter ego*	*Instrumental and timbral features*
0:00	Intro	Intro	Instrumental			Whoops
0:09	A	Verse 1	Rapped	I get it cracking like a bad back …	Roman Zolanski (RZ)	Whoops
0:18	B	Detached pre-chorus	Sing-song delivery on minor third and major second pitch dyad	Ice my wrist-es …	RZ	Squeaks
0:28	C	Verse 2	Rapped, ends with eight-bar hold on single word	Cause I pull up and I'm stuntin' …	RZ	Squeaks, bass riff, samba whistles, sirens
0:56	D	Chorus	Rapped	You a stupid hoe …	RZ	Whoops
1:16	A′	Verse 3	Rapped	Look Bubbles, go back to ya habitat …	RZ	Whoops
1:25	B	Detached pre-chorus	Sing-song delivery on minor third and major second pitch dyad	Ice my wrist-es …	RZ	Squeaks
1:35	C′	Verse 4	Rapped	Cause I pull up in the Porsche …	RZ	Squeaks, bass riff, samba whistles, sirens
2:04	D	Chorus	Rapped	You a stupid hoe …	RZ	Whoops
2:23	A″	Verse 5	Rapped	If you cute, then your crew can roll …	RZ	Whoops
2:33	E	Bridge 1	Sung, chest voice	Stupid hoes is my enemy …	Harajuku Barbie (HB)	Initially bass only, then bass riff, squeaks, and samba whistles
2:52	F	Bridge 2	Sung, goes into falsetto	You a stupid hoe …	HB	Squeaks, sirens
3:01	G	Bridge 3/Climactic segment?	Rapped	And I ain't hit that note …	Ambiguous; cues for both RZ and HB	Squeaks, heavy bass
3:11	Outro	Sample	Sample from Nicki Minaj's verse on "Y.U. Mad" by Birdman	I am the female Weezy	Female Weezy	Instrumental drops out, heavy reverb on voice

consolidation into a clear pitch and rhythmic pattern exhibits a pre-chorus' characteristic building up toward the "release" of a hook.[10] In this case, however, the pre-chorus dissipates back into a rapped verse (section C), thereby "detaching" it from the chorus that eventually follows in section D. The ambiguity of this section's function arises from the way the increase in predictability "tightens" the energy, producing a feeling of build that then dissipates when another verse starts. Initially, it could be retroactively understood as another verse, but its exact repetition between verses 3 and 4 (sections A′ and C′) helps solidify my hearing of the detached pre-chorus. In both instances, the pre-chorus section destabilizes formal expectations and makes the song's flow of energy harder to follow.

In addition to the formal ambiguity introduced by the detached pre-chorus, the build into the actual first chorus is accomplished by an eight-and-a-half bar hold on the word "would" (0:47–0:57), followed by a gasping intake of breath that overlaps with the onset of the chorus.[11] At its outset, the hold seems to stop the flow of the vocal performance, but as it continues an improbably long time, its energy begins to build. By holding a single syllable in a tense, nasalized tone, Minaj performs what I call an "anti-flow," in which she deliberately halts the forward momentum associated with rapping. In doing so, she ends up creating the kind of building intensification toward the chorus that is typically the domain of the pre-chorus by simply holding a single word for a full eight bars. The "chorus" that follows is marked by its repetition of the phrase "you a stupid hoe" sixteen times—once per measure, with minimal rhythmic variation. Its energy is aggressive and manic, and follows Bounce's hook conventions, with their predilection for extended repetition of raunchy phrases. Although the chorus' intensely repetitive structure distinguishes it from mainstream pop-rap hooks, its placement and repetition after a repetition of the verse–detached pre-chorus–verse pattern makes its formal role undeniable. What is unusual, however, is that the chorus does not return again in the last minute of the song, which instead comprises a string of bridge-like segments.

Following section A″ (2:23–2:33), the remainder of the song is through-composed without any repeated sections (though the lyrics remain organized around the phrase "stupid hoe"). Sections E and F are sung, and are the first of three bridge-like sections. Section F can be heard as a sung version of the chorus, as it too only repeats the phrase "you a stupid hoe," but this time on increasingly unsteady pitches and reaching from Minaj's chest voice well into her falsetto range. This dissolves into a final segment (G) in which she repeatedly growls, "fuck you stupid hoe." The track closes with a sample from Minaj's verse on "Y.U. MAD" (Birdman, feat. Nicki Minaj and Lil Wayne; released 2011), in which she states, "I am the female Weezy."[12]

Whereas the first two-and-a-half minutes of the song feature a playful variation on a verse-chorus paradigm, the song's final third pushes it toward a terminally climactic form, which produces "tension between the expected highpoint (repeated chorus) and the climactic ending, the song's actual highpoint" (Osborn 2013, 23). The closing sections of "Stupid Hoe" feature the core characteristics of Osborn's terminally climactic forms: "a repeated lyrical/melodic hook over a section that is marked by a dynamic, rhythmic, or harmonic change" (Osborn 2013, 23). The material of sections E and F exhibit a sudden melodic modulation to B-flat, shift performance modality from rapping to singing, and offer increasingly unstable renditions of the "you a stupid hoe" hook moving higher and higher into Minaj's vocal range. The final eight bars before the outro (section G) move back to an aggressive nasalized rap. Although the chorus is memorable in its repetitiveness, it is sections E, F, and G that give the song its paradigm-shifting dramatic shape by offering increasingly frantic performances of the song's central lyrical idea that shift the form's energetic balance toward the concluding 45 seconds.

Vocal Performance

Flow

Inextricable from the analysis of vocal performance is the song's connection with Nicki Minaj's "Roman Zolanski" alter ego, whom she describes as a homosexual man from London.[13] Roman is an aggressive and energetic character, and his lyrics can be offensive, unruly, boisterous, and even threatening.[14] Hanisha Kulothparan finds that when Minaj raps from the perspective of Roman, the flow tends to have a strict, tight structure with regular rhyme and rhythmic surface activity (2021, 5–7). Kulothparan contrasts these tendencies with Nicki Minaj's other frequently used alter ego, "Harajuku Barbie," whose flow tends to be more disjunct, higher in pitch, have less clear rhyme patterns, and may involve singing (2021, 7–9).[15] In "Stupid Hoe," the sung bridge segments (sections E and F) are characteristic of Harajuku Barbie, and visual cues from the music video—Minaj appears wearing a pink dress and wig and with enlarged eyes, standing on an oversize chair that gives her a doll-like appearance—support this interpretation.[16] As noted in Table 22.1, section G presents conflicting cues; the return to a fast, lower-pitched rapping style here suggests Roman Zolanski, while the visual cues from the music video are mixed, as Minaj is depicted styled alternately as Roman and Barbie in this section.

Rhythmically, most of the song's flow can be represented by eighths and quarters, with a few syncopations over beats and barlines. As with most rap transcriptions, the visual appearance of numerous "straight 8ths" fails to communicate the patterns contained in the rhyme and accent pattern; in many rap tracks, these patterns are far more complex and syncopated, as "syncopations often occur due to the placement of stressed syllables in flow" (Komaniecki 2019, 23). In "Stupid Hoe," Minaj reliably places rhymes at the ends of syntactic units, and frequently accents the third beat of measures, especially in the hook and C and C′ sections. Aside from the notated syncopations, the track generally follows the metrical accent pattern suggested by 4/4, albeit with emphasis on beat three frequently stronger than on beat one.

Interactions between the Vocal and Instrumental Layers

The song's instrumental track enhances the relentlessness of the flow, with every segment except for the first eight bars of the first sung bridge (E) backed by some combination of hand claps, whoops, squeaks, sirens, samba whistle, and bass. The whoops in particular, first heard at the opening of the song, tend to have an intensifying quality; in each section where they appear (Intro, A, D, A′, D, A″), their surface rhythm doubles from sounding on beats one and three to sounding on every beat in the last four measures of the segment. With the exception of the bass, the instrumental elements concentrate percussive energy in the mid and upper-frequency ranges, overwhelming the ear with high-pitched material. This intensely noisy version of a fast dance beat represents another of the song's borrowings from Bounce music.

Compared with the increasingly through-composed form explored by the vocal layer, the instrumental layer has a relatively strict repeating form that continues throughout the track. A spectrographic snapshot of the whole track shows that sections with bass activity alternate with sections in which bass is completely omitted (see Figure 22.1).[17] In the higher frequency ranges, the instrumental rotates three times through a fixed 56-measure pattern that includes sections A-D and their repetition; the third rotation is incomplete at only 40 measures (see Table 22.2).[18] Even when the vocal leaves behind the verse-chorus structure with the first sung bridge in section E, the instrumental rotation continues unchanged; the only exception comprises the first 8 measures of the E section, which departs from the rotation by thinning to bass only, before returning to where it "should" be in its pattern—the combination of squeaks, tresillo bass riff,

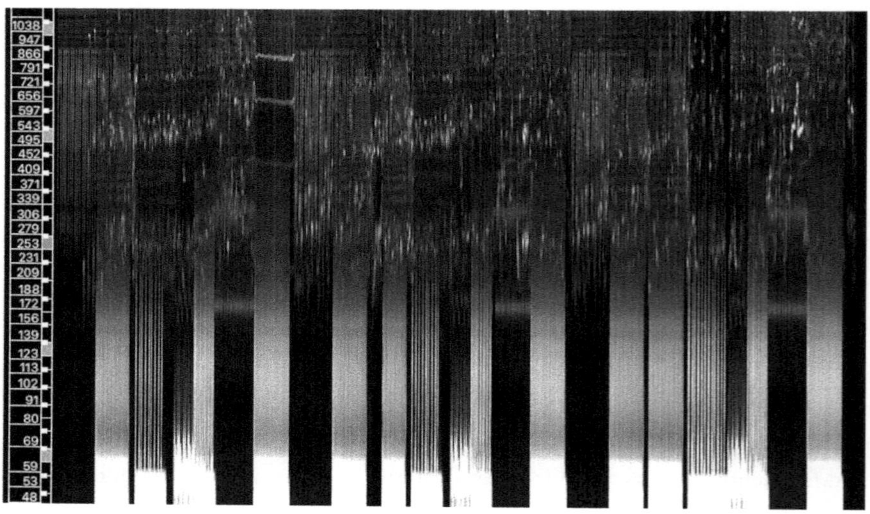

FIGURE 22.1 Spectrogram of "Stupid Hoe." Note how activity in the bass frequencies comes and goes throughout the track. Created with Sonic Visualizer.

TABLE 22.2 Form chart of instrumental layer details mapped against vocal structure

Time	*Section*	*Measure count*	*Instrumental and timbral features*
0:00	Intro	8	Whoops on beats 1 and 3
0:09	A	4	Whoops on beats 1 and 3
0:14		4	Whoops on every beat
0:18	B	8	Squeaks
0:28	C	8	Squeaks, tresillo bass riff, claps, samba whistle
0:38		8	Squeaks, claps, sirens
0:48		8	Squeaks and heavy bass
0:58	D	8	Whoops on beats 1 and 3
1:07		4	Whoops on beats 1 and 3, bass, backing vocals
1:13		4	Whoops on every beat, bass, backing vocals
1:17	[Instrumental repeats sections A–D]		
2:23	A″	4	Whoops on beats 1 and 3
2:28		4	Whoops on every beat
2:33	E	8	**Bass only, reverb on voice**
2:43		8	Squeaks, tresillo bass riff, claps, samba whistle
2:52	F	8	Squeaks, claps, sirens
3:01	G	8	Squeaks and heavy bass
3:11	Outro		No instrumental, heavy reverb on voice

claps, samba whistle that previously accompanied sections C and C′. This contrast between the continuity of the instrumental pattern and the vocal departure toward through-composition highlights the adventurousness of the last few vocal segments. The instrumental layer maintains its relentlessness, even as Minaj's vocal performance drifts further toward an impression of improvisational riffing on the song's title, signaling her abandonment of pop-rap conventions.

Pitch and Vocal Timbre

As Mike Barthel writes in his review of the song, "Nicki is … frighteningly good at communicating through tone, through timbre, through visuals, and through style" (2012). In adopting the Bounce aesthetics of raucous repetition, her vocal performance shines through as an expressive and experimental space that contributes enormously to the song's signature playful, deranged, and mocking tone. This comes through in obvious ways, such as the semantic satiation I hear generated by the relentless repetition of the chorus, and her enhancement of this effect by transforming the pronunciation of "hoe" into "how." It also comes through in more subtle ways, that become easier to grasp through close pitch and vocal production analysis, as her vocal strategies shape not only the local flow within verses but also the perception of the song's formal structure.

Spectral analysis indicates that Minaj's rapping voice hovers predominantly around C4 (see Figure 22.2), and the vocal energy of the song—apart from the melodic bridges—stays in this pitch area. In the rapped verses, Minaj tends to

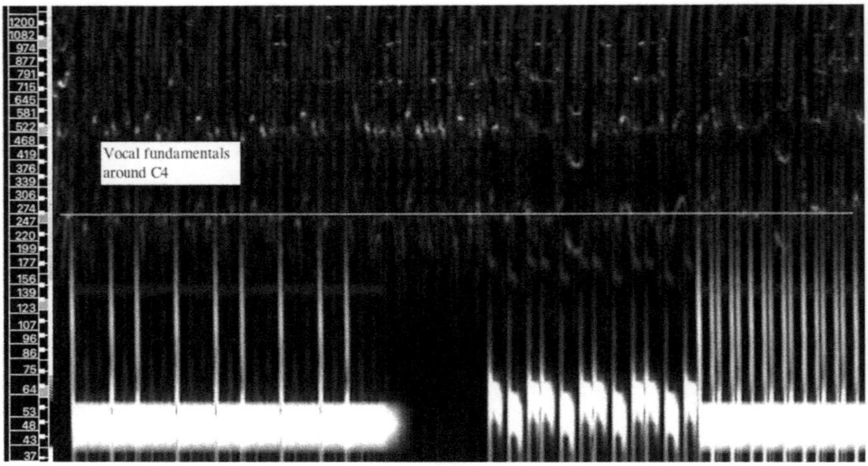

FIGURE 22.2 Spectrogram of 0:19–0:38, showing vocal fundamentals hovering around C4. Created with Sonic Visualizer.

gradually heighten intensity over the course of the segment by adding more sob tone,[19] pitch fluctuation, rapid register breaks between sob and vocal fry,[20] and/or by growing increasingly nasal. In the first verse (section A), her rapping moves toward a sob tone that fluctuates in register, culminating in her mocking question: "You see where Brad [Pitt] at?" After the first chorus, section A′ follows a similar pattern of increasing sob tone. The first eight bars of verse sections C and C′ are characterized by alternating rising and falling intonation. In section C, as she approaches the eight-bar hold (0:47), her voice grows more nasalized and pinched, with what sounds like a relatively high larynx and constricted throat, resulting in a twang-infused tone that she maintains through the hold.[21] The ending of section C′ (1:55–2:03) exhibits rapid shifts between sob tone and vocal fry, with the resulting registral instability creating a yodel-like effect. This foregrounded registral instability associates the lyrics' repeated insults with an increasing emotional intensity and instability, making her mockery of her counterparts clear as she builds toward rejecting "custody" of rappers who would follow in her footsteps.

Unlike the verses, with their continually morphing vocal styles, the detached pre-choruses (section B) and choruses (section D) exhibit vocal stability. The detached pre-chorus moves into a sing-song chanted delivery, mostly on C4, with an accented syncopation in each measure on E♭ or D. Here, the fundamental pitches are clearer, but her voice retains a spoken tone that makes some of the shorter durations less stable in pitch. Nevertheless, the salience of the motivic minor third (and occasional major second) and the relative pitch stability and repetitiveness give this segment the tight structure that makes it sound characteristic of pre-chorus material. The G2 in the bass—also present in the verse—also receives more of an implied harmonic context here, with all three vocal pitches stable over the bass. The chorus itself is characterized by its relentless repetition, with the phrase "you a stupid hoe" occurring sixteen times, once per measure, with consistent rhythmic placement. Although the vowel shift from /hō/ into /hou/ intensifies the loss of semantic meaning I hear in the repetition of the phrase, vocally, the chorus is stable and predictable, and the repetitive hook blends into the relentless whooping of the instrumental.[22] Up through the end of the second chorus, Minaj has alternated stable, predictable sections (pre-chorus, chorus) with unstable ones (verses), creating a balance between novelty and familiarity (or over-familiarity, in the case of the chorus).

The stability and tuning of the vocal pitches become more salient in the portions of the song that are clearly sung—sections E and F. Given the widespread availability of Autotune and other pitch correcting tools at the time this track was recorded, I assume that any instabilities in the pitch are deliberately created and/or retained. I should also make clear that my analyses of pitches as being "unstable," "out of tune," "flat," or "sharp" are not a judgment of vocal talent or ability, or of the quality of the song.[23] They are rather a documentation of facets of the vocal performance that contribute to the song's character and my enjoyment of it.[24]

The initial instability of the first sung bridge (section E) comes from the return of the G2 bass pattern first heard associated with the pre-chorus, with its melodic pitches of C4, D4, and E♭4. Here, the tonal center of the melody shifts down a whole step, to B♭3. The pairing of this new melodic material, with its emphasis on a lowered third scale degree (Db), sounds jarring over the return to the G in the bass, initially making the entire melody sound like it has gone flat. The effect is short-lived, however, with the G easily being recontextualized within the new B♭ tonal center.

Figure 22.3 shows my analysis of the sung pitches in sections E and F. Because I do not want to suggest that she is singing precise quarter-tones or other calculated microtonal pitches, I have opted to use simple + and – signs to indicate a pitch's audible sharpness or flatness, respectively.[25] In general, my notation preserves a somewhat idealized tonal version of the melody, and marks where the heard pitch audibly strays. Although pitch across this segment clearly outlines a melody based on B♭, one layer of instability derives from the way Minaj unpredictably uses the major and minor thirds of the scale, frequently alternating the two. In m. 125, the ambiguity regarding D4 vs. D♭4 reaches a peak—my analysis has her narrowly on the D♭4, but a very sharp version of it. The melodic high pitch of F4 is another site of melodic ambiguity. The first time she rises to the F, the pitch is flat, anticipating the drop back to B♭3. The second time, on "whack," the F4 is audibly sharp. Generally, when she hits F4 and quickly falls back to B♭ 3, the F4 is flat. There are a few places where the vocal pitch clearly renders the melody chromatic—for example, the somewhat sharp E-natural on the downbeats of m. 132 and 134—but even though the instabilities are clearly audible, they are never strong enough to cast doubt on the B♭ tonic to which the melody consistently returns.

FIGURE 22.3 Vocal transcription of sung bridge segments (sections E and F, 2:33–3:01). Plus (+) and minus (–) signs indicate audibly sharp and flat pitches, respectively.

Section F continues her exploration of B-flat, but the tessitura shifts upward into her mixed register and then breaks into her falsetto in m. 141, for a two-measure melisma on "stupid." Here, the surface rhythm slows, allowing her to scoop and drift pitches, with several drifting sharpward as she holds them. The final rapped segment of the song (G) seems to comment on her singing ("and I ain't hit that note"), although if she's referring to the song's high note of G♭5, she does in fact "hit it," sliding through it from flat to sharp before falling back to the D♭5 that ends the section. But the point is that the quality of her singing is immaterial—she wants her enemies to know that she can mock even the basic tenets of pop singing and still produce a hit.

Conclusion

In "Stupid Hoe," Minaj moves fluidly through vocal modalities and subjects each one to variations in register, production mode, resonance, and tuning. She begins with a loose verse-chorus form, and then abandons it to explore building toward a terminal climax. Harmonically, she employs a sudden modulation to a new tonal center, and flexibly moves between implied major and minor modes, while simultaneously exploring the limits of unstable intonation and tuning. She manipulates Bounce's raunchy repetition and hard four-on-the-floor beats into a parodic reinvention of the diss track formula.

"Stupid Hoe" was a hit, but not a chart-topper like "Anaconda" or "Moment 4 Life," both of which feature more mainstream pop approaches to rapping, singing, and instrumentals. Aside from Mike Barthel's nuanced positive review of the song and video, many critical responses were mixed to negative. Michael Cragg's review in *The Guardian* says the song "seems to forgo a chorus or even an actual tune in favour of cute one-liners ... it sounds more like she's run out of ideas," and that it sounds like "an early mixtape" (2011). Robbie Daw of *Idolator* gave a mixed review, calling her "rhyming acrobatics...interesting" but the track "monotonous" (2011). More recent public comments on the music video indicate that attitudes toward the song are shifting toward recognition that what was previously perceived as the song's failure to meet standards for "good" organization, singing, rapping, lyric-writing, and beat-making, were all deliberate choices meant to mock those very standards. Ten years after its release, commenters on the music video refer to the song as "experimental," "avant-garde," and "ahead of its time."[26]

Propelled in large part by its music video, it's difficult to disentangle the track as sonic artifact from the total audio-visual spectacle. In general, Minaj's visual aesthetics have received more attention than the sounds of her music; numerous articles (e.g., Roach 2019, Cosimini 2017, Hunter and Cuenca 2017, Schoppmeier 2015, and McMillan 2014) try their hand at analyzing her presentation of her body, her sexuality, her Blackness, and her rejection of hip-hop norms and conventions. McMillan in particular aligns Minaj's "deployment of a black camp sensibility" with Paul Gilroy's theorization of Black cultural resistance represented through "inauthentic" performances (McMillan 2014, 85, citing Gilroy 1995, 29). Because of this attention to what Minaj's performances *look* like, it's difficult to judge the polarizing nature of the song separately from its video. But stepping back from the video's manic editing and hotly debated imagery of Minaj gyrating on all fours in a cage (see Roach 2019), offers an opportunity to *hear* how these ideas of rebellion, experimentation, and deliberate camp come through sonically.

Notes

1 "Stupid Hoe" was written by Nicki Minaj (full name Onika Tanya Maraj) and Tina Dunham, and produced by DJ Diamond Kuts. It was released on December 20, 2011 on Cash Money Records as a promotional single for the album *Pink Friday: Roman Reloaded* (2012).

2 YouTube removed the like:dislike ratios on videos in late 2021, after I had already drafted the chapter (Perez 2021).

3 This record was broken a few months later by Rihanna's music video for "Where Have You Been" (Corner 2012b). In keeping with the theme of the book, this chapter will prioritize the analysis of the recorded track over the video, though the significance of the video to the track's overall reception should be noted. It is impossible to completely disentangle the controversial nature of the track from that of the video.

4 The colloquial usage of "hater" goes beyond simple dislike of an individual to connote a person who attacks an artist or performer at least in part out of jealousy over their success, and/or out of the belief that this success is undeserved. This connotation can be observed in the way individuals reject criticism, both valid and invalid, as the invective of mere "haters"—incompetent, jealous people whose opinions do

not matter. Varying shades of this idea can be surveyed on *Urban Dictionary* at https://www.urbandictionary.com/define.php?term=hater.

5 The idea of "defamiliarization" as a rhetorical tactic originates with Russian literary theorist Viktor Shklovsky in his 1917 essay "Art as Technique." The concept has been previously adapted into musical analysis by Rebecca Perry (2020).

6 One such fan-sourced history appears on the Nicki Minaj wiki at https://nickiminaj.fandom.com/wiki/Lil%27_Kim.

7 This lyric is used repetitively in "Where dey at?" (1991) by MC T. Tucker and DJ Irv, a track cited by Miller as foundational for the Bounce genre (2012, 81–83).

8 Miller also cites "Where dey at?" for popularizing this structure. The track moves through sections based on rhythmic variation of heavily repeated phrases (e.g., "shake that ass like a saltshaker," "fuck David Duke," and "bitch better have my money") (2012, 81–83).

9 The previously local label rose to national prominence with its launch of Lil Wayne's career (Miller 2012, 141–147).

10 Jay Summach (2011) describes the pre-chorus as a segment that builds momentum toward the chorus and writes that "changes in groove, lyric phrasing, and the length of formal units, as well as dynamic level, register, instrumentation, timbre, harmonic progression, and harmonic rhythm all have the potential to increase forward formal urgency" (Summach 2011, [3]). Studies of the pre-chorus (Summach 2011; Ensign 2015) have focused primarily on harmonic and melodic-motivic trends, which are difficult to apply to this track. However, for me, the change in vocal rhythm and pitch to a strictly repetitive pattern has the effect of "tightening" this segment's energy, producing a feeling of build that then dissipates when another verse starts.

11 Minaj performs a similar four-bar hold on the final syllable of "Waikiki" in Big Sean's "Dance (A$$) Remix" (2011), along with some whoops very similar to those heard in "Stupid Hoe."

12 "Weezy" is another name for rapper Lil Wayne. "The female Weezy" is one of Nicki Minaj's alter egos.

13 It can be difficult to definitively determine whether an alter ego is present in a song unless the lyrics name them (as in section C′ of "Stupid Hoe") or the artist has confirmed the alter ego's presence. There is consensus, however, that Roman Zolanski is the primary character throughout "Stupid Hoe" (see, for example, James 2012 and Hunter and Cuenca 2017, 41).

14 The Roman Zolanski alter ego began to appear in 2010 in Nicki Minaj's guest verses, including the Ludacris track "My Chick Bad" (2:14), Trey Songz's "Bottoms Up" (2:25), and Kanye West's "Monster" (3:40). He first appears on a Nicki Minaj album on *Pink Friday* (2010), perhaps most famously on the second track, "Roman's Revenge." This track also introduces Roman's mother, Martha Zolanski, as an additional alter ego; at the end of the song, Martha scolds and cautions Roman in an exaggerated, fake British accent. Additional fan-sourced history of the Roman character, his role in Minaj's music, and images of his appearance in music videos and live performances can be found at the Nicki Minaj wiki: https://nickiminaj.fandom.com/wiki/Roman_Zolanski.

15 The Harajuku Barbie alter ego first appears in Minaj's guest verse on the Young Money track "Bedrock" (2009); in the music video, she can be seen wearing a large "Barbie" pendant during her verse (1:31–1:57). Harajuku Barbie often appears in opposition to Roman Zolanski; Kulothparan's analysis of the dialogic appearances of both Roman Zolanski and Harajuku Barbie in "Bottoms Up" (Trey Songz) and "Monster" (Kanye West) clarifies the differences in their flow styles (2021, 4–14). Like Roman Zolanski, Harajuku Barbie appears frequently throughout Minaj's 2010 *Pink Friday* album. Additional fan-sourced history of the Harajuku Barbie character, her role in Minaj's music, and images of her appearance in music videos and live performances can be found at the Nicki Minaj wiki: https://nickiminaj.fandom.com/wiki/The_Harajuku_Barbie.

16 For a deeper analysis of Minaj's visual aesthetics, see Schoppmeier (2015).
17 For more on tracking pitch density and timbral balance in a track, see Moylan (2020, 252–54).
18 For more on timbre as a form-shaping element, see Moylan (2020, 258). My study of instrumental timbres in this track falls within Moylan's "middle dimensions" (261–62) and his study of timbre as "character" (rather than "content") (270, 283).
19 According to Kate Heidemann, sob tone is produced by means of a lowered larynx, widened pharynx, and raised velum, that when used in singing, creates a "darker" sound associated with crying (2016, par. 3.19). Minaj, however, incorporates sob tone into her rapping and combines it with rapid register breaks. Victoria Malawey associates such breaks and "abrupt register shifts" with various kinds of emotional expressiveness (2020, 54).
20 The "creakiness" of vocal fry uses the lowest vocal register and is produced by short, thick vocal folds, in contrast to sob tone's use of the falsetto register (Malawey 2020, 41, 52).
21 My understanding of twang as used here follows Heidemann (2016, par. 3.16).
22 See Moylan (2020, 183–84) for more on using phonetics and IPA in vocal analysis.
23 Where my analysis of the vocal line concerns precise pitch names, I proceed primarily by ear, as I wish to present an interpretation of the track that reflects how it might realistically be heard. I have assisted my aural analysis with the transcription software Transcribe!, by using its EQ feature to attenuate the track's bass and high end and slowed playback so that I can more easily pick vocal pitches out of the busy instrumental. Beyond this, all judgments about pitch—including flatness or sharpness—are made by ear. Although precise pitch analyses of fundamental pitches in rap are theoretically possible, Ohriner found that technological limitations restricted the feasibility of such analysis to capella vocals (2019, 418). Analysis of pitch in rap music by ear has precedent in Komaniecki (2021), who offers five levels of pitch usage in rap music, ranging from the vocal inflection routinely used for "rhyme strengthening" to "sung/chanted verses."
24 Moylan (2020, 178–85) explores these concepts of pitch, vocal performance, and vocal timbre as central to vocal expression and musical content.
25 This +/– system builds on the up/down arrow system suggested by Moylan (2020, 182) but differs in that it includes intervals of flatness and sharpness greater than the quartertone.
26 A brief review of top YouTube comments from the past year reveals a widespread understanding of the song as an expression of self-conscious mockery of mainstream rap. Examples of such comments include: "The fact that she purposely made this 'bad' and it still made mainstream shows she's the QUEEN OF RAP!!!" (from username "Idk Lol"); "I'm sorry but everybody hated this back in the day but it genuinely aged so well. Like this is an actual classic…I look back on this now and see how iconic and ahead of its time it was. Yes I'll admit, back then I thought it was bad, but I was one of those people who didn't appreciate it for what it really was. It was seriously ahead of its time and it just flew past all our heads. NICKI I AM SO SORRY" (from username "wutheringheights01"); "Someone on twitter said that this song is experimental rap and I couldn't agree more. Nowadays this isn't weird at all but this was so ahead of its time" (from username "antoniojpg"). Although the majority of recent comments are positive, the song still garners negative reactions, for example, "This song gave me a new brain tumor" (from username "ein lein"), and "I hope anyone who likes this song goes to prison for the crime that is this song's existence" (from username "Mxulin").

References

Barthel, Mike. 2012. "Nicki Minaj Masters the Art of the Diss Video with 'Stupid Hoe.'" *The Village Voice Blogs*. January 27. https://www.villagevoice.com/2012/01/27/nicki-minaj-masters-the-art-of-the-diss-video-with-stupid-hoe/

Casey, Christina Schoux. 2018. "New Orleans Bounce Music, Sexuality, and Affect." *Journal of Language and Sexuality* 7(1): 5–29.

Corner, Lewis. 2012a. "Nicki Minaj Breaks Vevo Record with New Music Video 'Stupid Hoe.'" *Digital Spy*. January 25, 2012. https://www.digitalspy.com/music/a361991/nicki-minaj-breaks-vevo-record-with-new-music-video-stupid-hoe/

———. 2012b. "Rihanna Breaks Nicki Minaj's Vevo Record with 'Where Have You Been?'" *Digital Spy*. April 5, 2012. https://www.digitalspy.com/music/a379957/rihanna-breaks-nicki-minajs-vevo-record-with-where-have-you-been/

Cosimini, Seth. 2017. "'I'm a Motherfuckin' Monster!': Play, Perversity, and Performance of Nicki Minaj." *Feminist Formations* 29(2): 47–68.

Cragg, Michael. 2011. "New Music: Nicki Minaj—Stupid Hoe." *The Guardian*. December 21, 2011. https://www.theguardian.com/music/2011/dec/21/new-music-nicki-minaj-stupid-hoe

Daw, Robbie. 2011. "Nicki Minaj Shoots Down A 'Stupid Hoe' with New *Roman Reloaded* Track." *Idolator*. December 20, 2011. https://www.idolator.com/6117302/nicki-minaj-stupid-hoe-roman-zolanski-track?safari=1

Ensign, Jeffrey S. 2015. "Form in Popular Song, 1990–2009." PhD Dissertation, University of North Texas.

Free, Christian. 2016. "Diss-Topia: Beef as Interdiscursive Cultural Policing in American Hiphop." MA Thesis, University of Chicago.

Gilroy, Paul. 1995. "'… to Be Real': The Dissident Forms of Black Expressive Culture." *In Let's Get It On: The Politics of Black Performance*, edited by Catherine Ugwu, 12–33. London: The Institute of Contemporary Art.

Heidemann, Kate. 2016. "A System for Describing Vocal Timbre in Popular Song." *Music Theory Online* 22(1): https://mtosmt.org/issues/mto.16.22.1/mto.16.22.1.heidemann.html

Hunter, Margaret, and Alhelí Cuenca. 2017. "Nicki Minaj and the Changing Politics of Hip-Hop: Real Blackness, Real Bodies, Real Feminism?" *Feminist Formations* 29(2): 26–46.

James, Robin. 2012. "What's Your Take on Roman Zolanski? Misogynist? Genius?" *It's Her Factory*, January 26, 2012. https://www.its-her-factory.com/2012/01/whats-your-take-on-roman-zolanski-misogynist-genius/

Komaniecki, Robert. 2019. "Analyzing the Parameters of Flow in Rap Music." PhD Dissertation, Indiana University.

Komaniecki, Robert. 2021. "Vocal Pitch in Rap Flow." *Intégral* 34: 25–45.

Kulothparan, Hanisha. 2021. "Flow in the Alter Egos of Nicki Minaj." Paper presented at the annual meeting of Music Theory Southeast, online conference, March 20–21.

Malawey, Victoria. 2020. *A Blaze of Light in Every Word: Analyzing the Popular Singing Voice*. New York: Oxford University Press.

McMillan, Uri. 2014. "Nicki-aesthetics: The Camp Performance of Nicki Minaj." *Women and Performance: A Journal of Feminist Theory* 24(1): 79–87.

Miller, Matt. 2012. *Bounce: Rap Music and Local Identity in New Orleans*. Amherst, MA: University of Massachusetts Press.

Moylan, William. 2020. *Recording Analysis: How the Record Shapes the Song*. New York: Routledge.

Ohriner, Mitchell. 2019. "Analysing the Pitch Content of the Rapping Voice." *Journal of New Music Research* 48(5): 413–33.

Osborn, Brad. 2013. "Subverting the Verse-Chorus Paradigm: Terminally Climactic Forums in Recent Rock Music." *Music Theory Spectrum* 35(1): 23–47.

Perez, Sarah. 2021. "YouTube Is Removing the Dislike Count on All Videos Across its Platform." *Tech Crunch*. November 10, 2021. https://techcrunch.com/2021/11/10/youtube-is-removing-the-dislike-count-on-all-videos-across-its-platform/

Perry, Rebecca. 2020. "Between the Signposts: Thematic Interpolation and Structural Defamiliarization in Prokofiev's Sonata Process." *Music Theory Spectrum* 42(2): 193–206.

Roach, Shoniqua. 2019. "Black Respectable Currency: Reading Black Feminism and Sexuality in Contemporary Performance." *The Journal of American Culture* 42(1): 10–20.

Schoppmeier, Sören. 2015. "'Hottentot Barbie' as a Multicultural Star: The Commodification of Race in Nicki Minaj's Music Videos." *Aspeers* 8: 49–70.

Summach, Jay. 2011. "The Structure, Function, and Genesis of the Prechorus." *Music Theory Online* 17(3). https://mtosmt.org/issues/mto.11.17.3/mto.11.17.3.summach.html

Tyree, Tia, and Melvin Williams. 2021. "Black Women Rap Battles: A Textual Analysis of U.S. Rap Diss Songs." *Women and Music: A Journal of Gender and Culture* 25: 64–86.

23

TRACKS AND TRANSFORMATIONS IN THE WAILERS' "CONCRETE JUNGLE" (1973)

Mike Alleyne

This chapter is supported by a webpage.

This chapter examines the sonic elements of "Concrete Jungle," discussing ways in which artistic consciousness and significant aspects of commercialization are simultaneously revealed in the recording. The analysis employs and adapts components of Moylan's approach to "Identifying Sound Sources and Materials" (2020, 159–62). Discussion of this reggae recording incorporates several cultural and commercial contexts inclusive of the song's meaning and historical significance. It aimed to sonically redesign reggae to appeal to rock audiences while maintaining its conceptual themes and performance dynamics, and this assessment considers the accompanying sociopolitical background before scrutinizing the aural details of "Concrete Jungle." By referencing the original studio track sheets and interviews with the overdub and mix engineer, this chapter addresses how the recording industry's resistance necessitated new sonic strategies in presenting the marginalized genre to the mainstream.

Contexts for *Catch a Fire* and "Concrete Jungle"

Jamaica had attained its political independence from Britain in August 1962, infusing the nation with fresh creative motivation and energy, but initial optimism was eventually tempered by harsh socioeconomic realities that ultimately characterized much of the emerging music (Alleyne 2012, 236). In 1973 when Island Records released The Wailers' international debut album *Catch a Fire*,[1] the group, featuring Bob Marley, Peter Tosh, and Bunny Wailer, had already spent a decade of dislocation both within and beyond Jamaica. Journeying through Jamaican music's creative transitions from ska to rock steady to reggae in the 1960s, they fought to be taken seriously by developing their songwriting and performance skills while they struggled to receive fair compensation in the

DOI: 10.4324/9781003089926-31

predatory recording industry environment of Kingston (Blackwell and Morley 2022, 157, 161–62; McCann 1993, 68; Masouri 2015, 147). The latter half of the 1960s intersected with the group's resentment of the country's oppressive neocolonial sociopolitical establishment, dissatisfaction with the restrictive creative conditions under which they toiled, and their embrace of the Afrocentric Rastafarian philosophy and spiritual practice. As a result of their outsider status, Marley, Tosh, and Wailer all had negative encounters with Jamaican law enforcement and prison to varying degrees between 1967 and 1968 (White 1992, 221), with other alarming incidents occurring later on (Masouri 2013, 226–31), though such experiences only toughened their musical resolve and sound.

By the end of the decade, the band appeared to be making little commercial progress despite advancing its overall creative chemistry and musical cohesion and scoring some hit singles in Jamaica. Producer Lee Perry's unauthorized licensing of Jamaican Wailers' recordings to British-based Trojan Records (eventually issued on the albums *Soul Rebels* in 1970, *Soul Revolution* in 1971, and *African Herbsman* in 1974) deepened their collective disillusionment with the music business (Alleyne 1999, 94; Alleyne 2012, 164, 206; Beaumont-Thomas 2021).

By 1972, Marley had recorded a Johnny Nash-produced solo single for CBS in Britain, "Reggae On Broadway."[2] Despite being an American singer with a fairly conservative recording history, Nash had become familiar with Jamaica's early reggae scene first-hand, scoring a major international pop hit with "Hold Me Tight" in 1968 (recorded in Jamaica), then repeating the feat by successfully covering the Marley song "Stir It Up" in 1972 (Alleyne 2012, 193).[3] Although the rock-inflected "Reggae On Broadway" single failed commercially (not being widely heard until it was reissued as the opening track on the posthumous *Chances Are* album in 1981), in retrospect it laid part of the reggae/rock crossover foundation for the Wailers' next phase. Island founder Chris Blackwell signed the group a week after internationally successful Jamaican singer-songwriter Jimmy Cliff left the label in the wake of his iconic movie debut in *The Harder They Come* (1972) and the release of its accompanying soundtrack album (Blackwell and Morley 2022, 160).[4] Sensing that further breakthroughs could be achieved beyond Cliff's mainstream transatlantic hit singles,[5] Blackwell was motivated to mold The Wailers into what he described as a black rock group to satisfy the label's larger commercial objectives (Blackwell and Morley 2022, 161; Bradley 2000, 413–15; Katz 2014, 54; Katz 2010, 62). As the *Catch a Fire* overdub recording and mix engineer Tony Platt recalls,

> I was one of the house engineers at Island Studios at Basing Street in London, and I got called into the studio office by the studio booker, and she said Chris (Blackwell) wants you to work with this Jamaican band. Some of them are coming over next week and I've put you down for these sessions. That was it basically, and then I had a conversation with Chris and he explained to me that what he wanted to do was to … bring in a couple

> of guys that we'd been using on various rock sessions, Wayne Perkins being one of those.[6]
>
> I, as one of the house engineers that happened to have worked on rock and reggae, was assigned to do those sessions. Probably the only sort of creative variable … was putting that combination of people together. The rest of it was whatever happened as we did it.
>
> The whole idea of the exercise was to bring this white rock music vibe into the reggae context. It was basically to pick up FM college airplay in the States. That was the big market everybody was aiming at, at that particular point in time, so that was the idea.[7]

The group's members were themselves equally hungry for wider success with the prospect of consistent financial backing, which had been a key component missing from their career development. Academic critic Paul Gilroy points to reggae's innate hybridity and how the genre "derived a different kind of cultural legitimacy" in Britain (1993, 82), where *Catch a Fire* arguably had its greatest initial impact on collective Black consciousness. Nonetheless, by the early 1970s the music's eclectic roots had become sonically congealed into a distinctive admixture from which significant commercial departures were readily recognizable. Blackwell's direct involvement with The Wailers and the engagement of Island's creative and business machineries also magnified racial reverberations surrounding financial control of the group's music and the paradoxes of harnessing an exploitative recording industry system to awaken audiences to the capitalist injustices articulated through the tracks.

Fundamentally, "Concrete Jungle" embodies a blend of contrasting cultural and recording spaces in both Kingston and London, and the recorded results from both cities demonstrate how resonances of place and lived experience can become integral components within the fabric of the finished work. "Concrete Jungle" provided an audible interconnection between exterior social spaces directly impacting upon the interior psychological landscapes informing tonality, color, structure, and dynamics in the recorded work, in conjunction with the spatial and human environments of the recording space. The "Concrete Jungle" friction and fusion between the divergent genres of reggae and rock was certainly not the first time that earthy Jamaican-recorded rhythm tracks interacted or collided with metropolitan production polish. In many earlier instances in the late 1960s, particularly on singles from the Trojan Records label (whose parent company, B&C, involved Blackwell[8]), the uneasy superimposition of orchestral strings to facilitate UK airplay on the BBC's radio airwaves was sonically and stylistically disconnected from the music's essence (Alleyne 2009, 254, 256). Moreover, the recordings comprising *Catch a Fire* targeted an audience conditioned to accept the credibility of artistry through the creation of integrated concept-oriented albums.[9] According to band member Bunny Wailer (d. 2021) a contributing distinctive factor was the decrepit condition of the instruments they used for rehearsal and recording, infusing the *Catch a Fire* album with a unique sound, while the adaptability of the Barrett brothers (Aston on bass and Carlton on drums) enabled them to creatively extract tones and timbres that might not otherwise have been available (Steffens 2017, 151–52).

By the early 1970s, reggae remained strongly associated with the singles market and was not seriously viewed within the mainstream industry as an album-oriented genre with appropriate musical credibility or potential market longevity (Alleyne 2009, 256). The *Catch a Fire* album was the first major step toward correcting this misperception, and "Concrete Jungle" was its opening track, written by Bob Marley.[10] Apart from setting all of the group's principal members firmly on their future respective solo paths to achieving differing degrees of international stardom, the album for which The Wailers received a £4,000 pound advance from Island (Blackwell and Morley 2022, 161) incorporates an emerging global postcolonial consciousness, and efforts to assert such a presence within the commercial industry sphere. The unison vocal that begins verse 2, "No chains around my feet/But I'm not free/I know I am bound here in captivity" speaks directly to ontological and collective consequences of residual imperialism directly impacting its victims. When interviewed during an American tour in support of the *Catch a Fire* album, Marley commented on reggae's relationship to Jamaica, remarking that "It's our blues" (Robinson 1973, 25). In so doing, he provided immediate context for the lyrical themes and the music's timbral substance shaped by harsh historical forces.

It is worth noting that despite poor initial sales amounting to only 14,000 copies in its first year (Blackwell and Morley 2022, 166; White 1991, R-17; Alleyne 2009, 256; Katz 2014, 54), the album eventually became a global million-seller and a significant influence on roots reggae artists and fans. The album's sleeper impact in America is partially evidenced by its *Billboard* 200 chart debut in 1975, two years after its release (and after the two studio albums that followed it), while in Britain it failed to chart at all (Whitburn 2018, 762; Roach 2009, 186). Now somewhat confusingly credited as a catalog release to Bob Marley & The Wailers (an entity not known in that form on Island until the release of the 1974 *Natty Dread* album),[11] the *Catch a Fire* 2001 Deluxe Edition utilizes the original Zippo lighter album cover identifying the performing act as The Wailers.[12] In addition, the Deluxe Edition crucially includes a separate disc featuring the previously unreleased Jamaican 8-track recordings that formed the album's basis prior to transformative overdubs taking place at Island's Basing Street studios in London, England (*Catch a Fire*, Deluxe Edition, 2001). According to engineer Sylvan Morris, the cohesion of those Jamaican recording session performances owed much to the extensive pre-production rehearsals that The Wailers had undertaken (*Classic Albums*), creating an intrinsically solid foundation upon which the subsequent enhancements were built.

Although this essay's principal concern is with the finished internationally known version of "Concrete Jungle," access to the record's basic pre-overdubbed version illuminates the critical aesthetic alterations that contributed to its final manifestation.[13] Moreover, the song had been previously recorded by The Wailers circa 1971 and released as a single with a markedly slower, more soulful rhythmic accentuation for Lee Perry's Upsetters label (while also released on an undated disc on Marley's Tuff Gong label, thought to have been issued circa 1972).[14] That version included horns but existed without any of the sonic accoutrements or the electric guitar solo space adorning the more quickly paced finished version issued by Island two years later. The spatialities and sinuous rhythms embedded within Perry's production prefigure his imminent pioneering sonic adventures in dub through which he and others such as King Tubby (real name: Osbourne Ruddock—d. 1989) comprehensively reframed sonic texts and influenced mainstream pop in the process.[15]

The sonic remolding of the constituent songs on *Catch a Fire* to appeal to rock audiences has been the subject of polarizing debate in the decades since the album's release. Some observers hail the transformation of reggae from the category of novelty singles to fully fledged albums by a self-contained group of unique identity rather than a loose compilation of songs featuring various vocalists. Alternatively, others have serious reservations about the album's commercial concessions in order to breach the hitherto resistant rock market, noting, for example, the deliberate reduction of typically characteristic reggae bass presence in favor of treble orientation and the inclusion of session musicians well-versed in rock, but having little or no prior exposure to reggae (Alleyne 1999, 95–96).[16] In his 2022 memoir, Blackwell argues that *Catch a Fire* "didn't seem a weakening or a pasteurization to market the Wailers as you would a rock act. There was no sellout" (Blackwell and Morley 2022, 166). However, cultural and aesthetic perspectives play critical roles in determining sonic perception, and the distinctly contrasting character of the record's aural identity within the reggae canon certainly demonstrates a conscious shift toward more commercially viable formulations.

The absence of credits in 1973 for additional American session guitarist Wayne Perkins from Alabama and Texan keyboardist John "Rabbit" Bundrick was probably strategic to focus critical attention on The Wailers.[17] Blackwell suggests that it was long incorrectly assumed that Peter Tosh had played the electric lead instead of Perkins or any non-Wailer (Blackwell and Morley 2022, 165), although as I have repeatedly argued (Alleyne 1999, 2009, 2012, 2018, 2020), the performative style and phrasing (of both Perkins' lead and melodic accompaniment) is immediately anomalous given the group's existing musical vocabulary.[18] The lead guitar on "Concrete Jungle" projected the contents of experiences from other musical and cultural spaces, and its tonal and emotional edges were filtered through a distinctly different aesthetic lens from that which informed The Wailers' expression on the recording. Notably, Perkins and Bundrick remained uncredited until the release of the 2001 *Catch a Fire* Deluxe Edition. These circumstances inevitably raise questions regarding authenticity and "production strategies focused on re-presenting the music" embodying "a recontextualization of reggae for the musical mainstream …." (Alleyne 2020, 26). Such concerns are intrinsic to analysis of the recording, although former Island engineer Tony Platt is adamant that "Bob (Marley) was part of the process" and was actively engaged in reshaping the material for mass market audiences (Alleyne 2020, 26).[19] Platt also underlines the unpremeditated, intuitive and instinctive approach to recording and production that characterized the early 1970s at Island and its effect on the Wailers' project:

> The *Catch a Fire* album sounded completely different from any other reggae record before. It still had the soul, but it didn't have quite the same extent of raw edges going on with it. And that's something that came out of the music. What we did was we left ourselves completely open to the music and what the music was doing. We let the music set the pace and we let the music set the direction. All that was left for us was to just kind of keep it in shape. It really was that simple and kind of organic.[20]

Notably, the printed label on the British album only specifically mentions the Jamaican recordings studios used for the project, while the American label states "Recorded in Jamaica." The exclusion of reference to any London recording sessions on either version and the absence of credit for the session musicians might be seen as authenticity strategies designed to firmly link the music to its cultural source, while conversely the sound of the LP yielded clear evidence of production intervention that steered the material in more commercial directions relative to the context of reggae in that era.

Sonic Staging and Political Perspectives

"Concrete Jungle" bore the informal name of a government housing project in Kingston's beleaguered Trench Town neighborhood occupied by The Wailers,[21] a social confinement from which lasting escape seemed improbable. The record's grim mood in which the protagonist displays both despair and determination is certainly affected and accented by the sonic hybridity of the Jamaican tracks with their low bass and midrange frequency emphasis against the brighter, more treble-oriented tone of the added electric guitar and other London overdubs, including keyboards.[22] Intentionally or otherwise, the soundscape reflects a sense of emotional polarization within the recording's sonic paradoxes.

As the opening song on *Catch a Fire*, literally introducing The Wailers to international audiences, "Concrete Jungle" was one of the songs featuring the greatest number of overdubs on the album, prominently adding electric lead guitar on 16-tracks in London to the 4-track and 8-track tapes that had been recorded in Kingston.[23] Released as a single by the summer of 1973, the record became an unofficial herald of resistance from a once largely muted British colonial empire (at least in its musical recordings, though far less so in its literature), striking back against imperialism not only with sonic aggression and intensity but also with pointed lyrical critique highlighting the historical situations that had brought sufferers from the developing world to a critical sociopolitical convergence. Ironically, the record's very existence negotiated a pathway through the capitalist commercial jungle of the recording industry to allow the postcolonial marginalized voices and sounds of The Wailers to be heard, replete with audible resonances of the mediating marketplace filters.

Tracking the Changes

Recalling his first listen to the raw tracks of "Concrete Jungle" in Kingston in 1972, Island Records founder Chris Blackwell emphasized that

> [I]t was just so far ahead of anything that had ever been made in Jamaica before – that one particular track, the structure of it, the whole thing. It was just unbelievable.
>
> *White 1991, R-16*

While that original recording contains an innate power, he still supplements the material with rock instrumental trimmings to gain the desired commercial traction in America. Despite Blackwell's ultimate financial vindication of his textual adjustments to The Wailers' Island debut, he nonetheless recognizes that the album's production involved forms of stylistic intervention:

> I consider the first record, *Catch a Fire*, to be the least reggae-like because it had overdubs from a rock guitarist I was working with named Wayne Perkins, and I edited the backing tracks in order to double the length of the songs and create space for solos, which reggae never had much of. These ideas were unique for reggae ....
>
> *White 1991, R-17*

The creation of sonic space for the guitar solo on "Concrete Jungle"—outside of the genre's norms—clearly signals an impulse to transform the reggae soundscape. The absence of any lead solo space in the Deluxe Edition's original Jamaican version speaks to Island's commercial crossover imperatives and the need to exert the rock presence (inclusive of guitar feedback) on the album's opening song to help configure its market identity. Platt recalled that the song "originally came from an 8-track tape from Jamaica and I copied it up to 16-track. I copied a chunk and just edited it with a razor blade."[24] Alleged in one account to have been captured on the third take (Salewicz 2009, 215), the solo's lasting sonic imprint also owes much to Platt's overall mix sensibility and addition of outboard effects, making it an aural spectacle.

Figure 23.1 illustrates the "Concrete Jungle" studio track sheet from the London overdub recording sessions at Island.[25] Despite the access to 16-track recording at Island's London studios, the soundscape of "Concrete Jungle" seems deceptively sparse, with Perkins' electric guitar overdubs accentuating an aural contrast to the deeply rooted centrality of the bass and drums. Platt recalled that the tapes of such songs from Jamaican studios frequently contained multiple

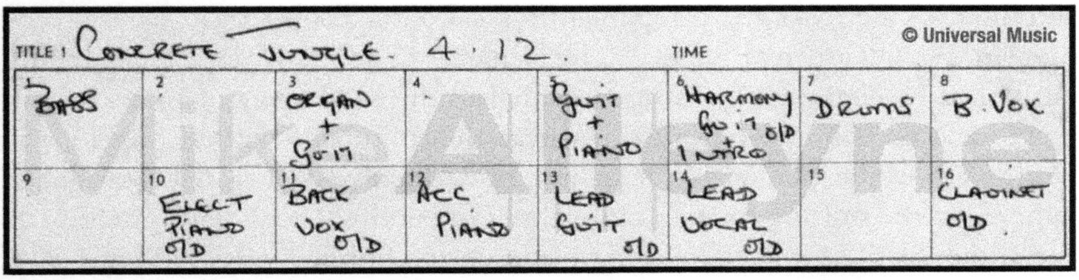
TITLE: CONCRETE JUNGLE. 4·12. TIME © Universal Music

1 BASS	2	3 ORGAN + GUIT	4	5 GUIT + PIANO	6 HARMONY GUIT O/D INTRO	7 DRUMS	8 B.VOX
9	10 ELECT PIANO O/D	11 BACK VOX O/D	12 ACC PIANO	13 LEAD GUIT O/D	14 LEAD VOCAL O/D	15	16 CLARINET O/D

FIGURE 23.1 The "Concrete Jungle" track sheet used at Island Studios, Basing Street, London, in 1972, indicating the recording's 16-track tape vocal and instrumental assignments. © Universal Music. All Rights Reserved. Reproduced with permission.

instruments mixed together on one track (Crane and Knoop 2012). Examination of the studio session track sheet for "Concrete Jungle" reveals that three tracks contained more than one instrument, while four of the sixteen tracks (2, 4, 9, and 15) were completely unused (Figure 23.1). In contrast, "Stir It Up," also included on the *Catch a Fire* album, utilized all sixteen tracks after overdubs were applied.

The potential technical drawback resulting from recording multiple simultaneously active instruments together on a single track is the inability to separate them later when mixing. All of the drums were dedicated to track 7 with the kick, snare, side stick, toms, and hi-hat all competing for space, while the bass guitar occupied track 1. In this instance (and indeed many others), the Jamaican studios capitalized on microphone leakage to create forms of cohesion fusing the otherwise separate aural spaces, accentuating the record's density. In so doing, the recording techniques applied to the basic tracks of "Concrete Jungle" mirror the protagonist's psychological internment articulated in the lyrics of urban annihilation. As the Island studio track sheet indicates (see Figure 23.1, track 14), Marley recorded a new lead vocal overdub in London,[26] and when he launches into the pre-chorus section singing "Darkness has covered my light/And has changed my day into night," the protagonist's acute existential anguish is achingly tangible in the final recording, with a deeper expressive edge than occurs on either the Lee Perry-produced version or the original Jamaican recording for the album. The emotional poignancy is powerfully accentuated here by the staccato monosyllabic harmony background vocals of Tosh and Wailer punctuated by mournful sustained notes at the end of the section. At such stages, the record communicates The Wailers' angst as a projection of lived reality through song and sound, embodying much more than a typical pop performative façade that artistically adopts narrative strategies to convey its concepts.

Mapping the Sonic Jungle

Exploration of the recording's sound stage facilitates navigation of its contents and understanding of how its aural spaces are utilized. With the exception of the exploratory overdubbed lead and melodic guitar parts, the overall sparseness and interlocking rhythmic components typical of the roots reggae era are readily evident in "Concrete Jungle." The elements of its sound stage are carefully positioned by engineer Platt to perhaps reflect the aural energy and movement of a rock mix. While the original Jamaican mix is imbued with its own kinetic fluidity, all of the elements in it remain in decidedly fixed spatial locations as opposed to the shifting stereo positionalities of Perkins' overdubbed guitar parts and Bundrick's clavinet and electric piano on the finished version. Figure 23.2 illustrates the perceived sound stage of the track.

The placement of the guitar solo initially center left of the sound stage (2:49–2:51) and then firmly right-panned for its remaining duration firmly draws overt attention to its presence, exploiting more of the record's spatial potential. The solo was in fact constructed from two separate takes and tracks, thereby facilitating and simplifying the stereo sound stage movement in an era before console automation.

In other writings, I've noted that "The opening bars of 'Concrete Jungle' … yield immediate evidence of intent to assimilate portions of the rock audience" (Alleyne 2018, 194) in an utterly deliberate manner, as Platt confirms. The record's first few seconds feature two guitars (one of which is from the Jamaican sessions) and an organ that collectively sound as though they are emerging from a rehearsal interaction or an unguarded contemplative moment, but this illusion is a resynchronization of studio happenstance. Perkins' opening guitar lines on the recording initially occurred earlier as he waited for the multitrack tape playback to begin. This necessitated manual editing to place them in appropriate context with the introduction. Platt noted that

> You'd get these wonderful things that happen. I always dropped into record right at the front of the track, right at the very, very beginning because you never knew what you were gonna get. And so you had Wayne (Perkins) … trying to find something out, and that stayed on the record. You have that kind of noise and these picky bits.[27]
>
> We had this twiddly bit on the front and Chris said 'That's great, we need that but it's too far in front.' So we put it on to another piece of tape and I recorded it back in and moved it. No Pro Tools. You couldn't just kind of drop it in. I had to physically move it on the tape. So it actually wasn't where it sounds. I moved it closer in ….[28]

Blackwell has suggested that Perkins' lack of exposure to reggae essentially meant that the recording captured "the sound of someone learning about the music he was inside of at the exact moment he understood it" and that the aural capture of this process in action made the music more compelling despite the guitarist's initial inability to establish a rhythm location within the track (Blackwell and Morley 2022, 164–65).

The introduction of the kick drum and hi-hat provide a rhythmic framework coupled with Perkins' exploratory overdubbed blues-rock guitar phrasing.[29] This

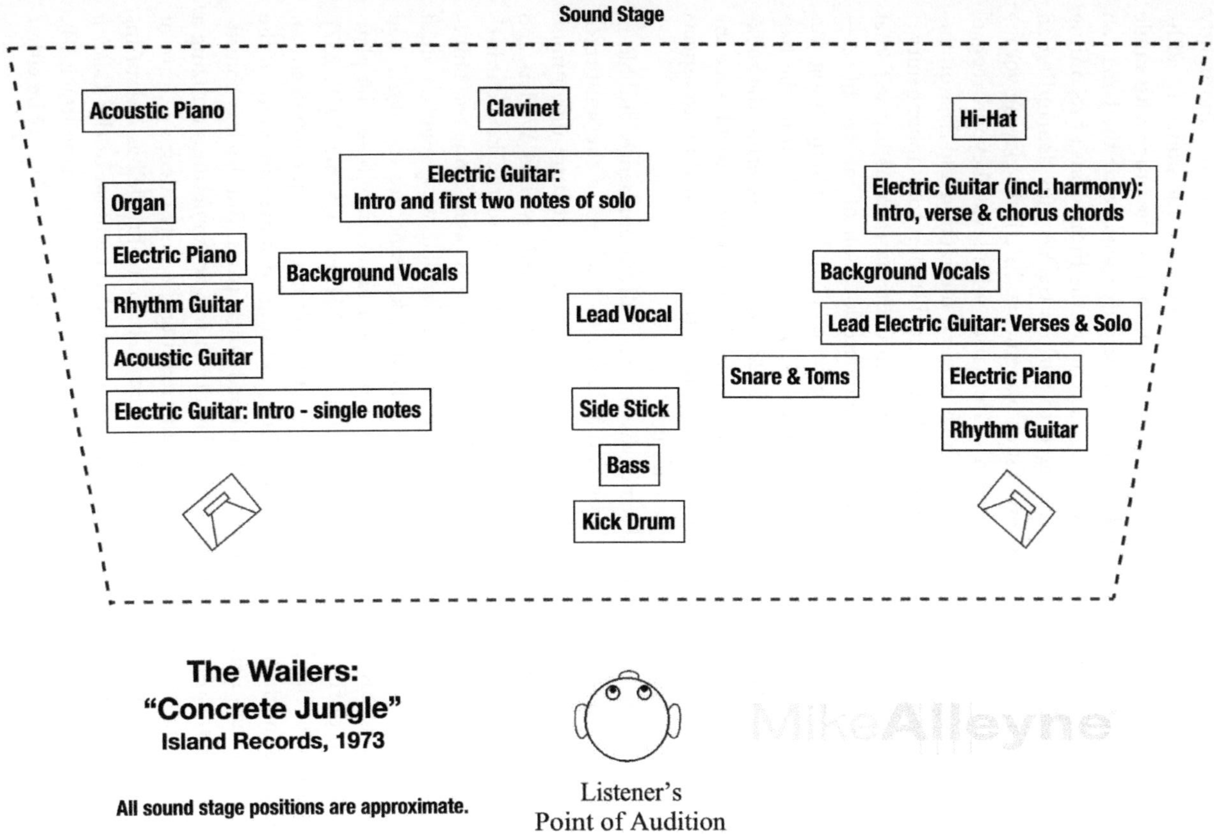

FIGURE 23.2 The stereo sound stage of "Concrete Jungle."

announcement of lead rock guitar presence barely eleven seconds into the record is especially important as a sonic signifier of both "Concrete Jungle" and the *Catch a Fire* album as new chapters in reggae history, a genre which had previously primarily presented the electric guitar in percussive rhythmic and lightly melodic roles without the same degree of pronounced tonal incision.[30] In addition, Perkins plays two overdubbed parts (one with chords, the other with single line phrases) accompanying the rhythm guitar (already included on the Jamaican session tapes) that emerges as verse 1 begins. Due to the limitations of recording on 16 tracks, guitar parts were relocated to various tracks, thus demanding an acute sound engineering awareness of such placements.[31] Perkins' playing vocabulary and tonalities certainly reveal his familiarity with styles beyond reggae, creating a peculiar sonic anomaly in which cultural interaction and friction fluidly coexist, though not without potentially alerting listeners to its incongruity.[32] It has been well documented, including in comments by Perkins himself, that he had problems finding "the one"—the standard location of the downbeat—on The Wailers' tracks including "Concrete Jungle," and that remedying this situation even involved an instruction from Blackwell not to listen to the bass (Alleyne 2009, 257; Blackwell and Morley 2022, 164).[33] Considering the inextricably central roles of bass in reggae, this suggestion—intended to advance the session progress—also highlights a creative dislocation that translates to aspects of "Concrete Jungle" and the wider album.

John "Rabbit" Bundrick's clavinet that is especially prominently audible in the "Concrete Jungle" introduction and assigned to track 16 of the multitrack tape is also a similarly atypical feature, introducing a sonic texture uncommon to reggae of that era but one that was readily embraced by The Wailers and very much a part of contemporary rock. Blackwell refers to the keyboardist's contribution as providing "more technological brightness" and "some few familiar points of reference" for rock and soul fans unfamiliar with reggae's aural terrain (Blackwell and Morley 2022, 166). Bundrick's exposure to reggae had been limited to his stint in Johnny Nash's band and rudimentary in-studio playing direction during the *Catch a Fire* sessions from Marley (Alleyne 2009, 257). Apart from the clarity of its first appearance, the clavinet subsequently forms an integral part of the record's rhythmic fabric, existing far more discretely in the overall mix than Perkins' multiple overdubbed guitar parts. Bundrick is considered to have introduced the use of the clavinet in reggae, sometimes utilizing a wah-wah pedal to enhance its presence and making the instrument a feature of reggae thereafter.[34] Interestingly, he has cited "Concrete Jungle" as one of his favorite Marley songs, emphasizing the experimental dimension of the *Catch a Fire* sessions (Bundrick, 2021). He also plays overdubbed electric piano (track 10), which is so subtly integrated into the recording's mix that it could easily be either missed or misidentified.

The introduction certainly sets the stage for the protagonist's philosophical contemplation of oppressive social and spiritual conditions, soon expressed through Marley's lead vocal and the urgent background vocals. The opening line, "No sun will shine in my day today," posits absence as the record's governing lyrical and psychological force, representing the spiritual darkness that looms over the song from start to finish. Perkins' telling five note blues-rock guitar phrase following that vocal line actively shifts the context of the urban desperation from local to global, although it remains somewhat out of context with the overall soundscape. The guitar then remains out of the mix until the first chorus, during which Perkins plays harmonic dual guitar parts (reminiscent of British rock group Wishbone Ash) woven around the lead and background vocals, an approach repeated in later choruses. His contributions in verse 2 feature interspersed bluesy phrases emphasizing the record's rock undercurrent that metamorphoses into central focus in the fiery solo.

Previously, I referenced the general centrality of bass on *Catch a Fire*, and "Concrete Jungle" provides a primary example. Session player Robbie Shakespeare (d. 2021), later of the internationally famous Sly & Robbie bass and drum rhythm section, was the actual bassist on "Concrete Jungle" rather than the established Wailers' player Aston "Family Man" Barrett who was Shakespeare's mentor. Barrett had also played on the previous Lee Perry production of "Concrete Jungle" as part of the Upsetters studio band (Blackwell and Morley 2022, 157). Although the bass presence on the original Jamaican recordings is audibly stronger, Shakespeare's powerfully kinetic performance is more than mere accompaniment and injects emphatic punctuation and rhythmic momentum, while his selective phrasing creates additional spaces for vocal engagement, inviting consideration of other aspects of the instrumental texture.[35] Even in mediated form, the bass depth on "Concrete Jungle" is more pronounced than, for example, on Paul Simon's debut transatlantic hit solo single "Mother and Child Reunion," released a year earlier in 1972 with its reggae rhythm tracks also recorded in Kingston.

Considering the bleakness of the ontological subject matter in "Concrete Jungle" and the starkness of the guitar solo and accompanying interjections, the record remarkably does not include any punctuating cymbal crashes to accentuate or magnify the protagonist's torment. Transitions are instead subtly signaled by a highly pitched snare and toms, actually sounding much closer to timbales.

Temporal Echoes

The dramatic ignition of lead electric guitar on the Island recording of "Concrete Jungle" is made all the more peculiar by the stark absence of anyone in Wayne Perkins' audibly present role in The Wailers' iconic British television debut appearance on BBC's *Old Grey Whistle Test* music show in 1973. The pre-recorded backing track inclusive of selected accompanying guitar phrases is played back under live vocals. Perhaps in recognition of the audiovisual contradiction, Perkins' solo is completely replaced by a clavinet competently played by Earl "Wire" Lindo (d. 2017), but the keyboard is unable to project the same degree of emotional urgency and sonic penetration. Since the guitar's inclusion on the record was intended as a key selling point for rock audiences who were the show's primary viewers, the semi-live performance BBC appearance represents a disconnection between the record, The Wailers, and their crossover audience targeted by "Concrete Jungle." It is intriguing that the televisual presence of a non-Wailer in a lead guitar role might have been perceived as an intrusion on the group's identity, while an even greater aural manifestation of the guitar work on the record was apparently not.

The song is also notably absent from the previously unreleased 2021 archival album and DVD, *The Capitol Session '73*, featuring live recordings in Los Angeles at the famed Capitol Records tower. Occurring very shortly after the release of The Wailers' second Island LP *Burnin'* (issued in mid-October), those recordings (minus Bunny Wailer, replaced by the group's vocal mentor Joe Higgs) constituting *The Capitol Session '73* still contain five other songs from the *Catch a Fire* album, but the exclusion of "Concrete Jungle" further suggests that key aspects of its distinctive sonic imprint including Wayne Perkins' guitar work were not readily reproducible live with the band's existing line-up. This later became a less challenging live situation when American blues-rock guitarists such as Donald Kinsey and Al Anderson performed with Bob Marley and The Wailers and also appeared on the group's studio albums.[36]

"Concrete Jungle" remains a crucial turning point for reggae despite some critical efforts to diminish its historical significance,[37] expanding its soundscape possibilities as the genre claimed more commercial and artistic space during a decade of remarkable stylistic innovation and production variety. The aesthetic validity of the sonic transformations on its tracks will remain a subject of continual debate, but the record irrevocably signals the emergence of album-oriented Caribbean music on its journey toward becoming an integral part of the global pop mainstream.

Notes

1 Although the given release date in the Deluxe Edition for the original LP is April 13, 1973, a January 27 Island Records advertisement aimed at retailers in *Billboard* magazine clearly implies that the product was already prepared for distribution and stocking. In addition, liner notes in the Trojan Records release *Trenchtown Rock: The Anthology* (2002) cite a UK release date of December 1972.

2 To facilitate the execution of the Island recording contract, Chris Blackwell had to extricate Marley from his CBS obligations by making financial concessions that granted singer Johnny Nash's business partner Danny Sims a two-percent override on the first six Wailers albums and a new songwriter's contract (Davis 1990, 105). Such circumstances may have generated greater commercial urgency on Island.

3 The UK release of Nash's single achieved hit status in 1972, but it only charted in America in 1973.

4 Blackwell and Morley (2022, 160) note that the continual "shooting, editing, and reshooting" carried out by the film's director Perry Henzell meant that wide American release was delayed until 1975, effectively stalling Jimmy Cliff's career momentum and hastening his departure from Island.

5 Cliff's "Wonderful World, Beautiful People" reached number 25 on the Hot 100 in America, debuting in December 1969. That record's overdubbed orchestral string arrangements were notably antithetical to The Wailers' creative approach and direction.
6 Tony Platt, interview by author, Cardiff, November 29, 2009.
7 Tony Platt, interview by author, August 9, 2021.
8 According to De Koningh and Cane-Honeysett (2003, 36, 47–48), Trojan Records Ltd. was formed in 1968 after Blackwell sold his interest in various subsidiary labels (often featuring Jamaican music) to concentrate on the rock market with Island Records. Blackwell recalls forming the enterprise in that year with his accountant/entrepreneur colleague Lee Gopthal (Blackwell and Morley 2022, 85) to whom he later sold his interest in the label.
9 Hoskyns (1995) includes comments from author Vivien Goldman who once worked for Island as a press officer, affirming that *Catch a Fire* "altered the landscape of reggae completely and irrevocably" in an era dominated by the singer/songwriter and the album format.
10 Maxine Stowe, the Jamaican attorney-at-law and manager of the late Bunny Wailer, has challenged Chris Blackwell's portrayals of his business and creative relationships with The Wailers and the fairness of his dealings with the group. She claims that much of the intellectual property assigned by Blackwell to Marley's authorship was instead the result of group interaction and that the perception of the Island founder's favoritism toward Marley was real rather than imagined (Gardner 2022).
11 Blackwell recalls that the band was previously known as "Bob Marley and The Wailers" during their years with Lee Perry's Upsetter Records in Jamaica and that he changed their identity to "The Wailers" for *Catch a Fire* before reverting to featuring Marley for the 1974 *Natty Dread* LP (White 1991, R-17) after the departures of Bunny Wailer and Peter Tosh.
12 Production on both the original LP release and the Deluxe Edition is credited to "Bob Marley and Chris Blackwell," possibly contributing to the perceptions of Peter Tosh and Bunny Wailer that Blackwell saw Marley as the band's central figure. Following various manufacturing issues with the original Zippo lighter cover, it was replaced by the less imaginative but more obviously rebellious tightly cropped photo of Marley smoking a "spliff." The original cover was congruent with the sonic adjustments made in attempting to reach the rock mainstream.
13 The version of "Concrete Jungle" on the first disc of the Deluxe Edition compiling the unreleased Jamaican versions does not appear to include the same lead vocal take, but it is nonetheless representative of The Wailers' concept of the record prior to overdubs.
14 The Lee Perry-produced recording appears on *Trenchtown Rock: The Anthology 1969–78* (Trojan Records); however, that release cites a copyright date of 1973. A pre-Island version of "Concrete Jungle" also appears on the JAD Records 1998 release, *The Complete Bob Marley & The Wailers 1967 To 1972, Part III.*
15 Mike Alleyne, 2012. *The Encyclopedia of Reggae: The Golden Age of Roots Reggae.* New York, Sterling: 73–75.
16 In *Jean-Michel Basquiat: Crossroads*, author and reggae insider Lee Jaffe (179) quotes lyrics from the song and recalls that the late visual artist queried whether Marley was present when the (lead) guitar was recorded or when the album was being mixed, and whether Bob thought he was selling out. While this chapter answers some of those questions in factual terms, Jaffe's account does not reveal whether Basquiat resolved his apparent creative misgivings.
17 In a series of interview comments included in Barney Hoskyns' (1995) article "Burnin': Bob Marley and the Wailers take Britain" (see References), Bundrick, described by the author as being "somewhat piqued," noted that Chris Blackwell had never mentioned the credit omission up to that time and suggested that the inclusion

of such information highlighting the participation of white rock musicians might have been contradictory to Island's marketing objectives. Hoskyns also commented on the peculiarity of "a reminiscence by Bundrick of the Stockholm sojourn" (during one of Marley's pre-fame transitory phases in Johnny Nash's entourage) being included in the 1992 *Songs of Freedom* compilation in which, once again, his role as an additional musician on *Catch a Fire* remained unmentioned.

18 Similarly, Eric Clapton's guitar performance on the 2005 posthumous Bob Marley single "Slogans" is readily identifiable as a stylistic overlay featuring a blues rock articulation (Alleyne 2018, 195).

19 Blackwell and Morley (2022, 161) indicate that Marley "was immediately intrigued" by the idea of The Wailers adopting the guise of a Black rock act, whereas Peter Tosh and Bunny Wailer were less enthusiastic. In Phill Brown's recording engineer memoir *Are We Still Rolling?* he recalls his work on The Wailers' second Island album (on which Tony Platt was also a major engineering participant), *Burnin'* (1973), noting that after mixing they were "speeding up some of the songs to make them more acceptable to a British audience" (2010, 94).

20 Tony Platt, interview by author, August 9, 2021.

21 Liner notes to *Songs of Freedom*, performed by Bob Marley, Island Records TGCBX 1 (512 280-2), 1992, CD.

22 Engineer Platt was highly complementary regarding the quality of the Kingston recordings, though certainly on "Concrete Jungle," the kick drum and bass appear to have more density than distinct clarity.

23 Recording engineer Tony Platt specifically mentions that some songs for *Catch a Fire* were only originally recorded in Jamaica on 4-track format while others appeared in 8-track form. This may be due to the use of three different recording studios in Kingston: Dynamic Sound Studios, Harry J Studios, and Randy's Studios.

24 Tony Platt, interview by author, August 9, 2021.

25 Many thanks are owed to Tony Platt, Paul Alger, and Dan Abela at Universal Music for invaluable assistance in making the track sheet available for reproduction in this chapter.

26 In Blackwell and Morley (2022, 163), Chris Blackwell notes that Tosh and Wailer "were not so keen on coming" and that Marley accompanied him and the tapes on the journey to Island's studios in London for mixing.

27 Tony Platt, interview by author, Cardiff, November 29, 2009.

28 Tony Platt, interview by author, August 9, 2021.

29 Perkins was at Island's London studios recording the second Smith, Perkins & Smith album for the label. Hailing from Birmingham, Alabama, he made his name as a session guitarist at the famous Muscle Shoals Studio, but he was totally unfamiliar with reggae. Chris Blackwell employed him to draw on his existing musical vocabulary to add rock touches to *Catch a Fire* (White 234). Perkins had previously backed Island artist Jimmy Cliff on recording sessions at Muscle Shoals Studios in Alabama (Blackwell and Morley 2022, 163), but the album emerging from those recordings, *Another Cycle*, was focused on R&B rather than reggae (Alleyne 2012, 46).

30 The ubiquitous session and featured artist performances of guitarists Ernest Ranglin and Lynn Taitt typify the instrument's roles in Jamaican popular music in the 1960s. See Alleyne (Forthcoming).

31 Tony Platt, interview by author, August 9, 2021.

32 It was this very anomaly on "Concrete Jungle" that eventually catalyzed my doctoral thesis.

33 Perkins mentions Blackwell's bass avoidance advisory in the *Classic Albums DVD* documentary and Platt confirms that Perkins had challenges finding "the one" in the songs. Masouri's contradictory account of the recording sessions (2008, 118) conversely suggests that "Blackwell advised him to be guided by the bass, rather than the melodies."

34 Tony Platt, interview by author, August 9, 2021. Also, see Blackwell and Morley (2022, 165).

35 The effect of Shakespeare's performance on "Concrete Jungle," on which he literally underpins the protagonist's inner turmoil, is particularly well-mirrored throughout his work on the brilliant, deeply introspective Grace Jones (2008) reggae-centered album *Hurricane* (her tenth studio album) and its remarkable spatially dynamic 2011 counterpart *Hurricane Dub*. Both records are Jones's least commercially successful releases (with neither album charting in America), and the scope of Shakespeare's bass versatility on the *Hurricane* iterations remains sadly overlooked.

36 Alleyne (Forthcoming). Al Anderson joined for the 1974 album *Natty Dread*, while Donald Kinsey appeared with both him and Earl "Chinna" Smith on parts of 1976's *Rastaman Vibration* LP.

37 See Bradley (2000, 414).

References

Alleyne, Mike. 1999. "Positive Vibration? Capitalist Textual Hegemony and Bob Marley." In *Caribbean Romances: The Politics of Regional Representation*. Ed. Belinda Edmondson 92–104. Charlottesville: University Press of Virginia.

———. 2009. "Globalization and Commercialization of Caribbean Music." *Popular Music History* 3 (3): 247–73. http://dx.doi.org/10.1558/pomh.v3i3.247

———. 2012. *The Encyclopedia of Reggae: The Golden Age of Roots Reggae*. New York: Sterling.

———. 2018. "Marketing Marley: Cultural & Commercial Consequences." *Anales del Caribe* 2017, 190–206. Havana: Casa de las Americas.

———. 2020. "Authenticity in Music Production." In *The Bloomsbury Handbook of Music Production*. Eds. Simon Zagorski-Thomas & Andrew Bourbon, 19–31. New York: Bloomsbury.

———. 2024. "Rhythm, Rasta, Rock & 'Electric Avenue': The Electric Guitar in Anglo-Caribbean Music." *The Cambridge Companion to the Electric Guitar*. Eds. Steve Waksman & Jan Herbst. Cambridge University Press.

Beaumont-Thomas, Ben. 2021. "Lee 'Scratch' Perry, visionary master of reggae, dies aged 85." *The Guardian*, August 29. https://www.theguardian.com/music/2021/aug/29/lee-scratch-perry-visionary-master-of-reggae-dies-aged-85

Blackwell, Chris, and Paul Morley. 2022. *The Islander: My Life in Music and Beyond*. New York: Gallery.

Bradley, Lloyd. 2000. *Bass Culture: When Reggae Was King*. London: Viking.

Brown, Phill. 2010. *Are We Still Rolling?* New York: Hal Leonard.

Bundrick, John. 2021. "Thoughts on Bob Marley's *Catch a Fire*." John Rabbit Bundrick Official Site. 2021. http://rabbitwho.com/pr_catch_a_fire.php

Crane, Larry, and Geoffrey Knoop. 2012. "Tony Platt: Reducing Creative Compromise." *Tape Op* 92. November/December 2012. https://tapeop.com/interviews/92/tony-platt/

Davis, Stephen. 1990. Bob Marley. Rochester, Vermont: Schenkman.

De Koningh, Michael, and Laurence Cane-Honeysett. 2003. *Young, Gifted and Black: The Story of Trojan Records*. London: Sanctuary.

Gardner, Claudia. 2022. "Maxine Stowe Calls Out Chris Blackwell's 'White Knight Storyline.'" *Dancehall Mag*, June 1, 2022. https://www.dancehallmag.com/2022/06/01/news/maxine-stowe-calls-out-chris-blackwells-white-knight-storyline.html

Gilroy, Paul. 1993. *The Black Atlantic: Modernity and Double Consciousness*. Cambridge, MA: Harvard.

Hoskyns, Barney. 1995. "Burnin': Bob Marley and the Wailers take Britain." *Mojo*. Bob Marley & the Wailers. Rock's Backpages. Accessed September 8, 2021. http://www.rocksbackpages.com.ezproxy.mtsu.edu/Library/Article/burnin-bob-marley-and-the-wailers-take-britain

"It's Island (Eye-Land) Month." 1973. *Billboard (Archive: 1963–2000)*, January 27, 12–13. https://ezproxy.mtsu.edu/login?url=https://www.proquest.com/magazines/island-eye-land-month/docview/1017436098/se-2?accountid=4886.

Jaffe, Lee. 2022. *Jean-Michel Basquiat: Crossroads*. New York: Rizzoli.

Katz, David. 2010. "Island Advocate." *Waxpoetics*. September/October, 54–64.

Katz, David. 2014. "Lively Up Yourself." *Mojo*, 251. October 2014, 52–57.

Liner notes to *Songs of Freedom*, performed by Bob Marley, Island Records TGCBX 1 (512 280-2), 1992, CD.

Masouri, John. 2008. *The Story of Bob Marley's Wailers: Wailing Blues*. London: Omnibus.

Masouri, John. 2013. *The Life of Peter Tosh: Steppin' Razor*. London: Omnibus.

Masouri, John. 2015. *Simmer Down: The Early Wailers' Story*. Brighton: Jook Joint.

McCann, Ian. 1993. *Bob Marley: In His Own Words*. London: Omnibus.

Moylan, William. 2020. *Recording Analysis: How the Record Shapes the Song*. New York: Routledge.

Platt, Tony. Interview by Mike Alleyne. Cardiff, Wales. November 29, 2009.

Platt, Tony. Interview by Mike Alleyne. Online. August 9, 2021.

Roach, Martin., ed. 2009. *The Virgin Book of British Hit Albums*. London: Virgin Books.

Robinson, Leroy. 1973. "Soul Sauce: Wailers Wooing U.S. Fans." *Billboard (Archive: 1963–2000)*, November 24, 25–25, 51. https://ezproxy.mtsu.edu/login?url=https://www.proquest.com/magazines/soul-sauce-wailers-wooing-u-s-fans/docview/1505874416/se-2?accountid=4886

Salewicz, Chris. 2009. *Bob Marley: The Untold Story*. New York: Farrar, Straus and Giroux.

Steffens, Roger. 2017. *So Much Things to Say: The Oral History of Bob Marley*. New York: W.W. Norton.

Whitburn, Joel. 2018. *Top Pop Albums 1955-2016*. Menomonee Falls, WI: Record Research.

White, Timothy. 1991. "1981–1991: The Bob Marley Legacy for the Reggae Revolution—Chris Blackwell: An Interview with the Founding Father of the Reggae Music Industry." *Billboard (Archive: 1963–2000)*, July 13, R2, R16, R17. https://ezproxy.mtsu.edu/login?url=https://www.proquest.com/magazines/1981-1991-bob-marley-legacy-reggae-revolution/docview/1505924948/se-2?accountid=4886.

———. 1992. *Catch a Fire: The Life of Bob Marley*. New York: Henry Holt.

Discography

Bob Marley & The Wailers. 2001. *Catch a Fire* (Deluxe Edition). Island 314 548 635-2.

Videography

Classic Albums: Bob Marley and The Wailers—Catch a Fire. Directed by Jeremy Marre. 1999. Eagle Rock Entertainment.

INDEX

Note: Page references in *italics* denote figures; in **bold** tables and with "n" endnotes.

For Product Safety Concerns and Information please contact our EU
representative GPSR@taylorandfrancis.com
Taylor & Francis Verlag GmbH, Kaufingerstraße 24, 80331 München, Germany

www.ingramcontent.com/pod-product-compliance
Ingram Content Group UK Ltd.
Pitfield, Milton Keynes, MK11 3LW, UK
UKHW020156200926
475702UK00016B/2738
* 9 7 8 0 3 6 7 5 4 6 3 1 1 *